Cybersecurity Governance
in Latin America

SUNY series in Ethics and the Challenges of Contemporary Warfare
—————
Steven C. Roach, editor

Cybersecurity Governance in Latin America

States, Threats, and Alliances

Carlos Solar

Published by State University of New York Press, Albany

© 2023 State University of New York

All rights reserved

Printed in the United States of America

No part of this book may be used or reproduced in any manner whatsoever without written permission. No part of this book may be stored in a retrieval system or transmitted in any form or by any means including electronic, electrostatic, magnetic tape, mechanical, photocopying, recording, or otherwise without the prior permission in writing of the publisher.

For information, contact State University of New York Press, Albany, NY
www.sunypress.edu

Library of Congress Cataloging-in-Publication Data

Name: Solar, Carlos, author.
Title: Cybersecurity governance in Latin America : states, threats, and
    alliances / Carlos Solar.
Description: Albany, NY : State University of New York Press, [2023] |
    Series: SUNY series in ethics and the challenges of contemporary warfare |
    Includes bibliographical references and index.
Identifiers: LCCN 2022017013 | ISBN 9781438491417 (hardcover : alk. paper) |
    ISBN 9781438491424 (ebook) | ISBN 9781438491400 (pbk. : alk. paper)
Subjects: LCSH: Cyberspace—Security measures—Latin America—Case studies. |
    National security—Latin America—Case studies. | Cyberspace
    operations (Military science) | Computer security—Government
    policy—Latin America—Case studies. | Cyberterrorism—Prevention—
    Government policy—Latin America.
Classification: LCC U167.5.C92 S6545 2022 | DDC 355.4/1—dc23/eng/20220713
LC record available at https://lccn.loc.gov/2022017013

10 9 8 7 6 5 4 3 2 1

For Josefina, Nicolás, and Manuel

Contents

List of Illustrations	ix
Introduction: Into the Cyber Realm	1
1 Cybersecurity Governance	23
2 Cyber Responses and the State	57
3 The Military Industry	81
4 Strategic Threats	107
5 US-China Rivalry	137
6 Alliances with the United States	165
7 Quantifying Cybersecurity	185
Conclusion: States, Governance, and Cybersecurity	203
Appendix	221
Notes	239
Bibliography	291
Index	333

Illustrations

Figures

1.1	ICT good exports (% total good exports) and ICT good imports in 2017 (% total goods imports), in selected regions	29
1.2	Scatterplot of Internet penetration and government effectiveness in Brazil and Colombia, 1996–2017	47
2.1	How often do you see political information on Facebook?	70
2.2	Policy and strategy levels of maturity for cybersecurity and cyber defense using the regional Internet average (percentage) as cutting point	77
3.1	World military expenditure and GDP	86
3.2	Latin American military expenditure and GDP	86
5.1	Huawei branches in the world	144
C.1	Average means of institutional support in the Americas, 2004–2018	214

Tables

I.1	Scores of selected countries in security ratings and GDP	13
1.1	Internet penetration (percentage of population)	50
1.2	Countries where governance indicators and Internet penetration are positively and statistically correlated	53

2.1	Bivariate correlation matrix selected variables in the Americas, 2018–19	63
2.2	Bivariate regression matrix selected variables in the Americas, 2018–19	63
2.3	Summary of cybersecurity and cyber defense policy developments	75
4.1	Military expenditure (% of central government expenditure)	112
4.2	Military expenditure (% of GDP)	112
4.3	Armed forces personnel	112
5.1	Best country as model for future development, valid percentages	158
5.2	Trust in foreign governments and multilateral organizations	160
5.3	Bivariate correlation matrix (a)	161
5.4	Bivariate correlation matrix (b)	162
A.1	Descriptive statistics, bivariate correlations coefficients, and adjusted r square for Internet penetration (predictor) and different governance variables (outcome)	222
A.2	Views on foreign governments, by Internet use, Mexico	224
A.3	Views on foreign governments by Internet use in Mexico, controlling for preferred model for development	225
A.4	Symmetric measures by model country for views on foreign governments by Internet use in Mexico, controlling for preferred model for development	227
A.5	Cybersecurity bilateral and multilateral cooperation in selected countries	228
A.6	Descriptive statistics for cybersecurity maturity by Internet penetration	229
A.7	Independent t-test output: Strategy development by Internet penetration	230
A.8	Correlation coefficients matrix	231

A.9	Linear regression results	231
A.10	Multivariate regression results	231
A.11	Means estimation	232
A.12	Bivariate regression results	232
A.13	Correlation matrix between all variables included in the factor analysis	233
A.14	Factor Analysis Output, total variance explained	236
A.15	Factor Analysis Output, component matrices	237
A.16	Bivariate regression results	238

Introduction

Into the Cyber Realm

Empirical research on cybersecurity has become highly relevant in this day and age. Although it is useful, we cannot learn enough about cybersecurity simply by reasoning descriptive patterns. Indeed, one of the great things about social sciences is that we can usually make the most out of observations and inferences to problematize and theorize about cybersecurity, for instance, by studying how countries protect and advance their national interest in cyberspace. Current cybersecurity research stems from many disciplines, such as political science, international relations, sociology, criminology, development, international law, psychology, and economics. It deals with various interconnected themes, including the rule of law, the spread of digital technologies, and democratization; most notably, it sheds light on what we can call the governance of cybersecurity. Scholars tend to ask broad questions including: How do states and citizens educate themselves to avoid hackers? How does the United States and China cyber rivalry cascade into the rest of the world? What is the relationship between Internet usage and antigovernment mobilization in developing democracies? Is the Internet increasing crime and terrorism?[1]

Most research acknowledges the fact that the proliferation of information and communication technologies (ICTs) has pushed the boundaries of the use of digital tools for good and bad. Between 2005 and 2018, there were more than 250 state-sponsored cyberattacks. Incidents transpired despite the setup of multilateral frameworks to monitor, deter, and respond to emerging threats in information security sponsored by the African Union, the League of Arab States, Shanghai Cooperation Organization, the Organization of American States, and adopted by intergovernmental entities, including the

Group of 7, the European Union, the North Atlantic Treaty Organization, and the United Nations.

The truth is that there is little common ground on how to enforce the measures required to address cybersecurity. Some Western groups of states tend to consider existing international law sufficient for guiding state behavior in cyberspace. Other industrialized countries, such as China and Russia, would prefer new normative guidance on state use and development of ICTs. International consensus is fragile, and cross-border cooperation does not suffice for real global collaboration.[2]

The outlook seems more gloomy when even global leaders, including UN Secretary-General António Guterres, are pessimistic about reaching a compromise. "When one looks at today's cyberspace, it is clear that we are witnessing, in a more or less disguised way, cyberwars between states. The fact is that we have not yet been able to discuss whether or not the Geneva Conventions apply to cyberwar or whether or not international humanitarian law applies to cyberwar," Guterres argued.[3] Undoubtedly, more intergovernmental, governmental, and nongovernmental discussion on cyberactivities is needed to strengthen international law, national law, and nonbinding cyber norms in different contexts.

In these early days in the history of cybersecurity, scholars have tried to address the implications for stakeholders both in- and outside of academia. Governments are interested in knowing more about how to build expertise to manage their cyber defenses, and civil society is highly responsive to relevant studies promoting open discussion and deliberation online. Private industry, as another major actor, might be the most updated and well-versed source of knowledge regarding cybersecurity issues. Not only do they use research to design and develop the various aspects of ICTs diffused globally, but, more importantly, they also seem to craft the pivotal channels of interactive information in the twenty-first century (i.e., email, blogs, vlogs, wikis, social sites, microblogs, etc.).

Public and private security logics meet at the melting point between government and corporate concerns toward cybersecurity. The health of ICTs and networks is many times dependent on the logic of markets, on profitability, and at the hands of the best buyer in today's competitive trade and commercial environment.[4] Does this mean public authorities have less to say in the governance of cybersecurity? Then why is cybersecurity a national security issue, as many in government have painted it?

As of the writing of this book, a different branch of research has consolidated in the subfield of cyberpsychology. This particular approach is understood as the "processes underlying and influencing the thinking, interpretation, and behavior" associated with human online interconnectivity.[5] What such a level of study has come to validate is that social, economic, and political forms of action using ICTs and Internet-based technologies ultimately shape new forms of socialization in at least three iterative ways: among citizens, between citizens and authorities, and among two or more nation-states. Mostly, empirical research has addressed these three aspects. Public opinion scholars have studied the application of technical developments and citizenship mobilization, including the equal opportunities presented by satellite communications, smartphones, cable television, and the Internet. Sociological and developmental studies have pointed out the potential use of e-governance and e-democracy toward community building and government information policy. Political scientists and international relations students have explored questions surrounding the idea of cyber conflict breaking out among rival states and issues of global norms among those believed to be allies. Criminologists have emphasized the role of cyberspace in fostering crime, deviance, and other forms of bending the law.

Much of the scholarship produced by international relations scholars and political scientists has involved the underlying meaning of cyberspace and why it raises security concerns. This consequently raises cybersecurity as a state-centered issue. In this literature, cyberspace is defined by Brandon Valeriano and Ryan Maness as the "networked system of microprocessors, mainframes, and basic computers that interact in digital space." More relevantly, they argue that "what happens on the physical layer of cyberspace is where we engage political questions."[6]

Political issues revolving around citizenship, society, and global norms in cyberspace have flourished from a vast array of standpoints, often too many to enumerate. What concerns this book the most is the extent to which practitioners and scholars have underscored the links between digital technologies, cybersecurity, and the military:

> Derived from a realist theory of world politics in which states compete with each other for survival and relative advantage, the principal cybersecurity threats are conceived as those affecting sovereign states, such as damage to critical infrastructure within their territorial jurisdictions. This approach delegates responsibility for the security of cyberspace to military, intelligence, and law enforcement agencies, which together constitute the state's

national security apparatus. These agencies tend to operate with limited public accountability, oversight, and transparency.[7]

Cyberspace provides even more opportunities or disruptive technological transformation that could provide a decisive advantage, on the one hand, but might also rick uncontrolled escalation on the other.[8]

ICTs are said to have integrated almost fully into how modern states manage their military affairs:

Unlike the infantry or artillery revolution, the information revolution didn't just create information warriors, it informationized all conventional warriors. These are infantry soldiers, munitions experts, forward air controllers, pilots, and war-fighting staffs that are dependent on digital technologies and information to conduct conventional operations. They are the front-line combatants, armed with M-16s, radios, and combat iPads. It is virtually impossible to separate modern warfare from digital capabilities.[9]

We might have decided to view cybersecurity as a purely civilian issue and conceptualized both current and future cyber "attacks" as criminal matters. But as with humanitarian assistance, rule of law programs, strategic communication, and so much else, the military has jumped in to fill the vacuum created by squabbling and under-resourced civilian agencies.[10]

The military itself recognizes that the continued advancement in technology has changed the conduct of warfare:

Today's commanders must drive integration of lethal and non-lethal effects across Land, Air, Sea, Space, and Cyberspace to create unity of action while maintaining our competitive military advantage on the battlefield. Our failure to operationalize and normalize the cyberspace domain effectively cedes it to our adversaries, gives them a competitive edge, and ultimately, creates an increased attack vector against our objectives.[11]

States around the world have emphasized the need to invest in the development of cyber military capabilities, and examples of armed forces adopting new doctrines for such principles are becoming the norm. Experience from battlegrounds in land, sea, and air serve as useful roadmaps for future cyber military operations. However, to some observers, "Tactical terms used in continental warfare such as vital ground and in-contact are not consistent with an environment where bits and bytes are proxies for warriors."[12] For some scholars, the hype behind cyberwarfare is overblown:

> [I]t is highly unlikely that cyber war will occur in the future. Instead, all past and present political cyber-attacks are merely sophisticated versions of three activities that are as old as warfare itself: subversion, espionage, and sabotage. That is improbable to change in the years ahead.[13]

> Even if cyberspace exists primarily for war, it does not follow that war would itself become the primary means by which cyberspace would be governed. Unlike land, air, and sea, cyberspace is a sociotechnical institution rather than a natural, physical domain.[14]

There is a dividing line among perspectives:

The cyber hype perspective would suggest that we are seeing a revolution in military affairs with the advent of new military technologies. The moderate view is guided by careful consideration of what the real dangers are, as well as the costs of the overreaction.[15]

Cybersecurity expertise has grown exponentially. The literature increases so rapidly that the knowledge we think we possess on governance, legislation, and state interactions becomes obsolete very quickly. "Cyber expertise research is like astronomy: what we perceive is actually a snapshot of the past," argued Robert Thompson of the Army Cyber Institute at the U.S. Military Academy.[16] More than ever, studying the cyber domain thus demands innovative claims, rethinking previous theoretical paradigms, marrying sometimes distanced research methodologies, and, more notably, bringing forward new case studies in a nuanced and empirically grounded way that can defy the test of time.

Take the following example. Hacker attacks are deemed the "new normal" in geopolitics, reaching both state and subnational levels.[17] On August 22, 2019, the *New York Times* ran a story about municipalities across the United States held hostage by ransomware attacks. More than forty local authorities had been victims of unknown hackers that infiltrated their systems and encrypted their data.[18] In ransomware attacks, systems are locked and later unlocked after the victims pay a sum of money, usually in cryptocurrency, to the captors. Victims will say that paying the ransom is cheaper than rebuilding the systems all over again. In cases like this, higher-up authorities are called. Intelligence agencies, the armed forces, police, and public prosecutors will in theory align their resources to try and identify the source of the hacking. Aiding citizens and victims of these types of attacks demands sophisticated countermeasures from trained and well-resourced national agencies. Incidents targeting a municipal library can quickly escalate to threaten other services such as water, power, hospitals, communications, intellectual designs, military secrets, and, what is much feared, the denial of service or destruction of critical national infrastructure. Concurrently, the U.S. justice system has overseen the prosecutions of hacking groups led by nationals from Iran and China for infiltrating domestic computer networks, including those of critical institutions such as the Pentagon.

Researchers in both developed and emerging countries use small-N case-selection or large-N contexts (i.e., systematic surveys) to highlight the way states deal with cyberincidents. Research, either qualitative or quantitative in style, derives inferences that go beyond the particular observations collected.[19] Put simply, cybersecurity scholars observe incidents and states' responses that help us learn about governing the cyber realm. In a large-N pooled time-series analysis, for example, scholars found that in 170 countries surveyed between 1996 and 2010, higher Internet penetration usually came with reports of more stable governance, regardless of regime type.[20] From this type of research, we usually learn about transparency, accountability, and good governance themes, although less about policy action. In contrast, from the small-N example taken from ransomware incidents in municipalities, we learn that some authorities prefer to pay the ransom and that only a few communities in the United States have cybersecurity protocols to counter incidents such as these.

If we were to collect facts systematically, let's say across a vast number of federal states, we might arrive at the conclusion that local authorities are usually cash strapped in their attempts to access appropriate cybersecurity for their systems. We might then infer that, beyond the observations collected, this phenomenon might be common in other countries, as similarly ill-resourced municipalities abound in countries affected by austerity measures. If we had the data to infer that based on a sample from developing countries, cities are poorer than those in the United States, we might assume that if hackers were to target them, they would be eventually less prepared to avoid such an infiltration in their systems—although maybe, as less able to pay ransom, less appealing to greedy hackers! Nevertheless, the future of cybersecurity in not-so-rich countries is difficult to forecast. Much of cybersecurity's nature is said to depend heavily upon the (good and bad) users themselves.[21]

Cybersecurity Governance

Academic studies surrounding cybersecurity governance have burgeoned in the advanced democracies, notably in English-speaking countries. Little research originates in or uses case studies from developing societies where cybersecurity "uncertainty" seems more significant. This book acknowledges how unclear cybersecurity governance can be and seeks to untangle a few pertaining complexities.

The main concern of the book is the fact that states struggle when dealing with perceived threats that are too difficult to measure and scenarios too complicated to assess.[22] Cyberthreats easily exploit uncertainty. We are more likely to overstate or misjudge the actual danger posed by the threats if we understudy them.

One way to deal scientifically with uncertainty and complexity is to search for generalizations and make inferences. Across the book, I look for such interpretations in the developing world, and explore cybersecurity governance beyond what we know from the advanced democracies (i.e., United States, United Kingdom, European Union, and Australia, among others) and other industrialized states (China and Russia).

Martin Libicki has defined cybersecurity as the "efforts to prevent systems from being compromised."[23] I take this approach a step farther, and define cybersecurity governance as the actions and policies adopted by civilians, the military, industry, and the private sector to safeguard digital space. However, the focus of this book is directed on the military.

The book deals with three much-hyped and highly debated concepts: cyber, security, and governance.[24] I intend to theorize over practical efforts (*governance*) to solving complex problems (*cyber*) that affect the state (*security*). From this standpoint, I shall approach the policy perspective that shapes cyberspace security as a national security issue. While some scholars question the value of "national securitizing" cyberspace—because it militarizes civilian problems, subjects freedom to vaguely defined controls, and incentivizes rivalry among states—I take into consideration the established trend that many countries have already adopted. A glance at the UN Institute for Disarmament Research (UNIDIR) cyber policy portal confirms that rich and emerging states have constructed, or are in the process of building, vastly prepared cyber military commands and that legislation has been adopted that allows these groups to plan and act against threats concerning cyberspace. Both the Pentagon and NATO have declared cyber a "domain," just like the sea, air, land, and space.

This book is not a justification for national security doctrines. On the contrary, it aims to create awareness of the issues presented when militarizing cyberspace by acknowledging that this has become a standard feature among some nation-states. As others have argued, the "militaries must be very careful about what missions they accept in cyberspace and must circumscribe their forays into cyberspace lest they are overwhelmed by the sheer scope of the domain."[25] My point is that by focusing on actual governing practice across levels, space, and jurisdictions, we are more likely to grasp the cyber, security, and governance arenas without losing touch with reality.

While this book is mainly diagnostic in identifying trends and problems, its contribution is to challenge prescriptive assumptions regarding cybersecurity governance happening in and beyond the industrialized North. I argue that sample bias and the overemphasis by international relations scholars on a few selected cases often distort the analysis and understanding of the broader dynamics of cyber, governance, and security as both separate and combined concepts.

Regarding case studies, it is not my intention to cry wolf and ask social science researchers to select diverse case studies from beyond the rich countries. Methodologies such as most similar cases, most different cases, or single case-study analysis are frequently used to explore new trends in world politics, and most of the time, they are correctly executed. However, by incorporating a sample with understudied diverse case studies—Gerring defines these as "a set of cases that encompasses a range of high and low values on relevant dimensions"[26]—we can account for a full array of variations on the selected variables of interest. That is, if we study cases that rank high and low in different cybersecurity governance dimensions and the role of the military in them, better and stronger representativeness can be claimed.

Entering the cybersecurity realm of research requires, thus, describing and explaining differences. Whether we study many phenomena (i.e., cyberwar, cybercrime, cyberterrorism, cyber espionage, etc.), or just one, describing and explaining demand recollection of many observable implications that make sense of our theories.[27] More specific contributions to the scholarly literature on cybersecurity increase our ability to engage in scientific explanation. Much of the cybersecurity background literature in this book reflects on the fields of political science, international affairs, and criminology but also on other equally relevant subjects (i.e., computer science or cyberpsychology). In sum, to make an interdisciplinary contribution to the study of cybersecurity, I argue that important topics, as observed in advanced democracies, might garner equal explanatory success in developing countries.

The Argument

The book has three main pillars of research: (1) to assess the cybersecurity scenario in developed countries and illustrate current events in the developing Western Hemisphere; (2) to explore the governance of cybersecurity in comparison to other perceived threats to national security in the region; and (3) to illustrate the militarization of cybersecurity policy.

In the first pillar, I consider that states in the Western Hemisphere find themselves in a scenario where global norms call for cybersecurity capacity building. In Latin America, the risk of interstate conflict among countries has diminished, and nations put their limited financial resources into addressing nontraditional and human security threats, including cybersecurity, that call for adequate military resources. The ongoing cybersecurity developments have pushed states to rethink their traditional warfare roles and the equipment their armed forces use against the threats posed by state and nonstate actors such as drug cartels, paramilitaries, organized crime groups, gangs, hackers, and terrorists.

Countries can opt to develop the means of achieving cybersecurity by considering the changing character of war and extreme criminal violence, yet political and economic limitations influence how nations recapitalize their military budgets. Leaders respond differently in a complex and uncertain world. Some craft strategies across a range of contingencies, others adopt a more focused approach. I argue that in times of rapid technological change, some states might move quickly to utilize resources and skills. In contrast, others stall and postpone any investment in new security contingencies.[28]

My primary goal is to deal with a real-world problem of great social significance: cybersecurity governance as characterized in the industrialized countries and the spillover effect it exerts on developing nations. The focus of the book is on the developing countries, specifically in Latin America. Theoretically, I engage with the prior literature on cybersecurity governance, international relations, comparative politics, and criminology. I have chosen theories from these subfields to illustrate that they might be wrong in their explanation of how developing states craft cybersecurity governance and the role of the military in administering it. I am adamant about the need to engage with the United States' cyber history to tell the case of Latin America. It is not enough to focus just on the history of cybersecurity in Latin America without analyzing events transpiring in North America. Focusing on the history of cyberspace militarization in the most technologically advanced nation in the world opens a door to discussing worldwide events and their cascading effect across the hemisphere. Many in Latin America and elsewhere might not be familiar with such a history, and that is why I write this book.

The book presents in-depth country-by-country research while constantly referring to the industrialized countries' cyber history to contextualize global outcomes. A set of empirical events explained in the book led me to generate observable implications from the theory. In particular, I focus on what I call the first wave of cybersecurity governance in Latin America, which began in the late 2000s. I take Brazil's 2008 National Defense Strategy, which named cybersecurity as one of the critical strategic domains for national security and which consequently launched a full-scale effort to institutionalize a policy designed to unite different sectors and multiple stakeholders across the public and private governance ecosystem controlling national security, defense, and information security.[29]

Brazil's strategy came at the time of the first systematic uses of cyberattacks against one state by another (accompanying a military campaign), when Russia targeted Georgia and, later, Ukraine with cyberattacks and propaganda campaigns during the short wars of 2008 and 2009, respectively. Moscow's raids, together with the Chinese People's Liberation Army's (PLA) and Beijing's "cyber-militia's" conspicuous cyberattacks, led observers to argue that cyberspace had finally turned into a medium for conflict and strategic warfare. Earlier, in 2007, Estonia was the target of a significant cyber hit that originated from inside Russia, which, to some observers, marked a tipping point comparable to what the Hiroshima and Nagasaki bombings did for the nuclear age.[30] Marina Kaljurand, the Estonian ambassador to Moscow during the crippling 2007 cyberattacks against her country, described retrospectively the daunting effects of the assault:

> Those were the first explicitly political cyberattacks against an independent, sovereign state in history. If put into today's context, the attacks were not very sophisticated—even primitive. But back then, they were very disturbing. By that time, Estonia already had widely established Internet and e-services, and an e-lifestyle; when those services were interrupted—mainly in the banking sector—it was highly disruptive. As to the effects of the attacks? They did not kill anybody, they were not destructive. They were highly disruptive to our lives though.[31]

By the early 2000s, cybersecurity was a matter of policy discourse based on little or no tangible evidence (i.e., the United States launched its first cyber executive review in the *National Strategy to Secure Cyberspace* in 2003). Less than a decade later, more severe political actions unfolded (i.e., NATO placing and resourcing its cyber center command in Estonia soon after the Russian attacks), and policy researchers called for more attention to "cyber defense capabilities" and "cyber deterrence doctrines." Both concepts quickly gained traction among those studying and leading the armed forces' operational realities.[32] "Computer security" no longer sufficed to characterize the cyber domain. Modern "cybersecurity," understood in this context, came to integrate aspects of computer security plus an array of national security elements, henceforth operationalizing the term at the highest levels of policy and politics.[33]

In July 2010, it was publicly disclosed that the Stuxnet computer virus, the product of a joint cyberweapons operation started in 2005 by the U.S. Defense Department and Israel, had infiltrated Iran's nuclear facility Natanz, destroying uranium enriching centrifuges. Today, this kind of malware keeps being disseminated in more sophisticated versions, thus calling for more cybersecurity measures.

Cyberweapons are the multifold tools and technologies used to disrupt or destroy computer network operations.[34] Cyberweapons can also be, as Thomas Rid and Peter McBurney put it, the "computer code that is used, or design to be used, with the aim of threatening or causing physical, functional, or mental harm to structures, systems and living beings."[35] The worm Regin, for example, a so-called "cousin" of Stuxnet, according to the private firms that have tracked it, is a powerful spyware created by some government that has targeted public and private organizations in developing countries.[36]

Another example came after the 2016 United States presidential election, won by the Republican candidate Donald Trump, which became known as the most politically significant use of digital technology to that time. No critical infrastructure was brought down; instead hundreds of thousands of low-tech fake accounts, stories, tweets, and other social media platforms were used to influence the political discourse and the election outcome. The national security and intelligence bodies of the United States stepped in and uncovered plausible evidence that Russia was behind the attack. In a "textbook information-warfare operation," Moscow was able to hack the Democratic National Committee, publicizing e-mails from Hillary Clinton's top aides.[37]

Low-level but intense attacks continue to flood the Internet. In June 2019, officials from the White House announced that Donald Trump had ordered cyber retaliation attacks on Iranian military computers as a direct response to Teheran shooting down a U.S. surveillance drone. In August, cybersecurity company Anomali uncovered suspected North Korean hackers as responsible for a phishing campaign that targeted foreign ministries and multiple research centers in the United States and Europe. Hackers mimicked the French Ministry for Europe and Foreign Affairs, tricking users into entering their credentials onto the malicious website, so they could later use the information to spy on the affected inboxes. That same week, Twitter and Facebook announced measures to delete state-led disinformation campaigns disseminated by a spammy network of approximately two hundred thousand accounts originating from mainland China, which they accused of "deliberately and specifically attempting to sow political discord in Hong Kong."[38]

Cyberincidents have led us to rethink what we know about computer network attacks, their scale, and whether they are used in combination with conventional armed conflict scenarios, launched on their own, or used in conjunction with a type of armed conflict that does not qualify as an act of war.[39] More relevantly, it leads us to think about the role of the military in such scenarios.

Table I.1 shows the Global Peace Index (GPI) results for a subsample of Latin American countries. The GPI reviews the state of peace in the world, ranking countries according to three thematic categories: ongoing

Table I.1. Scores of selected countries in security ratings and GDP

Global Peace Index, 2019 Rank	Country	Score	Rank in 2016	Intentional homicides, 2016 (per 100,00 people)	GDP, 2018 (per capita, in US$)
27	Chile	1.635	27	3	15,923
75	Argentina	1.989	67	6	11,652
116	Brazil	2.176	105	30	8,920
140	Mexico	2.600	140	19	9,698
143	Colombia	2.661	147	56	6,651
144	Venezuela	2.671	143	26	16,054*

Note: Venezuela's GDP value is from 2014. An estimate from the IMF puts the 2018 value in US$ 3,373.
Source: Institute of Economics & Peace (2019), IMF (2019), and World Bank (2019).

domestic and international conflict, social safety and security, and militarization. A score closer to 1 estimates higher levels of peace. We see that only Chile ranks in the top thirty most peaceful countries globally. Argentina and Brazil have levels of peace in the middle of the index (where most countries are clustered). Meanwhile, Mexico, Colombia, and Venezuela scored at the bottom of the table.

Indexes such as the GPI frequently show proof of Latin American states entering the perils of the digital age while dragging many other security issues, most notably rising levels of violent crime, along with them. The region remains the world's highest in homicide rates (especially in Central American and the Caribbean), high incarceration rates, and organized crime, all of which is highly detrimental to state and human development.[40] The main question here is whether cyber insecurity is different from these other threats to peace. For some observers, threatening activities in cyberspace also go against human development as they "undermine people's trust in ICTs as well as their wellbeing in cyberspace."[41]

Mapping states' efforts to deal with both international security and domestic violence include cybersecurity capacity building. As they have with terrorism, transnational crime, and human trafficking, among other issues, states have implemented several initiatives to improve drafting strategies, processes, guidance, and laws that typically strengthen the creation of dedicated agencies and response teams. For these purposes, they rely heavily on international cooperation. One or more countries can make themselves more cybersecure, promoting safer Internet governance between them.[42] When ICTs establish more secure networks, more people can engage in more activities in cyberspace (from electronic commerce to e-government services). The Organization of Americas States (OAS), for example, supports cyber-related capacity building programs for its members via training, crisis management, and exchange of best practices through its Inter-American Committee against Terrorism (CICTE) and Cyber Security Program.[43]

Another way to escape single-handling cybersecurity is through intergovernmental alliances. Most notably, the United States has been keen on facilitating bilateral cooperation on security, including cyber capacity building. Realist scholars argue that the United States has been successful in the role of central authority when it comes to international security in the Western Hemisphere, moderating the chances of interstate conflict, and shaping regional dynamics in response to perceived threats to peace.[44] It is understood that part and parcel of the great powers is to deter conflict and maintain peace by knowing which regional states are likely to initiate conflict, and why, so that they can anticipate and intervene with deterrence mechanisms. Of course, more theoretical development is needed from competing views, as conflict can happen despite deterrence efforts by a regional powerhouse.

By deterrence is understood as, "the means of dissuading someone from doing something by making them believe that the costs to them will exceed their expected benefit."[45] The goal of deterrence is to create disincentives and discourage the onset of hostile actions. In cyberspace, a deterrence stance warns other states against any seriously hostile act.[46] For this purpose, states dedicate resources not only to defensive but also toward building retaliatory offensive systems. Cyber deterrence mechanisms need not act solely in the cyber domain. Deterrence and retaliation to a cyberattack can derive from a broad range of tools (trade policy, foreign policy, military responses) and sectors (land, air, sea, space).[47]

A few questions then arise. First, can the United States deter traditional conflict in the so-called hot spots in Latin America for much longer? Second, and more relevant, can the United States deter cyber crisis escalation between regional states? How does Latin America fit in the so-called cyber problem[48] in U.S.-China relations, and what are the most critical military problems? What guides the United States in the Western Hemisphere in light of the dangerous tit-for-tat in cyberspace with Moscow and Beijing that could bring them into conflict? On the other hand, will the global economic revolution trump cyber rivalries in the long run?

These questions are the crux of this book. I ask them mostly because, beyond any other country's efforts, the U.S. military and civilian experts at the Pentagon have put long-term resources into organizing the state around cyber affairs. Ronald Deibert at the University of Toronto put it this way:

> The recognition that cyberspace is a warfighting domain led to the creation of the US Cyber Command, which centralizes command of cyberspace operations across the US military. That the world's largest military defines cyberspace as a domain within which to project power and to fight and win wars, inevitably has system-wide repercussions, both materially and ideationally. In basic terms, the reorganization of the US military prompts changes among allied armed forces who need to be synched up operationally to cooperate.[49]

It is reasonable to expect that U.S.-made cyber knowledge is traveling abroad through military-to-military diplomacy. Despite the perceived declining U.S. multilateralism on the global stage and little actionable policy toward Latin America in recent years, Washington has not rescinded its commitments regarding enforcing treaties and supporting allies when it comes to security affairs. This strategic demand is what the U.S. government has called building partner capacity, or as former Secretary of Defense Robert Gates put it, "helping other countries defend themselves, or, if necessary, fight alongside the U.S. forces by providing them with equipment, training, or other forms of security assistance."[50] Building the military and security forces of allied countries has been in Washington's repertoire of actions since the Cold War.[51]

However, and considering costly and controversial examples beyond Latin America, including Western Europe, Greece, Philippines, South Korea, and more recently in Afghanistan, Pakistan, Yemen, and Iraq, the United States has receded and picked battles more carefully. "Helping other countries better provide for their security will be a key and enduring test of U.S. global leadership and a critical part of protecting U.S. security, as well," explained Gates.

In the Western Hemisphere, security cooperation has occurred most notably through civil and military bureaucrats pushing policy through the channels of the Pentagon, State Department, and the U.S. Southern Command stationed in Florida. As an example, I recall that the U.S.-Chile Executive Cyber Consultation mechanism focused on bilateral cooperation, collaboration, and the protection of critical infrastructure, incident response, data security, information, and communication technology procurement, and military and law enforcement cooperation. The consultation mechanism is attended by senior-level officials from the United States, including representatives from the Department of State, the National Security Council, the Department of Homeland Security, the Department of Justice, the Department of Defense, the Department of the Treasury, and the Department of Commerce. The Chilean delegation is led by a senior official from the Ministry of Defense. It includes representatives from the General Secretariat of the Presidency, Ministry of Foreign Affairs, the Public Prosecutor's Office, the National Intelligence Agency, and the Ministry of Interior.[52] A similar mechanism was established in 2017 between the United States and Argentina, another of Washington's cyber allies in South America.

This multitude of state and nonstate bodies engaged around the cybersecurity issue leads me to discuss matters of governance. I do this in consideration of my second pillar. I argue that cybersecurity governance has come to push forward the growing agenda of human security that has captivated policy- and decision makers in the post–Cold War theatre, where traditional conflict has become a lesser priority. This way, the governance for cybersecurity has been set up among a web of decision makers, as conventional and new state resources are used toward countering risks both to the population and to the state. Scholars Lennon Chang and Peter Grabosky theorized on these matters, writing that "the governance of cyberspace is no less a pluralistic endeavor than is the governance of the physical territory," where democratic institutions overlap "to help secure cyberspace."[53]

New governance interpretations have come to challenge the folk versions of politics and policy. Centralized bureaucracies do not rule the policy arena any longer. The expansion of public policy matters to include new actors has led people to believe there is a failure of bureaucracies to solve complex social problems. State actors now try to involve new stakeholders for more extensive and widespread action. These aspects of governance are, by far, well identified, and a large body of literature has discussed their relevance. The governmental "do-it-alone" mode of thinking has given scholars the space to propose new networked, interactive, multilevel, and collaborative forms of governance.[54]

In Brazil, for example, after the 2008 National Defense Strategy was published, the government set up interactive policy communities to review the risks to their critical infrastructure, involving the state-owned Petrobras, and the ministries of defense, external affairs, health, science, and technology, plus other institutions, such as the central bank and the federal government's IT and information and security departments. Consultations on best practices, official guidelines, and monitoring standards revealed evidence of a wide range of vulnerabilities, cybersecurity holes, and network exposure that needed patching. Brazil's growing network of stakeholders now also includes public utilities, private companies, and telecom providers, all collaborating to improve regulation and expand cybersecurity measures on interconnected computer systems.[55]

I also review how countries in Latin America followed Brazil's cybersecurity endeavor by mimicking much of these in the name of national security. I treat the term *national security* in the book as a military matter. Still, I also include other issues, such as economic, climate, energy, and cyber affairs that mandate defense and foreign affairs actions from the government.[56] Brazil's Cyber Defense Command (CDCiber, its acronym in Portuguese), a unit within army, acts as the country's national center and cybersecurity-responsible agency. It is responsible for planning, coordinating, directing, integrating, and supervising cyber operations in the defense area. The CDCiber coordinates with two other agencies, the department of information and communications security, and the federal police's unit for combating cybercrime (URCC). Although the critical position rests with the head of CDCiber, which leads me to my third and final takeaway.

For the militarization of cybersecurity research pillar, I focus on the mushrooming of security networks to argue that what should hold our attention the most insistently is the gradually increasing presence of the armed forces, as steering nodes have started to take over. For example, Brazil's 2013 Defense White Paper identifies cybersecurity as a "fundamental strategic sector for national defense." It goes on to say that "efforts in the cyber sector aim to ensure confidentiality, availability, integrity, and authenticity of data circulating in Brazil's networks, which are processed and saved." It also elevates the CDCiber mandate to match "those of other existing government organizations, including through protection against cyber-attacks." It gives the Navy a role in developing technologies necessary "in particular in the area of cyber warfare," and among its priority projects it includes "the acquisition of the supporting infrastructure, and acquisition of cyber defense hardware and software solutions (to be implemented in 2010–2023)."[57] Brazil's 2008 defense strategy has a similar rationale, as it identifies the cyber issue "as one of the three fundamental sectors for national defense of strategic importance"; it grants the Navy autonomy "in cyber technologies that guide submarines and their weapons systems, and enable them to work in network with other naval, land and air forces," while it also seeks to enhance "cyber capabilities through the development of cyber training in industrial and military fields."[58] In the United States, the Cyber Command and the National Security Agency (NSA) are under the authority of the same military officer in charge of priming both cyber defense and offense capacity. In Latin America, the same principle reigns: the military is preparing for digital operations for information and control.

Overmilitarization of cyberspace is risky. Express cyber norms of engagement at the international level are unclear despite the initial effort marked by the publication of the *Tallinn Manual 2.0 on the International Law Applicable to Cyber Operations*. As stated by legal scholar William Banks, the *Tallinn Manual* provided "much-needed confidence for states that international law applies in the cyber domain and supplied a framework for applying to cyberspace widely understood norms from kinetic conflict." However, it is considered that the manual "is not a treatise on international cyber law, nor does it establish new international law or represent the views of any states on their cyber operations. There could be no such treatise at this time because of insufficient state practice, a paucity of official state legal views, and a lack of consensus on norms."[59]

In sum, military and civilian actors enter the global cyber theatre under quite generic, still blurry, and mostly unsettled legal circumstances. In post-transitional democracies, the military's interventionist role in politics has not strengthened democratic stability.[60] Marrying the digital space to highly politicized armed forces is at least troubling.

Plan of the Book

The book presents another eight chapters. Chapter 1 builds my theoretical foundation. It introduces the debate on the case studies and the themes of cybersecurity governance to lay the ground for the networked governance approach to studying state policy and decision making. Chapter 2 reviews the digital pax Latin Americana, a term used to describe the state of peace in the Americas; this is the chapter wherein cyberthreats are pinned to transnational nonstate parties acting outside international law. I then move on to provide some examples of the risks and opportunities for the Americas' states when they join up cybersecurity governance with the strategic agendas of defense and national security, in other words, the plausible militarization of cybersecurity. I conclude by arguing that in this new scenario of global insecurity, states are at a sensitive stage. As other scholars have highlighted, it is *how* and *when* nations decide to join the cyber era that will mark any nation's potential to use the cyber resource as an advantage.[61] I theorize on the idea that due to the militarization of cybersecurity, the next steps in the digital realm for emerging democracies will be to define their mid- to long-term strategic security dynamics, whether against threats originated in the advanced democracies, from other developing countries that pose today as traditional rivals, or from nonstate organizations with sufficiently developed cyber technologies to constitute serious risks to them.[62]

Chapter 3 debates the changing aspects of military missions and the security and defense industry in the Western Hemisphere. It discusses the political economy of arms industrialization and different policy choices adopted to develop the sector (export liberalization, protectionism, and wealth creation). It argues that cybersecurity governance has as prerequisites both goals and means that involve technologies to manage capacity and capabilities. Chapter 4 analyzes the underlying changing character of war and the strategic threats perceived in the region (interstate conflict, transnational crime, terrorism, paramilitaries and insurgencies, social and ethnic tensions, natural disasters and climate change, and cybersecurity), and how these relate to current military-industrial and economic progress. I use six comparative case studies to review my arguments: Argentina, Brazil, Chile, Colombia, Mexico, and Venezuela.

Chapter 5 surveys the U.S. and Chinese approaches in the remaking of international order and how it has carved new political and security scenarios in Latin America. The issues of cybersecurity lie at the heart of the chapter's discussion. Still, more, and broader, perspectives on commerce, international security, and military diplomacy are brought into question to provide a full picture. By the end of the chapter, I return to the cases of Mexico and Brazil to draw examples of these countries' Internet usage and belief in either the Chinese or American model of development.

Chapter 6 explores in more detail the prospects for U.S.-Latin American cyber partnerships. It argues that U.S.-led cybersecurity efforts in the region come at a moment of the overall revamping of the military's role and mission, which has brought forward, among other issues, the creation of cybersecurity forces and cyber commands. I review three positive cases where cyber partnerships have happened (Argentina, Brazil, Chile) and three negative cases (Colombia, Mexico, and Venezuela). Chapter 7 explains different dimensions for cybersecurity maturity currently measured by OAS, which provide a review of the "first wave" of cybersecurity measures adopted by Latin American and Caribbean countries.

In writing this book, I collected empirical data from academic literature, official memos and white papers, legislation, surveys, databases, and various other public sources to build robust evidence for my primary research queries. This plan includes the use of qualitative and quantitative sources of information. For the latter, I assume no prior knowledge of mathematics or statistics. I have tried to keep the train of ideas flowing despite the usual repetition of some statistical concepts. Some of the quantitative exercises in the book are supplemented in the appendix, which includes further methodological explanations and statistical estimations.

Integrating quantitative methods to the study of cybersecurity is helpful in at least three meaningful ways. It provides a fundamental approach to understanding the large quantity of the literature being published. Second, it gives the reader a better understanding of research practices in the social sciences. Third, it relates to the current use of big data analytics using statistical methods becoming more relevant today. The idea, nonetheless, is to have first a solid grasp of the theory and then to use it intelligently to guide some exercises of hypothesis testing. The main concern here is the "relationship between theory and empirical work, not the relative merits of quantitative or qualitative approaches."[63]

In the concluding chapter, I discuss relevant results that offer a new interpretation of the theory and practice of cybersecurity governance. The arguments and conclusions will challenge scholars and policymakers to understand cybersecurity's new international and national perspectives as well as the recent policy efforts made by individual countries.

Implications for Policy

Political and economic risks during the Cold War were identified in part by the superpower rivalry between the United States and the Soviet Union. Twenty-first-century political risk is no longer split into blocs but instead crowded with uncertainty rising from state and nonstate actors across the globe.[64] Responses to cybersecurity interweave with international politics and global economic issues. I argue that governing cybersecurity is thus an interactive and dynamic exercise. It demands that political, economic, and social systems come together and produce dyadic relations between national and supranational actors. Studying governance, as a means of collaboration in goal-directed networks, sheds light on the internal mechanics that allow organizations to function together. This approach merits new questions: Is cybersecurity governance a typical case of networked collaboration? How are actors interconnected? How do they communicate, share responsibility, and make decisions?

Across the globe, specific and compelling state agencies lead cybersecurity governance—namely, Cyber Command in the United States, National Cybersecurity Centre in Great Britain, Signals Directorate in Australia, Federal Office for Information Security in Germany, Cyberspace Administration of China, and Security Council in Russia. A closer look into Latin America's reality might allow us to better know how regional cybersecurity agencies arrange collaborative agendas, how important decisions are taken, and how power and legitimate authority interact. However, one may always doubt such an argument and ask: Do the affairs of cybersecurity encourage or hinder collaboration, decision making, and lawful use of authority? Complex policy issues, even in liberal and participatory forms of democracy, pose serious questions about governance.[65] In the name of national security, many governing processes remain secret and under little or zero external accountability beyond a close group of agencies. Further, I argue that cybersecurity poses a challenge for governance themes of democratic interest. Considering lead organizations taking over cybersecurity, other stakeholders might restrain their willingness to participate if not treated equally. Overt militarization, for instance, might capture expertise around cybersecurity above civilian expertise. Vertical government structures might lead to fragmentation and constrain policy communities in smaller subnetworks from speaking to each other, for instance, among those that prosecute cybercrime, those in charge of cyber defense, and those dealing with espionage. Prosecutors, military personnel, and spies are each one of a kind and not always prone to overlay under interactive forms of governing.

Is, then, cybersecurity "governance" a misnomer? This is a question that is causing much perplexity in both policy and academic circles.[66] By identifying national levels of interaction between government agencies and organized interests, this book aims to explore further how cybersecurity is planned, sustained, and supported by the state.

One

Cybersecurity Governance

States around the globe share a great concern about the idea that a significant life-threatening cyberattack will hit them at any given moment. To prevent a Web 2.0 catastrophe, governments ratchet up national security agendas to coordinate public resources against online threats coming from hostile state actors and cybercriminal groups. Most of the time, the authorities work behind the scenes during and after a cyberattack, because of the covert nature of the military, intelligence, and law enforcement agencies in charge.

In October 2018, the National Cyber Security Centre (NCSC) in the UK carried out a review that showed that, on average, the country had received ten cyberattacks per week since 2016. Events were categorized under five types: incidents affecting individuals most likely to require a response from local police force; threats to small organizations; attacks on larger organizations or broader government entities; attacks on large populations, the economy, or the government; and, finally, the most severe attacks, causing sustained disruption of essential services, leading to severe economic and social consequences or to loss of life.[1] Although it is publicly suspected that at least twenty countries have sponsored cyber operations (including other highly militarized Western countries, such as the United States), the British government argued that the majority of the attacks hitting the UK were perpetrated by computer hackers directed, sponsored, or tolerated by Russia, North Korea, and China.

Nations have used cyber operations (i.e., distributed denial of service, espionage, defacement, data destruction, sabotage, and doxing)[2] to cause incidents, for example, power outages, as Russia is suspected to have done in Ukraine in 2014, 2016, and 2022.³ In response to cyberincidents, states now use sanctions to punish or censor alleged attackers and their government sponsors.

According to the U.S. National Security Strategy, for instance, the protection of critical infrastructure from "malicious cyber actors," is considered as necessary as the control of borders, or defense against missile attacks and jihadist terrorism. The United States, the strategy goes on to say, seeks to "preserve peace through strength," reinforcing its capabilities, including those in space and cyberspace, and is meant to "hold countries accountable for harboring these (cyber) criminals."⁴ The document adds a particularly self-centered idea about the nature of cyberspace: "The Internet is an American invention, and it should reflect our values as it continues to transform the future for all nations and all generations. A strong, defensible cyber infrastructure fosters economic growth, protects our liberties, and advances our national security."⁵

U.S. General Joseph Dunford, chairman of the Joint Chiefs of Staff, told a Senate Armed Services Committee hearing that in the next decade Russia, China, and North Korea posed the greatest threats to the United States (in that order of severity). Dunford, who was by then the highest-ranked American military official, believed that Moscow presented the top military threat based on its nuclear, cyber, and electronic warfare capabilities, some of them tested in the military interventions in Crimea and Ukraine.⁶ Paradoxically, a few months earlier, during a meeting at the G20 with his Russian counterpart Vladimir Putin, President Donald Trump had suggested that it was time for possibly working together on a cybersecurity initiative to combat election hacking. The suggestion had lasted less than twelve hours following harsh criticism from the Republican Party and the Washington security establishment, which made Trump backtrack on the idea.⁷

Cyberattacks can have a severe impact on a large population, the economy, and the government. As things stand, both developed and developing nations sustain little doubt that their defenses will be tested to the full in the years to come. Critical national infrastructure (CNI), such as telecommunication systems, the civil nuclear sector, air navigation services, energy, the oil and gas industries, and government domains, is increasingly digital, which raises severe issues of cybersecurity. Most of these networks are interconnected and spread out to hundreds of organizations across the territory. With the inclusion of ultrafast broadband across public and private digital services, maintaining security and resilience is said to be at the heart of national cybersecurity. Cyber-resilience, or the capacity to "withstand, recover from and adapt to the external shocks caused by cyber risks," is nowadays determinant to complex organizations operating in computerized and interconnected fashion.[8]

What is puzzling about current cybersecurity operations is the overarching authority vested in central governments over the domain of the Internet and network information systems. In the United States, Canada, Europe, Australia, and in emerging democracies too, such as those in Latin America and Southeast Asia, cybersecurity has become a perceived top-tier component of economic and national security affairs.

Since Sir Tim Berners Lee created the World Wide Web, governments have moved at a different speed to secure their telecommunications and digital networks. In 1988, the Morris Worm, a computerized virus launched by a graduate student from the computer systems of the Massachusetts Institute of Technology, forced the United States to create an emergency team after it caused widespread infection of public and private networks. Nations began to spend hundreds of millions of dollars in delivering cybersecurity programs and later national strategies that led to the creation of cybercrime and cybersecurity units during the early 2000s. National computer emergency response teams, or CERTs, mushroomed across the globe to unite separate parts of central governments that had a role in protecting cyberspace. Today we find functioning CERTs from Austria to Vietnam.

In this chapter, I bring forward theoretical approaches to explore how the ongoing wave of cyber-incidents among the advanced democracies has echoed across the globe. I argue that the continuing cyber engagement between major Eastern and Western powers have exposed the high-tech threats looming over the world and the vulnerability of less-protected information systems. The consequences of a cyber hit in the developing world might be at least disastrous. Due to various factors, the lack of precious human and physical resources in nonindustrialized countries results in the exposure of considerable portions of their national digital interests. Latin American states, worried at their lack of preparedness, have tasked the states' security institutions to respond to the threats of cyberspace.

I begin discussing what the "cyber" has brought forward for Latin American countries (and theorize further in Chapter 2). I identify those countries in the American hemisphere that are not yet advanced industrialized democracies like Canada and the United States. However, their strategic conditions, more than those of other developing states, have allowed them to enhance their diplomatic and security priorities in a regional system less dominated by U.S. hegemony. These states include Brazil, Mexico, Colombia, Argentina, Chile, and Venezuela. I also group them by their having set up domestic policy communities partnering with the multiple bodies that deal daily with cybersecurity. These new security networks combine the armed forces, homeland security agencies, diplomatic corps, intelligence community, prosecutorial bodies, private-sector companies, NGOs, and academic institutions. The experiences of other states that are not superpowers, for instance among NATO members (i.e., Denmark, Belgium, the Czech Republic, The Netherlands, and Norway), is that networked cybersecurity governance is a hard case, posing, therefore, a strategic challenge for policymakers. I make the case that cybersecurity is networked, just as war, diplomacy, business, media, society, and even religion are believed to constitute valuable networked forms of connectedness.[9]

In 2017, for example, the Chilean executive joined the global cyberthreat-mitigation effort, rolling out its first-ever national cybersecurity policy. The gradually installed strategy, meant to be completed and in place by 2022, aims to mitigate risk at the micro level (people's privacy and property) and the macro level (the states' information systems), filling a void in security mechanisms in sensitive areas such as defense, intelligence, criminal justice, emergency systems, financial trade, and foreign affairs. Chilean authorities have set up a security policy community comprised of various institutions dealing daily with the risks found in cyberspace. It is partly because of this sophisticated networked governance approach that it is difficult to predict the extent to which cybersecurity outputs will be manageable and sustainable. The integration of cybersecurity above, within, and beneath state level, requires an enormous amount of foundational work.

Emerging democracies usually struggle when it comes to governing via interinstitutional policies. Cybersecurity requires the government to operate in unfamiliar ways that include sharing power with experts in IT commercial ventures and with other nonprofit organizations. As Zoe Baird, president of the Markle Foundation, put it: "To achieve an Internet that reflects a commitment to the public good as well as to commercial interests, we have to create more pluralistic models for Internet governance, models in which government, industry, and nonprofit organizations craft policy—balancing each other and working together in transparent processes that earn public's trust."[10]

In the Americas, for example, questions regarding political meddling, institutional underdevelopment, interagency rivalry, overmilitarization, and issues of human rights neglect shed light on the obstacles that surface when steering security governance among a plethora of capable bodies. Given such a scenario, the prospects for managing Internet and cybersecurity governance in the long term are, to say at least, hard to predict.

The "Cyber" Prefix

For some observers, the cyber (short for cybernetics) prefix is a useful shortcut for describing computer or digital interactions.[11] Cyber is therefore the essential way in which we socialize. Digital technology in modern life is holistic and full of overwhelmingly positive outcomes, although it also brings a dark side. The evolution of what we can call cyberincidents depicts the highly organized, politically or financially motivated, technologically sophisticated, and transnational presence of potentially criminal activity.

Threats to cybersecurity, whether to individuals or organizations, might be cyber dependent and cyber enabled. Regarding the former, the literature cites our universal dependency on computers, computers networks, or other forms of ICTs, which marks the technology itself as the target of the criminal activity (i.e., hacking, malware, and DDoS attacks).[12] *Cybercrime* refers to more traditional crimes that are increased in their scale or reach by using computers, computer networks, or other ICTs, including child pornography, stalking, copyright infringement, and fraud. "Computer-supported crimes," as a third category, are crimes in which "the use of computers is an incidental aspect of the commission of the crime but may afford evidence of the crime."[13]

As evidence of the many abilities attributed to ICTs to facilitate activities against the rule of law, countries have fallen into a problem characterized by the lack of consensus on the meaning of *cyber*, including when the term is linked to incidents, attacks, crime, warfare, and terrorism; the lack of distinction between offenses that are perpetrated by means of technology; the poor statistical reporting; the lack of human and resource capabilities among security, defense, and intelligence enforcement agencies; methodological problems in reporting cyberincidents; and, finally, media sensationalism and the social distortion of computer and ICT security.[14]

To solve many of the issues presented above, the advanced democracies have undergone significant reform in order to adjust their national jurisdictions, develop new areas of the law, create governance structures to assess and reduce risks, build regional and global collaboration and information-sharing mechanisms, systematically formulate and execute plans to mitigate and prepare long-term national cybersecurity capacity. On the other hand, in the developing world cybersecurity capacity building is catching up to the global community struggling to find human resources, funding, and technological capabilities to build security resilience. Political factors, nonetheless, play another role in driving progress forward.

Today the nonindustrialized world bears a legacy of peace, conflict, and rule of law problems that blur nations' capacity to politically influence elites to agree on many responses to risk, including the most pressing traditional (i.e., armed conflict) and nontraditional (i.e., transnational organized crime, global warming, and cyberwarfare) sources of insecurity. The wave of cybersecurity preparedness comes to the developing world at a time when superpowers are reconfiguring the international system, with, more clearly, the United States balancing forces against Russia and China. With the inclusion of some Western European nations, these superplayers are the significant forces in international security, translating their physical arsenals into modern digital weapons of disruption. In exploring such a global scenario, we can theorize on the state of the art of Latin America's cybersecurity regime.

Cybersecurity policymaking in the advanced and developing nations demands an overall response from agencies inside and outside the state's remit, but most notably including the criminal justice, military, and intelligence bodies. Latin American nations have adopted governing arrangements to counter cyberthreats in the last decade in part because of CERTs spreading quickly throughout the Western Hemisphere, Africa, and Asia, through peer-to-peer knowledge sharing.

The private ICTs and the Internet platform industry have also encouraged countries to adopt new forms of cyber protection, replacing the global discourse on the utopian notion of the Internet as liberalizing the flow of information. Figure 1.1 shows the total of ICT goods exports and imports in world terms and five regions. From the figures, we can see that Latin America is behind the three main industrial centers of North America, Europe, and East Asia, and below the world's average.

ICTs are seen by institutions such as the World Bank as a viable way to "offer vast opportunities for progress in all aspects of life, from economic growth, improved health, and better service delivery to learning through distance education and social and cultural advances."[15] However, one might also argue that nations and corporations are by now extremely conscious of the Internet's exponential power in promoting both freedom and censorship.

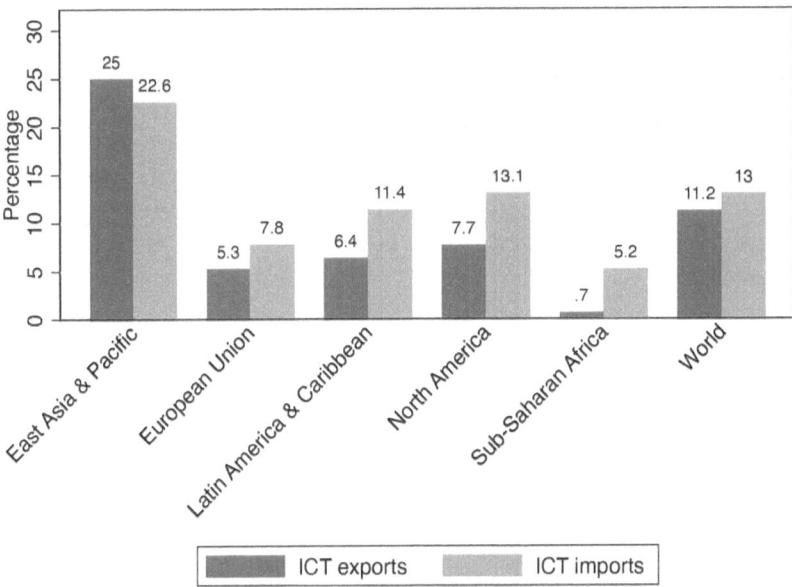

Figure 1.1. ICT good exports (% total good exports) and ICT good imports in 2017 (% total goods imports), in selected regions

Google, the American multinational, is said to have worked with the Chinese authorities to create an overreaching version of its search engine in mainland China. The authorities in Beijing are known for prying into the lives of their citizens, especially those with views on issues of freedom of expression and thought, minorities, and human rights. As Suzanne Nossel, CEO of Pen America, put it, by submitting to China's intrusive and coercive policies, Google "will deal a huge victory to Beijing and its campaign to entrench cyber sovereignty in the global order."[16] In Egypt, a recent cybersecurity law allowed the authorities to legally block any website considered "a threat to national security" or the "national economy."[17] After the disappearance and murder of reporter Jamal Khashoggi, Saudi Arabia's modern and resourceful authoritarian state decided not to silence the entire Internet, but rather to overflood it with patriotic messages, "designed to drown out critical or credible information" against the interest of the government, a technique also used on a grand scale by Russia.[18] During the civil war in Syria, many believed that the pro-regime Syrian Electronic Army (SEA) had received technological assistance from both Iran and the Lebanese Shiite militia Hezbollah in conducting a crackdown on Internet-based opposition and conducting other cyber operations.[19] The military junta established in 2014 in Thailand tightened its control over the 2007 Computer Crimes Act, which has been used to lock up journalists, rights activists, and other government critics, further encroaching on Internet freedom and privacy.[20] Reporters in Cambodia sounded the alarms after a massive cyber-espionage operation was uncovered preceding the country's national elections in 2018. Cybersecurity specialists believed the attacks aimed at the Cambodian National Election Commission, Interior Ministry, diplomats, and opposition lawmakers were launched by TEMP.Periscope, a well-known hacking group, working on behalf of the Chinese government.[21]

In the Western Hemisphere, recent attempts to meddle with the regulation of the Internet in the United States, the so-called net neutrality debate,[22] and other attempts across Latin America in general, give proof of powerful forces behind elaborate laws aiming to regulate, filter, and remove content to create a more "secure" Internet. As scholar Edwin Grohe argues, the use of cyber operations (from regulation to pro-government hacking groups), as well as the financial and economic interaction among other elements of power used by governments, represent different methods by which nations and nonstate actors might attempt to achieve political ends.[23]

A United Nations report on freedom of expression, for example, notes that Latin American and Caribbean regulators sometimes serve and enforce political agendas rather than the public interest by failing to renew media licenses, using government ministries for regulation actions, and lacking transparency and independence on matters related to regulating Internet companies such as connectivity providers, whether domestic or foreign based. Regionally, Latin America is the second-deadliest area in the world for journalists practicing their trade. There, 125 reporters and members of their crews were assassinated between 2012 and 2016, a total exceeded only by Northern Africa, which tops the list with 191 murdered professionals. According to the data, there are growing perceived threats to digital safety, which includes cyberattacks, surveillance, hacking, and online harassment, especially of women journalists. During the same time frame, in Mexico, 37 reporters were killed, followed in the region by Brazil (29), Honduras (19), Colombia (12), Paraguay (6), and Peru (4). The most likely sources of violence involved in these killings were political groups and military officials.[24]

War and Cyberwar

In the last century, democracy has taken root worldwide, with autocratic regimes becoming the least common form of government. The world's total number of nuclear weapons has also been in decline since 1986, and the quantity of military personnel has dropped since 1968 to a level of around 58 percent. The threat of all-out war has diminished since World War II. Interstate conflict gave way to civil wars, terrorism, and rising violence in the Middle East, North Africa, and Latin America. In 1958, the number of countries engaging in internal armed conflict passed the number involved in external armed conflict. Among the consequences of the shift from external to internal armed conflict are (1) more than one-third of armed conflicts worldwide are civil wars with international powers involved, and (2) nearly 1 percent of the global population are displaced people, comprising a record high in modern history.[25] According to a report by the Institute for Economics & Peace:

> The re-emergence of extra-state war in the early 2000s has been primarily driven by an increase in conflicts in the Middle East and North Africa, with a smaller increase in sub-Saharan Africa, and Central and Southern America. The most striking trend of the past decade has been the rise in internationalized internal conflicts, which made up 36 percent of total conflicts in 2016, compared to just 3 percent in 1991.[26]

Armed conflicts involving state actors can be classified in four main ways: extrasystemic or extrastate, interstate, internal or intrastate, and internationalized internal armed conflict.[27] Africa, the Middle East, and Latin America record higher levels of extrastate and internal conflict, which have persisted into the present day in the forms of transnational crime, terrorism, insurgencies, and organized criminality. Younger political institutions in the global South are more likely to experience internal conflict, social unrest, and corruption of state forces after dealing with these issues. In Nicaragua, most recently, the government has been accused of human rights abuses after a wave of protests and civil unrest started in early 2018 against the government's social security taxes and cuts in pensions. More than three hundred people have died in the disturbances, the majority at the hands of the government, it is believed.[28] Nicaragua experienced a civil war from 1970 until 1990, which provided hundreds of thousands of people with military experience. Another escalation of violence might bring the country to the brink of chaos and civil war, destabilizing all of Central America.[29]

Placing the security and privacy of the Web 2.0 in the hands of the state is at least worrying, considering current conditions of insecurity and underdevelopment in Latin America. The region is the world's most violent (when measured in murders), despite not having any active interstate war or major civil conflicts such as those currently being fought in the Middle East, North Africa, and South Asia. Violence in Brazil, Mexico, El Salvador, Honduras, Jamaica, St. Kitts, and Venezuela, plus a rising trend in numbers of disappeared, speak of states unable to enforce the rule of law.[30] Since the disastrous chapter of the war on drugs launched in 2005 under President Vicente Fox and later reinforced by Felipe Calderón, the sad reality is that Mexico joined the twenty nations with the highest homicide rates in the globe. Colombia, on the other hand, has shown a downward trend in homicides and growing overall stability regarding violence and crime, and has succeeded in adopting an initially unpopular peace accord with the guerrilla group Fuerzas Armadas Revolucionarias de Colombia (FARC). Colombia still has to prove successful in avoiding a large-scale return to arms and peacefully dismantling the remnants of other paramilitary groups and other criminal organizations, known as BACRIM. Venezuela is a peculiar case, as it has stopped submitting homicide data to international bodies. However, day-by-day reports give proof of uncontrolled violence, unofficial killings, and state killings. In Brazil, on the other hand, rural and urban violence has led to a spike in lethal force, with state institutions struggling to keep a hold on the wave of overall insecurity as Brazil has topped the list of the fifteen most murderous countries around the globe.[31]

The scholarship in conflict and warfare studies argues that the dominant scenario for military operations in this century is the urban environment. In the urban areas, war, criminality, terrorism, humanitarian interventions, and environmental degradation meet. The intersection between armed actors and military police has dominated urban life in many cities across the globe, most notably in Latin America megalopoli.[32] Stephen Graham, for example, has argued critically against the increased militarization of urban space. In his view, the securitization of the urban element has allowed governments to control what happens in metropolitan areas.[33] Also, within conflict-torn cities, there is growing evidence of the sophisticated use of technological weapons, intelligence, and a combination of armed forces doing police work and police forces overmilitarizing their practices in enforcing civilian law. Urban countercriminality and counterinsurgency, especially in the slums in the global South, have diminished democratic practices that enforce the rule of law. Scholars would also argue that political violence in urban life operates through surveillance, technology, and foreign wars. Drones have promoted high-tech breakthroughs,[34] as flying machines today can be commanded from faraway locations (e.g., pilots of U.S. unmanned aircraft in the skies of Afghanistan were based at the drone command site in Nevada). A new way of waging warfare is nowadays deeply entrenched along with the intelligence environment, civilian-military authorities, and corporate manufacturers in the military-industrial complex who can provide top-of-the-line technological weapons. On the other hand, modern wars also include the presence of pseudo-militaries, mercenaries, paramilitary forces, gangs, and other irregular armed factions.[35]

The list of conflicts worldwide worsening under the patterns mentioned above includes the wave of criminal violence in Mexico, civil war in South Sudan, violence in the Central African Republic, Al-Shabab in Somalia, humanitarian crises in Yemen and Afghanistan, war against the Islamic State in Iraq, the Rohingya crisis in Myanmar, conflict between Turkey and armed Kurdish groups, and war between Russia and Ukraine.

Despite the sizeable military, intelligence, and humanitarian responses to these conflicts, we hear hardly any contemporary stories of cybersecurity being applied in Africa, the Middle East, or Southeast Asia, other than drone strikes employed as a proxy to the use of highly capable technologies.

Although Latin America has been war-free since Colombia's five decades of civil conflict came to an end, threats to the rule of law are moving swiftly to accommodate other dimensions of nontraditional security, such as cybersecurity. The global fight conducted by ambitious states to dominate and exploit ICTs in pursuit of national interests not only directly affects the Western Hemisphere but suggests that nations all around the world enjoy a fragile state of peace in which new threats can rapidly collapse in all-out war triggered by covert information warfare against the newly established democratic societies.[36]

Human and traditional security approaches demand an infinite number of alternative solution paths that link science, technology, and warfare. As William Astore put it, "[S]cience may be defined as organized knowledge; technology, as applied knowledge; and warfare, as organized violence"; however, currently, the newest manifestations of warfare are difficult to predict with any degree of certainty.[37]

Cybersecurity stresses science, technology, and warfare and thereby reveals new linkages between social structures, logistics, and material considerations. Through an iterative relation, technology influences the military, and the military influences technology. In this vein, in the nineteenth and twentieth centuries, the scientific movements in the industrialized nations fostered technological diffusion which impacted heavily in warfare, revolutionizing military affairs.[38] The early computer age in World War II and the vast growth during the Cold War of the electronics industry profoundly influenced warfare, leading to the development and deployment by both East and West of air defense and precision warfare technologies. Robotics and ICTs in the twenty-first century have further revamped military-related industry and recast the future of warfare and cyber conflict.[39]

Valeriano and Mannes, nevertheless, have aptly identified the misleading interpretations used by the media, policymakers, and some academics of the term *cyber conflict*, thus unveiling the lack of consensus on the proper lexicon to apply. They opt for discarding "the popular terms cyberwar and cyber-attack and replace them with the terms cyber conflict, cyber incidents, and cyber disputes, suggesting that this more grounded terminology should become the standard for research on cyber conflict in the international system."[40] By *warfare*, they mean the "destructive operations that seek to maim, kill, or wound physical individuals and/or to damage or destroy property." War implies a military conflict in which one side is a state, which results in at least one thousand battle deaths of military personal. "This is a key point to remember, as some have noted that there is really no such thing as a cyberwar if there are no battle deaths," they argue.[41] Strategic cyber operations, instead, seem focused on campaigns that use cyber weapons to achieve strategic outcomes without resorting to armed attacks. These new varieties of war require different explanations and policy prescriptions that no doubt necessarily fall within the traditional model that consists of a militarized crisis leading to kinetic war.[42]

Data Weaponized

In the 1996 Hollywood movie *Independence Day*, the United States leads an assault to repel an alien invasion, using the best of its post–Cold War arsenal. Fighting off the extraterrestrials with strategic bombers, tactical fighters, and even nuclear weapons, ultimately the planet is saved thanks to an ex-scientist who develops a computer virus that is loaded into the alien's mothership.[43] Fact or fiction, "weaponizing" ICTs by infecting machines in order to cause havoc is today a frequently used technique across the globe.

In a cyber operation branded "Machete," a single Java file from an unknown source was able to target intelligence services, military institutions, utility providers (telecommunications and power), embassies, and government facilities in Spanish-speaking countries such as Ecuador, Venezuela, Peru, Argentina, and Colombia, as well as, reportedly, Korea, the United States, the Dominican Republic, Cuba, Bolivia, Guatemala, Nicaragua, Mexico, England, Canada, Germany, Russia, and Ukraine.[44] The malware was available in the public domain and had undergone minimal changes to avoid detection at least twice during its attacks in 2012 and 2014. It is believed to be still actively storing stolen information in hidden servers. "Machete" is capable of cyberespionage operations that include logging keystrokes, capturing audio from a computer's microphone and screenshots, and geolocation data and photos from the computer's Web camera, copying files to a remote server and a unique USB device, when inserted, hijacking the clipboard, and capturing information from the target machine.[45]

To stop "data being weaponized," corporate firms have called for strengthening privacy laws. Tim Cook, the CEO of Apple, has endorsed privacy laws and regulators tightening their scrutiny in the United States and the European Union. Cook argues that everyday personal information is used by companies to enrich themselves by building a massive commercial complex around user data. Apple sells mostly hardware, unlike Facebook or Google, which sell ads tailored by individuals' personal information and patterns of Internet use.[46] These last two are especially susceptible. In September 2018, Facebook detected a massive breach of the social network that compromised between at least thirty to fifty million users' access tokens, thereby allowing the attackers to get into people's accounts.[47] Facebook was involved in the Cambridge Analytica scandal, when the data consulting firm that worked with Donald Trump's election team harvested millions of Facebook profiles of U.S. voters without authorization in order to target them with personalized political advertisements.[48] Facebook reportedly has 2.23 billion monthly active users worldwide.

In October 2018, the *New York Times* reported that President Trump used an unsecured personal cellphone to take calls whenever he did not want to have a call logged into the White House call records. Although, the article added, his aides repeatedly warned him that his phone was vulnerable to infiltration by Russian and Chinese spies, it seemed that Trump kept using his iPhone. The U.S. intelligence community corroborated information with other foreign governments that groups in China and Russia were intercepting communications between foreign officials, including the president's calls.[49]

Days before the *Times*'s revelations, the U.S. Government Accountability Office (GAO) reported that many of the weapons systems controlled by the Pentagon were vulnerable to hacking from the exterior. A group of investigators put to the test the Defense Department's cybersecurity measures adopted over the previous five years on their cutting-edge weapons systems. The report found that "using relatively simple tools and techniques, testers were able to take control of systems and largely operate undetected, due in part to basic issues such as poor password management and unencrypted communications. Also, vulnerabilities that DOD is aware of likely represent a fraction of total vulnerabilities due to testing limitations."[50]

Observers have identified the dangerous effects digital manipulation might have in the event of war. Elias Groll argues that "an attacker with such access to U.S. military systems could conceivably alter what U.S. sailors aboard a warship see on their radar screens. They could manipulate computer systems to cause navigational errors or make enemy ships disappear from sensors." It is believed that only "sophisticated adversaries" have the ability to hack the U.S. systems so thoroughly.[51]

Performing cybersecurity, like performing diplomacy, forces all interested parties to face emerging complexities, which can push into marginalization and irrelevance those who lack the capacity to adapt. Just like diplomats, government officials require nowadays new types of solving mechanisms to meet the challenges the world faces, including global warming, epidemics beyond borders, and cybersecurity.[52] Cybersecurity is deemed of primary importance for government officials, as communication technologies require modifications not only in state apparatus but also in the operational practices of defense and foreign ministries. Military diplomacy, for instance, in both war and peacetime, entails cooperation between armed forces and their civilian counterparts that reflect not only civil-military relations but also the executive functioning of a given state. In the Western Hemisphere, military collaboration and engagement in conflict situations require responsible conduct in exchanging those communications related to the planning of military activities.

Since the end of the Cold War, defense diplomacy has gained momentum as countries have built bridges by employing military cooperation as a means to prevent a resurgence of conflicts between former enemies.[53] North-South military cooperation, for instance, between the United States and other nations in Latin America, goes beyond the traditional military establishment and usually invokes broader social, political, and economic consequences that extend the purpose of defense diplomacy.[54] Diplomats, military personnel, intelligence officers, and members of other security agencies rely on forms and degrees of interaction and communication that have been profoundly affected by changes in technologies. Policy groups within countries and even the governments of the countries themselves do not adapt to these changes at the same pace, or use the same tech-oriented strategies. The impact of the information revolution has nevertheless pushed most states to go "virtual" in reaction to changing political events and adapt rapidly to twenty-first-century reality.[55]

Internet Penetration

Internet usage has penetrated 55 percent of the world's population (in other words, there are 4,208,571,287 Internet users). Asia leads with 49 percent of the total number of Internet users, followed by Europe (16 percent), Africa (11 percent), Latin America (10.4 percent), North America (8.2 percent), the Middle East (3.9 percent), and Oceania/Australia (0,7 percent). In the Western Hemisphere, Internet penetration rates based on population are 95 percent in North America, and 67.2 percent in Latin America and the Caribbean. Europe shows 85.2 percent, Australia/Oceania had 68.9 percent, and the Middle East up to 64.5 percent. Only Asia and Africa are below the world average, with 49 percent and 36.1 percent, respectively.[56] Internet traffic flows without interruption around the world, as reflected in an evaluation measured between zero and 100, where higher values indicate faster and more reliable connections, with North America (97), Asia (95), and Europe (92) on top, followed by Australia (86), and South America (84). No data are available for Africa because not enough routers are available to test the significant paths of the Internet connecting the continent.[57]

It is commonly accepted in advanced democracies that virtually all adults are Internet users. In the UK, for example, 99 percent of adults aged sixteen to thirty-four years are Internet users, compared with 44 percent of adults aged seventy-five years and over. The proportion of men who have recently used the Internet is higher than women, by 91 to 89 percent. In 2018, at least 8.4 percent of adults in the UK had never used the Internet, representing some 4.5 million adults, of whom more than one-half (2.6 million) were aged seventy-five years and over.[58] In the case of Latin American states, Mexico, records 71.3 million Internet users representing 63.9 percent of the population aged six and over. Women account for 50.8 percent, while men represent 49.2 percent. The age group with the highest proportion of users is sixteen to thirty-four years (85 percent), while the age group with the least use consists of women over fifty-five years old. Mexicans use the Internet mostly to seek information (96.9 percent), and entertainment (91.4 percent), to communicate (90.0 percent), to access audio-visual content (78.1 percent), and to use social networks (76.6 percent). Either via land or mobile connection, 17.4 million Mexican homes have Internet access (50.9 percent of the national aggregate), making Internet usage an urban phenomenon with 86 percent of the users located in cities.[59]

Under such massive permeation by the Web, understanding developing states' adoption of cybersecurity capabilities proves equally essential for both policymakers and civil society.

Proliferation of Cyber Military Units

At the Council on Foreign Relations, a think tank based in New York, researcher Anthony Craig developed a dataset on national cyber capabilities for countries observed since 2000, in which he assessed the causes and consequences of cyber capability build-ups. Craig defined cyber capabilities as the resources and assets available to a state that it can draw on or use to resist or project influence through cyberspace. His dataset explored these capabilities from two perspectives: latent and active. The former was understood to indicate the "broad societal-based resources such as computer science knowledge and IT industry that governments can draw on to achieve their strategic interests in cyberspace." Active cyber capabilities, on the other hand, referred to the "operational capabilities that governments directly control and can deploy in computer network operations." Through his research, Craig noted,

> One example of an observable development in cyber capability is the creation of military computer network operations (CNO) units, which I define as government entities located within a state's military structure that are tasked to engage in operations involving computer networks. An obvious example would be U.S. Cyber Command created in 2010. Using this variable, the first thing that can be confirmed by the data is the rapid acquisition of cyber capabilities.[60]

The Council's cyber operation tracker lists publicly known state-sponsored incidents that have occurred since 2005. Data show an increase in military CNO units from five in 2000, to sixty-three in 2017 among ninety-five countries listed as victims of cyberattacks. "When compared to previous military innovations such as nuclear weapons, battleships, or aircraft carriers, the rate at which military cyber capabilities have been adopted is notably high," Craig mentions.

States that have experienced cyberincidents should be more likely to develop military cyber capabilities, since defense build-ups are a response to external security threats. Exploring the relationship between whether a country created a military CNO unit and whether that country had experienced a cyberincident against it in any of the three years previous to that creation, Craig notes:

> A cyber incident provides clear evidence to the state of its vulnerability, thus motivating it to develop the operational capacity to defend against and respond to future aggression. Nevertheless, there are more factors involved in this process that are yet to be uncovered, given that most CNO units are created in the absence of cyberattacks.[61]

Potential explanations for the increase of CNO units include the escalation of a cyber arms race. Although countries, for example in the global South, have tended to create military cyber units, this does not equate to substantial technological capacities in terms of how their cyber domains operate. For research and policy purposes, focusing on state-sponsored incidents, one the one hand, helps "identify when states and their proxies conduct cyber operations in pursuit of their foreign policy interests."[62] Also, state-sponsored incidents are believed to facilitate reporting when compared to those carried out by nonstate actors such as hackers. Events caused by unknown cyber groups, on the other hand, are equally telling as they reveal sophisticated attacks against organizations and governments across the world.

The hacking group called Sowbug, for example, has focused on illegally accessing and stealing information on foreign policy and diplomacy from South American and Asian governments (e.g., Argentina, Brazil, Peru, Ecuador, Brunei, Malaysia) since at least early 2015. According to some observers, this wrongdoing "is an exceptional event because this type of espionage is relatively rare in those regions, compared with North America, Europe, and other areas of Asia."[63] Although who is behind Sowbug is unknown, some have pointed out that the targets represent greater value to nation-states than to criminals. In 2018, different cyber operations incidents involved Latin American countries. For example, a threat actor believed to have been sponsored by China targeted multiple organizations with trade ties to that Asian country, including some connected with China's Belt and Road Initiative such as United States, Kenya, Mongolia, and Brazil. In Operation Parliament, an unknown threat actor used "spear-phishing" techniques to target parliaments, government ministries, and academic and media organizations for espionage, primarily in the Middle East but also in Chile. Finally, Operation GhostSecret, believed to be the work of the Lazarus Group and sponsored by North Korea, targeted critical infrastructure and organizations in the entertainment, finance, healthcare, education, and telecommunications sectors, with victims reported in the United States, United Kingdom, India, Thailand, Guatemala, Germany, France, Canada, Brazil, China, Bangladesh, Thailand, Hong Kong, Japan, and Australia.[64]

In Chapter 2, I argue that cyberincidents are carried out not only by state actors but many times through proxies using technologies that only some states have been able to pull off. In the case of Sowbug, the attackers stole information only after office hours to avoid detection, and they moved through the target network unviolated. Attackers were also able to stay in some systems for more than six months by hiding their tracks, blocking antivirus software, installing backdoors into the hosts, and impersonating other software commands. After some of the infected networks were spotted, Sowbug shut down and was believed to have moved to other critical infrastructures within the same systems.

Cyberespionage among nations is just as common nowadays as cyber-retaliation. In the UK, for instance, the authorities reported that after Western military action was deployed in response to a chemical strike by the Syrian regime of Bashar Al-Assad, Russia ramped up a cyber response flooding the Internet with fake news. Only hours following Great Britain's missile strike on Syrian targets, the NCSC elevated its defenses to the "highest possible level," according to Whitehall sources.[65] More often than not, Western governments blame Russia and China for supporting proxy groups that spy on sensitive political and military targets, including NATO, the EU, and government ministries. In both Beijing and Moscow the authorities have systematically denied any support for hackers.

Security analysts often trace the digital DNA of hacking groups to track back their incidents. Hackers from the group called ATP 28, for instance, with a suspected connection to Russia, have pursued long-term campaigns of cyberespionage against the Polish, Hungarian, and Georgian governments, NATO, the European Commission, the Organization for Security and Co-operation in Europe (OSCE), and dozens of defense attachés, including individuals working in the United States, UK, Canada, and Norway. There are many ways hackers can lure an official's attention to allow infected files to access private networks. Most commonly, this can happen through phishing attacks hiding malware inside fake emails.

Before the February 2022 Russian invasion of Ukraine, Moscow was blamed for striking both NATO and its allies in tandem with severe cyberattacks on Kiev. Moscow attacked Ukraine with two more first-of-their-kind cyberattacks, similar to the time when Snake malware hit the Ukrainian government's computer systems in March 2018. Kremlin-linked hackers were blamed earlier for attacks on the Ukrainian energy grid in 2015 and 2016, which caused temporary blackouts. In 2014, Energetic Bear malware was suspected of having penetrated various European energy companies, also with cooperation from the Russian state.[66] In the European Union, the continental authorities have responded to these incidents by setting up protocols of resilience and adaptability under novel cybersecurity policymaking processes.[67] For example, the EU and NATO have reconvened expertise within their organizations to include strategies countering so-called hybrid threats. In 2016, they concluded the Technical Arrangement on Cyber Defence. This program facilitates information exchange between emergency response teams strengthening cybersecurity cooperation. They listed seven areas for increased cooperation, including that of cybersecurity and cyber defense, such as countering hybrid threats, operational cooperation in the Mediterranean Sea area, defense capability development, defense research, defense and security capacity building in third countries, and the organization of joint exercises of various kinds.[68] At the OSCE forum, European states have applied confidence and security-building measures (CSBMs) since the Ukraine conflict began in 2014, although with mixed results.[69]

A cyberattack on the U.S. power grid in 2016 in which hackers sought to penetrate multiple critical infrastructure sectors, including energy, nuclear, commercial facilities, water, aviation, and manufacturing, led the Department of Homeland Security and the FBI to condemn the "multi-stage intrusion campaign by Russian government cyber actors." Weeks before, the U.S. Treasury Department imposed sanctions on nineteen Russian individuals and five groups, including Moscow's intelligence services, for meddling in the 2016 presidential election.[70]

In February 2018, Washington released its new Nuclear Posture Review (NPR), which expanded scenarios for the possible use of nuclear weapons against nonnuclear threats. These include cyberthreats, under the assumption that these capabilities would broaden the range of nuclear deterrence required against potential adversaries, including Russia.[71] The use of nuclear capabilities is worrying at least when it is positioned as a "hedge against the potential rapid growth or emergence of nuclear and non-nuclear strategic threats, including chemical, biological, cyber, and large-scale conventional aggression."[72] Going back as far as a decade ago, the United States has reportedly rebuffed hundreds of attempted attacks each day on computer systems that affect national security. In 2007, an unknown hacker managed to enter the Pentagon's email systems, leading officials to take off line up to 1,500 accounts from which information had been downloaded, including the secretary's office. The Pentagon believes that attacks on the U.S. defense come from a variety of sources, including recreational hackers, self-styled cyber-vigilantes, groups with nationalistic and ideological agendas, transnational actors, and nation-states.[73] The Pentagon cyber raid was later attributed to the Chinese PLA. It is believed that the U.S. and Beijing's military regularly probe each other's networks, raising the scale of disruption to serious concern more recently. China's foreign ministry has repeatedly denied these accusations and said the claims reflect a "Cold War mentality."[74] Finally, reports show that at least three other cases of hacking, affecting the Western Hemisphere, were allegedly sponsored by China (i.e., Nitro attacks in 2011, NetTraveler in 2013, APT10 in 2017). Setting cyber and security issues aside, I direct the theoretical focus next onto the affairs of governance.

Issues of Democratic Governing

Challenges to the liberal model of democracy and effective governance have uncovered a fair amount of tension in recent sociopolitical developments in Latin America. While most countries in the region have established functioning institutions, and a return to military authoritarianism seems unlikely, liberal and representative attributes usually associated with democracy always seem at peril. In many countries, informal mechanisms whose aim is to concentrate power in the hands of the executive have severed the traditional linkages between citizens and authority. Party systems have collapsed, and other types of cultural forms now represent civil society.

Conventional, representative liberal-democratic institutions are confronted by the rise of new social movements, popular protests, and cyber-activism, reshaping public space and the deliberative construction of democracy.[75]

Digital technologies are believed not only to enhance democracy and the mechanics of government through affecting the ways politicians and citizens accomplish political tasks. Some authors are keen to address the expansion of digital peer-to-peer communications technologies and the way these will democratize governments. Another camp argues that the digitization of social life is deemed to have a damaging effect on citizenship by promoting only partisan interests and undermining the democratizing impact of the World Wide Web.[76] While studying politicians, public sector reform, and the tools available to government in the information age, Christopher Hood argues that we are at the door "of a new information-industrial complex with large corporate interests at stake in the outsourcing and computerization of government's once-distinctive information-collecting, filing, and case-handling operations."[77]

Cybersecurity is a problem in the Western Hemisphere's young democratic regimes, mainly in places where pseudo-authoritarian politicians have continuous access to public office. In different parts of the continent, the voices of political opposition and civil society have been diminished to near-silence, and there is little accountability over the instruments used by the government to put together information on its people. As well, not every government today has a philanthropic vision that involves ICTs enhancing the way it interacts with citizens.[78] The quality of democracy, civil liberties, checks and balances, the rule of law, and popular participation can profoundly affect the establishment of information governance. The concentration of power in the executive and weak mechanisms of horizontal accountability between the branches of government, especially those in charge of defense, security, and intelligence affairs, hinder democratic performance.

In Latin America, agenda-setting in the areas of security and military affairs is usually deeply rooted in a tradeoff between formal and informal rules that affect democratic performance. Political scientists Gretchen Helmke and Steven Levitsky argue that everyday governing practices interact with formal institutions, accommodating, complementing, competing, or directly substituting the ways of ineffective, but formal, governance.[79]

After looking earlier in the chapter at the way cybersecurity unfolds in the more advanced democracies, I argue that its governance rests in many constitutional intricacies (such as updated legislation) that enhance state capacity to solve sociopolitical problems and institutional responsiveness to complex policy issues. Constructing such fundamental administrative and institutional capabilities for governance have been a challenge in Latin America.

Since the third wave of democratization, the mushrooming of what Guillermo O'Donnell called the brown areas of governability (geographies that escape the state's reach) has created ample room for questioning the provision of security.[80] For some of the states democratized by the third wave that are experiencing crises of security in the region (e.g., Mexico, Brazil, Argentina, and Venezuela), the main problem is that the police, the judiciary, the military, and the intelligence bodies seem conditioned to patronage, clientelism, corruption, and other inherited authoritarian traits that have outlived many attempts to run government in a more democratic way. The wave of cybersecurity preparedness, therefore, hits Latin America while antidemocratic pockets continue to erode the capacity of the state to perform adequately. Political authority over security governance seems fragile, and cybersecurity opens a new challenge.

I argue that capacity-building efforts and democratic responsiveness to "wicked" policy issues such as crime, security, and warfare can create offsetting balances of power among state institutions.[81] Eduardo Dargent, for example, argues that some state capacity-building efforts have created a class of technocrats that have become experts in running certain areas of government, which can lead to practices with little democratic accountability.[82] Security has undoubtedly created policy communities across the public sector that deal in the form of networks of governance.[83] For some scholars, network governance as a theory helps explain why and how private and public actors interact in the planning and execution of national and regional security policies.[84] The network security governance approach has provided critical ontological grounds to develop security governance theory building through the study of regional, national, and subnational interaction contexts and the systemic and subsystemic components of political, social, and economic aspects that affect security outcomes.[85] I deal with these issues further in chapter 3.

Security networks can be subdivided into policy communities broader in scope (for example, national security) and policy networks in more limited policy areas (for instance, drug trafficking, money laundering, and cybersecurity), which deal with more narrowly defined issues of the rule of law, such as prosecution, intelligence, national security, public security, and justice. Because policy issues such as cybersecurity are considered holistic to many areas of the state, these issues tend to fall into an in-between arena where authority is blurred, capacities are mixed, and remits overlap.

For example, the armed forces care about cybersecurity from a national security perspective, public prosecutors from a criminal lens, and the intelligence community from an espionage and information perspective. Of course, one should say that cybersecurity in all these areas has significantly impacted the understanding of the very meaning of each field. For example, current policies and practices of cyberintelligence have prompted many to ask more generally what intelligence consists of under such a scenario.[86] Intelligence agencies have used cybersecurity in their own favor, becoming "major international actors," mostly by expanding their techniques to cyberspace and generating, in many cases, what is "tolerable," even if it is not officially accepted by the international community. Cyberintelligence remains largely unregulated and hidden in the tangled substate apparatus.[87]

More widely, nonstate actors, such as those in the private sector (i.e., ranging from individuals to large corporations), think of cybersecurity from an asset-protection perspective. Approaching cybersecurity thus demands that we visualize and understand where our knowledge gap is situated regarding why and how cybersecurity is constructed among differentstakeholders. Although every country in Latin America reacts idiosyncratically to perceived security threats, some common patterns provide an excellent place to start testing some hypotheses, beginning in the next section, for further review in the ensuing chapters.

Governance and Internet Penetration

At this early stage in the book, I find it necessary to debunk some myths about cyber affairs and governance. In this final section, I set out to explore whether Internet penetration has any effect on governance, considering the theoretical discussion presented at the beginning of this chapter, which will continue as I make progress in the book. My curiosity stems from previous research establishing a positive (and statistically significant) relationship between Internet use and governance across different regime types.[88] I hypothesize that countries with higher Internet internet penetration rates will have higher governance indicators scores. I use Internet penetration as the independent variable and six different governance indicators as dependent variables over a sample of Latin American countries. My results show that

Internet penetration does not correlate positively with higher estimates of governance. When looked at separately, only some nations with midlevel rates of Internet penetration have statistically positive and significant relationships with the governance indicators selected. The results speak in favor of those arguing on behalf of better quality and use of the Internet to improving governance, instead of strategies based solely on growing the number of users.[89] The section contributes new insights to the bulk of the literature reviewed so far on digital technology, security policymaking, and democratic governance.[90]

It is helpful to start with a visual aid to get an idea of how the relation between Internet penetration and governance unfolds. We would assume this relationship is linear.

The scatterplot in Figure 1.2 shows two correlation lines between Internet penetration (understood as the percentage of the population with Internet access) and one dimension of governance (government effectiveness) as measured by the Worldwide Governance Indicators (WGI). The WGI dimensions are reported in scores from −2.5 to +2.5, with higher scores meaning better governance.[91] I want to begin by examining how the variables covary with each other. If we assume that the correlation between Internet penetration and governance is positive, more Internet users will be associated with a higher governance score. The way the metric variables relate to each other will show if there is any visible hint of a correlation.

While the evident increasing tendency in Internet penetration worldwide is a constant, the direction of the scores on governance once plotted is not so straightforward. For that, we will benefit from statistical tests for two purposes. First, we will be able to challenge the rest of the literature scientifically. Second, we will be able to generate new results.

Figure 1.2 shows data points and fits a linear model. The results diverge. For the case of Brazil, the Internet penetration and government effectiveness variables covary negatively. For the case of Colombia, the variables show a positive covariance. Later I will measure how Internet penetration and other governance indicators covary with each other. For now, I want to glance over the grand theories presented earlier in the chapter and the more empirically driven approaches that can test logically the hypothesis that the higher the value of Internet penetration, the higher the value of governance.

Governance means different things to different authors. To Robert I. Rotberg, governance is the performance of government in delivering political goods to citizens. Governance can thus be measured through a clear set of

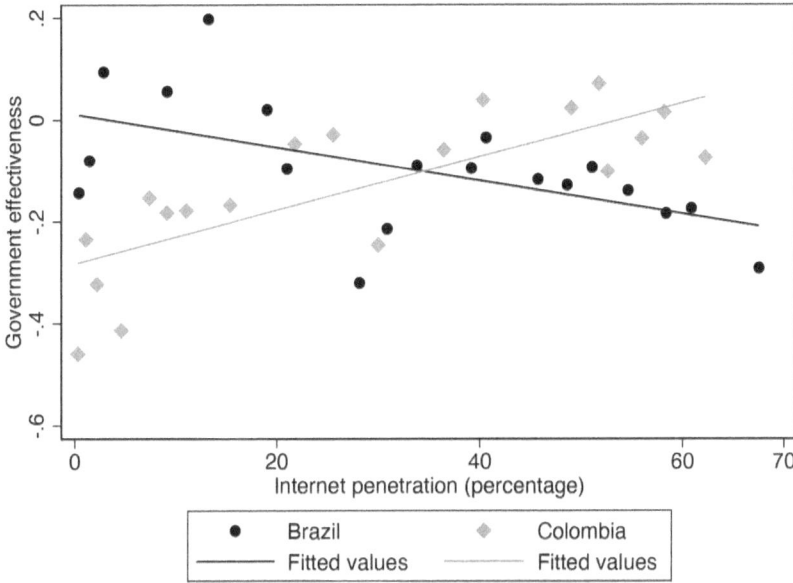

Figure 1.2. Scatterplot of Internet penetration and government effectiveness in Brazil and Colombia, 1996–2017

specific criteria, that is, an objective examination of performance results.[92] To Francis Fukuyama, governance is a bureaucratic capability.[93] He defines it as a "government's ability to make and enforce rules, and to deliver services, regardless of whether that government is democratic or not."[94] What is clear from many theories of governance is that governance can be examined by disaggregating different criteria related to state performance through subjective measures (what others think of such performance), or objective results (by measuring actual outputs).

In these two ways, the literature on state performance as governance circles on the idea that procedural measures of bureaucratic quality, capacity measures to deliver goods, and the outputs of such delivery, will define how states accumulate power through their governing institutions.[95] Measures of the quality of such bureaucratic features is, for some observers, better compiled by using impartial views of the state.[96] In theoretical respects, we want governance quality to reflect general, independent, and nonpartisan aspects of the rule of law. In practice, we want to be able to independently verify how given variables covariate with a particular measure of governance.

For the remainder of this section, I stick to the WGI definition of governance, which has an emphasis on the "traditions and institutions by which authority in a country is exercised."[97] Irrespective of whether countries are flawed or consolidated democracies, general and impartial governance measures should give us useful and comparable measures of state quality and thus bring us back to theoretical approaches.

The WGI disaggregates governance aspects of state capacity in six dimensions (government effectiveness, control of corruption, political stability, the rule of law, regulatory quality, and voice and accountability). The data are based on bureaucratic quality according to impartial expert surveys and aggregates of other existing measures based on more than thirty underlying data sources. Because many variables will contribute to any measure of governance, I say that Internet penetration may be "a" cause rather than "the" cause of better governance. Statisticians call all predictors covariates. This terminology avoids making causality claims that it would be hard to justify,[98] while it also implies that several other covariates of a dependent variable might be found in the political, economic, and social dimensions of life.[99]

For most developing democracies, such as those in Latin America, Sub-Saharan Africa, and East and Southeast Asia, good governance, in conjunction with well-developed ICTs, is essential to springboard into higher levels of development. For optimistic observers, developing ICTs and promoting greater Internet penetration favor transparency and accountability in governments while they also fuel social participation in civic life.[100] Presumably, we can expect higher Internet penetration to show some positive relation to the WGI's voice and accountability dimension (I define what each WGI governance category means shortly hereafter in the chapter).

On the other hand, less-optimistic observers argue that greater Internet access has misfired against the same people it was supposed to help. The openness that has allowed entrepreneurs, consumers, and political organizers to foster innovation through the Internet's global network has also enabled abuses on user data, propaganda, extremist views, targeted ads, false content, censored information, and suppressed dissidents. In the opinion of some people, the Internet is increasingly being "weaponized" on the global stage, as I explained earlier in the chapter. To others, this is an exaggeration.[101] Presumably, we can expect higher Internet penetration to have some relation to the WGI categories that oversee corruption, political stability, rule of law, and regulatory quality. Although in these cases, the relationships might be either positive or negative.

Finally, the Internet is essential for the delivery of many government services. A more developed e-government, together with greater Internet

Cybersecurity Governance 49

access by the citizenry, can enable better policy impact, making bureaucracies more accessible, usable, and affordable. Central executives worldwide have launched e-government initiatives to allow the online use of public services. The digitization of democracy supports policymaking and improved overall management of government processes involving documents and data.[102] Presumably, we can expect increased Internet penetration to have some relation to the WGI's measuring government effectiveness. In this case, the relationship might be positive.

The Dataset

The WGI database has proven a useful tool for cross-country comparison over a long period. Data on Internet penetration was taken from the International Telecommunications Union (ITU), aggregated and made available online at the World Bank's website. The WDI data measured Internet penetration as a percentage of the population. Internet users are defined as individuals who have used the Internet (from any location) in the last three months. The Internet can be used via a computer, mobile phone, personal digital assistant, games machine, digital TV, etcetera.

The countries selected for the comparative analysis were Argentina, Brazil, Chile, Colombia, Mexico, and Venezuela. The cases in the sample were selected based on two criteria. First, the country had an Internet penetration estimation between 1996 and 2017. Second, the country has information for all six governance indicators. The case selection strategy followed a diverse-case study methodology. Diversity refers to a range of variations in the variables (for this case, the (y) governance dimensions, and the (x) Internet penetration), which are useful for hypothesis testing.[103] The countries in the sample have extreme values of internet penetration, according to the most recent values, and have experimented with serious increases over the last ten years (see Table 1.1). A detailed methodology is indicated in the chapter's Appendix file.

Empirical Findings

According to the Worldwide Governance Indicators, the government effectiveness category "captures perceptions of the quality of public services, the quality of the civil service and the degree of its independence from political pressures, the quality of policy formulation and implementation,

Table 1.1. Internet penetration (percentage of population)

Country	Value in Year 2007	Value in Year 2017	Ten-year increase
Argentina	26%	74%	284%
Brazil	31%	67%	216%
Chile	36%	82%	227%
Colombia	22%	62%	281%
Mexico	21%	66%	300%
Venezuela	21%	72%	342%

Note: Values for Mexico and Peru correspond to 2018.
Source: International Telecommunications Union (2019) and World Bank (2020).

and the credibility of the government's commitment to such policies."[104] We are going to look at the relationship between government effectiveness and Internet penetration. I would hypothesize that countries with higher Internet penetration will have highly rated government effectiveness. Using regression analysis, I will be able to test the hypothesis that Internet penetration (x) is a predictor of the WGI variables (y).

Empirically, we can see from the descriptive statistics Table A.1 in the Appendix that in the dimension of government effectiveness, means for each country in the last twenty years vary from a low −1.08 in Venezuela to a high 1.17 in Chile. We see that the values for Pearson's r correlations range from a negative .837 in Chile to a positive .769 in Colombia. All cases, apart from Colombia, show a negative relationship between Internet penetration and government effectiveness (when, as here, the sign is negative, it means that if x increases, y decreases). Those with a p-value below the .05 cutoff mark indicate a moderate (Chile) to strong significance (Colombia and Venezuela). Variance in N, the number of cases, is due to Mexico's reporting their Internet penetration scores for 2018. Finally, the Adjusted R-squared value is given after a linear regression analysis using Internet penetration as a predictor of government effectiveness, or the outcome. Values indicate the independent variable predicting the outcome with a modest 23 percent in Chile, and a robust 58 and 68 percent for Colombia and Venezuela, respectively. The control of corruption dimension "captures perceptions of the extent to which public power is exercised for private gain, including both petty and grand forms of corruption, as well as 'capture' of the state by elites and private interests."[105] I would hypothesize that countries with higher Internet penetration will have highly rated control of corruption. We can see that in the dimension of control of corruption, the recorded means over the last twenty years vary from a low −1.15 in Venezuela to a high 1.39 in Chile. The correlation coefficients show a negative relationship in all the statistically significant cases (Brazil, Chile, Mexico, and Venezuela).

The dimensions in political stability and absence of violence/terrorism measures "perceptions of the likelihood of political instability and/or politically-motivated violence, including terrorism."[106] We can see that in the dimension of political stability, means over the last twenty years vary from a low −1.56 in Colombia to a high .53 in Chile. The correlation coefficients show a positive relationship in the statistically significant cases of Argentina's values are moderate at .47.

The rule of law indicator captures perceptions of the "extent to which agents have confidence in and abide by the rules of society, and in particular the quality of contract enforcement, property rights, the police, and the courts, as well as the likelihood of crime and violence."[107] The mean over the last twenty years of values in the rule of law dimension range from −1.55 in Venezuela to 1.27 in Chile. In two cases, the rule of law and Internet penetration have a relationship that is significant: Colombia and Venezuela. For Colombia and Venezuela, the r values are very strong, although, in Venezuela, the relationship is negative.

Regulatory quality captures perceptions of the "ability of the government to formulate and implement sound policies and regulations that permit and promote private sector development."[108] I would hypothesize that countries with higher Internet penetration will have highly rated regulatory quality. The means of the countries in the study for the regulatory quality dimension of governance vary from −1.34 in Venezuela to 1.43 in Chile. Unlike the previous WGI dimensions, this time more than half of the countries in the sample have positive mean values. Strong correlations exist between Internet penetration and regulatory quality in Argentina, and very strong correlations in Brazil, Colombia, and Venezuela. Only Colombia reports a positive relationship between the two variables.

The voice and accountability dimension captured the perception of the "extent to which a country's citizens can participate in selecting their government, as well as freedom of expression, freedom of association, and a free media."[109] The mean of the Voice and accountability dimension varies with values ranging from −.737 in Venezuela to 1.02 in Chile. For this dimension, all but two countries scored positively, making it the best performance of the countries in the sample. The bivariate correlations between Internet penetration and voice and accountability report and moderate positive relationships in Brazil, and a strong correlation in Colombia; while negative relationships and modest, moderate, and very strong correlations are seen in Mexico, and Venezuela, respectively.

Brief Analysis

So far, I have looked at whether high scores on Internet penetration go together with high scores or rankings on each of the six governance dimensions for eight countries in Latin America. For the hypothesis that countries with higher Internet penetration will have highly rated government effectiveness, results showed that only in the case of Colombia did this turn out to be statistically significant. Until 2018, Colombia did not have high Internet penetration. Instead, it lagged behind several other nations, for instance, twenty percentage points below the leader, Chile. From the results, we can conclude that government effectiveness, as measured by the WGI, does not necessarily correlate with higher levels of Internet penetration.

Regarding the hypothesis that countries with higher Internet penetration have highly rated control of corruption, this did not hold at all. None of the countries in the sample unveiled a positive relationship between the two variables. Rather strong negative correlations were reported, particularly in the cases of Brazil, Mexico, and Venezuela.

My next hypothesis was that countries with higher Internet penetration would have highly rated political stability. This was confirmed for the cases of Argentina and Peru.

For the hypothesis that countries with higher Internet penetration will have highly rated rule of law, the results were ambiguous, showing a positive and strong correlation in Colombia, and a negative and strong correlation in Venezuela.

For the hypothesis that countries with higher Internet penetration will have highly rated regulatory quality, only Colombia showed a very strong positive correlation. Still, again, the second subset of countries showed strong, robust negative correlations. Thus, the hypothesis only holds in the minimum of cases.

Finally, I proposed the hypothesis that countries with higher Internet penetration will have a highly rated voice and accountability dimension. This turned out to be true in Brazil and Colombia. In sum, Table 1.2

Table 1.2. Countries where governance indicators and Internet penetration are positively and statistically correlated

Governance Dimension	Country		
Government effectiveness	Colombia		
Control of corruption	—		
Political stability	Argentina		
Rule of law	Colombia		
Regulatory quality	Colombia		
Voice and accountability	Brazil	Colombia	Venezuela

shows the countries where governance indicators and Internet penetration were positively and statistically correlated. Chile (with the highest level of Internet penetration in the sample) turned out to have no statistically significant results.

Discussion and Conclusion

Law and technology scholar Susan W. Brenner raised a critical idea this book seeks to build upon in the forthcoming chapters. Brenner says that as new communications technologies are developed, and the limitless boundaries of cyberspace are enlarged, anyone can come up with new methods of launching a threat in cyberspace. The latter is not only an excellent place to hide from any authority, but it also demands an adequate response; not all states are capable or willing to make it possible. Once cyberincidents occur, several questions are raised: Who should respond, and how? But more, how might states prepare military and law enforcement personnel to respond to incidents or attacks?[110]

The current knowledge gap in cybersecurity governance in Latin America is significantly driven by the main theoretical implications generated in this chapter. Latin America has remained under the radar of the current cyber readiness unfolding globally. The subcontinent is sidelined by the fast-paced technologies and responses from the United States and its allies considering the cyber disputes with Russia and China. The role of states in Latin America in the overall security governance of the region is currently downplayed because of the emerging democracies' inability to deal with security issues and the undefined roles they should play when tasked with protecting the Web or, by default, their own ICTs. There are a few ways to break with this central orthodoxy, which I present in the ensuing chapters.

Much of the available literature misses the case of Latin America and relies on events occurring in the advanced democracies in North America, Europe, and Asia. Recently, Tim Maurer aptly warned on the risk of cyber conflict escalation among the rich democracies, yet with little attention to the role played by the emerging economies.[111] One might build upon such an argument to incorporate the most relevant cyber dynamics evidenced among the developing countries and how they compare to those evidenced in the advanced democracies. In chapter 2, I start exploring more deeply the regional cybersecurity geopolitics and the actors embedded in it, from state operatives to hackers. Some authors have focused more specifically on China (I present this discussion in chapters 5 and 6). The emerging Asian power provides a doorway to growing cyber capabilities and the balance of power with the Western Hemisphere. Chinese cyberactivity combines the realms of espionage, military institutions, and national policy. One can take the Chinese case study as a counterpoint and ask what Latin American countries can learn from the Eastern powerhouse and its relationship with the rest of the world players in today's international regime.[112]

Second, the recent cyberincident episodes involving Eastern and Western powers are said to be changing the nature of war, conflict, and security from how we knew it in the twentieth century. In this vein, P. W. Singer and Allan Friedman acknowledge that there are distinctions between cybersecurity and cyberwarfare that draw boundaries around the most powerful states' policy responses.[113] A second group of authors, for example, represented by Valeriano and Maness, challenge such common knowledge on the cyberthreat. They argue that there is very little evidence that cyberwar is or is likely to become a severe threat.[114] Cyber deterrence, rather than warfare, has occurred through restraint and regionalism. Of course, both perspectives are well worth debating to shed light on deterrence capabilities in the developing American states and the extent to which regionalism and security play a part in such affairs. I argue that there are plenty of ways in which the Latin American countries take care of their international security, one of them by building up a deterrence mechanism through more armaments. A firm deterrence stance policy "will threaten to punish bad behavior but also to withhold punishment if there are no bad acts or at least none that meet some threshold"[115] (I review some these ideas in chapters 3 and 4.) If cyber deterrence complements rather than replaces traditional instruments of statecraft and power, then cyber conflict can be a useful and novel way to achieve powerful effects in international affairs.[116] How do we include these ideas in the Latin American scenario?

Third, and in response to the above question, cyber conflict is said to be dictating relationships in security and technology between the superpowers, but what about the nonindustrialized nations? Because cyber deterrence led by military organisms is expanding among the middle-sized and small powers globally,[117] we might also argue that the impact of cyberspace upon the most central aspects of statehood and the state system—"power, sovereignty, war, and dominion"[118]—is also under pressure to change in developing regions, although there is still a need to find concrete proof of such dynamics. More military discretion in the conduct of cyber operations (defensive and offensive) is one plausible explanation.

For the United States, for example, cyber conflict is said to be creating new practices of national security and their strategic implications with the international community. This is a good starting point to trace down what has happened in the rest of the Western Hemisphere. What needs to be questioned, then, is what role cybersecurity fulfills in the actual agenda of the United States in Latin America, and what are the pros and cons of foreign supervision over the region's preparedness for cyber conflict.[119] Brian Mazanec and Bradley Thayer, for example, claim that the United States has produced deterrence cyberwarfare by bolstering strategic stability in cyberspace.[120] Capabilities for deterrence are worth pinpointing when discussing the challenges of building policy toward preventing and preempting cyberincidents (I aim to do this in chapters 6 and 7). In the next chapter, I argue for a new view of governance coming from the global South, which also demands rethinking new approaches to cybersecurity.[121]

Finally, at the closing section of the chapter, I overlapped many of the ideas above in a short statistical exercise designed to locate my interest in the theoretical and empirical relationship between Internet penetration and governance indicators. Based on previous theoretical claims, I argued that the advantages of more Internet penetration were beneficial to better governance. When we suggest a relationship such as this, we immediately tend to question how "perfect" it is going to be. I conclude that the previously assumed hypothesis that higher Internet penetration correlates positively with higher levels of governance does not always hold.

Two

Cyber Responses and the State

The wave of cybersecurity policies rolling across the globe has found its momentum in the emerging democracies, as more and more governments copy and adapt plans to ring-fence themselves from the risks found in cyberspace.[1] The core executives in the developing Western Hemisphere, for instance, have joined the global effort to mitigate cyber risks at both the micro-(people's privacy and property), and the macrolevels (states' digital information systems and infrastructures). For matters of strategic relevance, the move has come to fill a void in highly sensitive areas for the state's security and development, including defense, intelligence, criminal justice, emergency systems, financial trade, and foreign affairs.

To make cybersecurity comprehensive enough, countries such as Brazil, Argentina, Colombia, Chile, Panama, and Uruguay have set up domestic policy communities partnering with the multiple bodies that deal daily with cybersecurity. These new security networks combine the armed forces, homeland security agencies, diplomats, the intelligence community, prosecutorial authorities, private sector companies, NGOs, and academic institutions. Such a type of networked cybersecurity governance mirrors the policymaking tested in more advanced cyber domains such as Canada, the United States, Europe, and Australia. In these corners of the globe, and according to scholars, cybersecurity policymaking has not come free of difficulties, despite the level of institutional development and democratic anchorage expected in more affluent nations.[2] Networked cybersecurity governance undoubtedly poses a strategic challenge for policymakers.

As I mentioned, emerging democracies usually struggle when it comes to governing via inter-institutional policies. In this vein, I aim to advance the following argument. In the mushrooming of cybersecurity networks, what should hold our attention most is the gradual presence of the armed forces as steering nodes, which have started to take over, with Brazil at the pole position. The move mirrors what has happened in advanced democracies over the last decade.[3] The plausible idea that the military now sits atop powerful cyber centers or commands (thus enjoying holistic defensive and offensive capabilities for planning and executing cyber strategies) is the main theme explored next.

States and Cybersecurity Governance

States around the world are currently caught within what the superpowers have made out of so-called twenty-first-century cyberwarfare.[4] Over the last seven decades, Washington, Beijing, London, Moscow, Pyongyang, and Seoul have raced to best control their computer-based capacities to inflict intimidation or straightforward sabotage and espionage onto one another.[5] The question here is, What can the less-influential, but equally globalized nations do under such a strategic scenario, where access, functionality, and activity in the cyber world responds to no single actor's control?[6]

Thomas Rid proposed that the recent hype over cyberwar has prompted changes in the way political violence occurs. States now avoid bloody conflicts, instead engaging through the meticulous work of very skilled people using indirect adversarial means that prevent collateral damage, operating instead behind a powerful battery of information technologies. Rid's case for this new form of warfare sheds light on what the middle powers (with a tradition of avoiding conventional warfare) ought to ask themselves: Should they even be worried about this particular type of cyberwar?[7] Daniel Abebe, assessing such ideas for the case of the United States' strategic challenges coming ahead, put it this way:

> The potential for cyberoperations and cyberwar generates an interesting set of questions about the suitability of the traditional foreign affairs understanding of war for a new, technologically sophisticated type of warfare[8]

For the emerging democracies in various regions of the globe, the chances of returning to the old expensive ways of preparing for or deterring an attack from a potential enemy become less every day. The volatile conflict hot spots requiring the hardest militarized responses are narrowing down to zones of underdevelopment (maybe with the exceptions of Ukraine and Mexico).[9] For these states, the prospects of switching from traditional to nontraditional strategic capabilities and becoming skilled at setting up cybersecurity mechanisms are thus higher (I return to this idea in chapters 3 and 4). That is particularly the case in those states where austerity-driven officials try to manage public resources more efficiently.

Uneven Competences and Human Security

The interconnected dimensions of global security have brought consequences on many overlapping policy areas in the developing democracies, where the security-development riddle is now assessed through interagency policymaking.[10] Human security approaches have reinterpreted security as an encompassing phenomenon that holds equal relevance to a country's population as it does to its state governance, two aspects previously addressed in separate ways, through internal and external security.[11] Cybersecurity, as Nazli Choucri proposes, flourishes in a state in which "all dimensions of security are strong."[12] I borrow from international relations theory on the so-called middle powers to argue that emerging states in the global scenario are burdened by the challenge of building holistic and functional cybersecurity governance, especially those countries with humbler developmental standards.[13] Some nations do not have the critical infrastructure and knowledge to develop a cybersecurity structure, as evidenced in the superpowers or a handful of well-developed emerging powers.[14]

Take, for example, the growing cyber readiness in Australia and New Zealand, compared to other countries in Asia, as well as Israel in contrast to other Middle Eastern and North African nations or affluent Western Europe in relation to weaker states in Southeast Europe.[15] In Latin America, some developing countries have a lower technological intensity and are expected to be less engaged in cyberspace activities (e.g., Bolivia and Ecuador). On the more advanced spectrum, we might foresee a set of states whose national resources are stressed due to keeping up with a technologically intense environment driven by the northernmost advanced democracies (Canada and the United States).[16] In the in-between category, some states are more likely to have certain advanced ICTs than others, while still in the maturing stages of the domain of cyber policy.

In Chile, for instance, a formative stage has allowed public and private sector actors to minimize threats to national security, with the armed forces playing an important role, such as happened in Brazil, Colombia, Mexico, and

Peru. As one of the two Latin American nations in the OECD,[17] Chile has benefited from substantial public expenditure on technological connectivity. Cybersecurity efforts have come hand in hand with an entrepreneurship movement aiming to set up a startup hub in the Southern Cone. The move imitates what some nations have done in other critical areas of the globe, for instance, Israel's economic-scientific cooperation that strengthened the country's cyber diplomacy as a "soft power instrument" stretching to Arab, African, and BRICS nations.[18]

In 2017, the Chilean government launched its first national cybersecurity strategy with technical support from the OAS Cyber Security Program of the Inter-American Committee against Terrorism (CICTE), similar to what other countries (e.g, Colombia, Paraguay, Trinidad and Tobago, and Jamaica) have recently adopted.[19] The national policy understands cybersecurity as a "cross-cutting and multi-factorial concept." It underscores the creation of a networked policy community to strengthen its information infrastructure and better respond to cyberattacks. Chile intends to create its first Computer Security Incident Response Team (CSIRT, elsewhere also known as CERT), with a specific area dedicated to defense and military affairs to cover security risks and threats including internal leakages, the destruction of critical information infrastructures, espionage and surveillance carried out by other state actors, and cybercrime.[20]

Argentina, on the other hand, is said to have little "cyber resilience" for protecting its national and defense infrastructure, such as exists in Bolivia, Ecuador, Paraguay, Venezuela, and Nicaragua. Despite an early awareness of cybersecurity, Argentina has lagged in the cyber effort. The country was one of the first in Latin America with an up and running CSIRT and has produced national programs and cybersecurity plans oriented mostly to protect the critical infrastructure.[21] In 2013, the government decided to start a bilateral cooperation forum with its more advanced partner, Brazil, to catch up on shared cybersecurity tasks.

Finally, in Uruguay a series of strategic analytical capabilities put together over the last decade have supported the coordination and allocation of national cybersecurity resources.[22] In 2009, the country passed a decree requiring all government agencies to develop a cybersecurity policy according to matching needs. Its indigenous CSIRT is responsible for coordinating incident responses, processing cyberintelligence data, and coordinating with foreign governments and international organizations. Its various partners include, for instance, the U.S. Department of Homeland Security through the STOP.THINK.CONNECT campaign for raising awareness of safe Internet connections in both the public and private spheres.[23] Despite most efforts in these countries, cyberattacks are still a threat. In 2016, according to the country's CSIRT, Uruguay only received fifteen "high severity" intrusions into its cyber infrastructure. This is far from the totality of cases as institutions with traditional security measures tend to have limited powers to detect everything that is happening in their networks.[24]

For some Latin American states, ensuring responsive and overall cyber capabilities through internal and external security measures seems to be the best way forward.[25] The development of a human security agenda, now including cybersecurity, through a lens of networked governance, has meant the setting up of interconnected nodes for stakeholders, where the affairs of security are met through combining public and private interests to address shared goals.[26] The move embraces the use of state and nonstate actors' unique capacities to alleviate complex policy issues on the broad security spectrum. As an example, the long battle against drug trafficking in the Americas has meant that the states need to reexamine their skills through foreign and internal security amalgamation, of course, with military resources sometimes taken to a dangerous extreme, considering weak accountable and democratic institutions.[27]

It should be mentioned, nevertheless, that it is common for some factions to play the human security card in perverse ways, especially those states pursuing less democratic ways of ensuring the rule of law. In the developing nations, these types of semiauthoritarian states abound (e.g., Brazil in the Americas or Nigeria in Africa).[28] The higher levels of suspicion and interagency rivalry have limited the networked effort for crime, justice, and other security policies. Moreover, the intelligence agencies in Latin America have, for years, been used to spy on both political rivals and allies by the incumbent government. Policing bodies in the region also suffer from high levels of compartmentalization in light of federal politics (Brazil, Mexico, Argentina), on top of a tremendous problem with corrupt practices and political patronage.[29] The military, on the other hand, has a record of authoritarian practices, human rights abuses, and partial subordination to the elected authorities. However, this does not mean that networked security is not happening at all, since interconnected architectures can be formed in countries, or in specific policy areas, where the institutions do not behave in these ways.[30]

Cybersecurity Networked Governance

In an ideal world, networked governance over cybersecurity can benefit decision- and policymakers alike. Developing states can profit from smaller policy arenas that need to be linked together to become operative; in contrast, for instance, to the superpowers, where major silo agencies struggle to find consensus when dealing with security governance.

On the downside, for network governance to be effective it requires durable, democratic, and demanding meta governing and steering of policy.[31] Most of the time, such a role falls to a central state actor with leading capabilities to ensure that policy purpose is attained. For matters regarding security affairs, emerging democracies tend to put such a burden in the hands of one or two core executive institutions. These usually cause uncertainty or mistrust in relation to other bodies, such as the corporate ones (policing authorities, regulatory and overseeing agencies, and the armed forces), or, among autonomous institutions that provide solutions to different types of security demands (independent criminal justice actors, private security companies, organized civil society, among others).

Using such an approach, states confront the ongoing militarization of cybersecurity. Countries tend to correlate cybersecurity with a higher defense capability, at the end of the day equating cyber responses to national security and military terminology. In one way, this has secluded cybersecurity, turning it into a semi-independent institution with relatively high corporate powers and the usual disdain for civil authority.[32] Without proper democratic steering, any institution can direct cybersecurity to its means of securitization,[33] endangering its accountability and democratic control, and finally allowing it to overshadow other equally resourceful actors with better expertise in the setup and operationalization of cyberspace, including, for instance, the private telecommunications sector. However, because the military is seen as capable, resourceful, strategic, and solution-oriented, many of human security's most challenging roles (now including cybersecurity) are left to them as the ultimate state crisis solvers.[34]

The question of support for the armed forces is a hard one in Latin America. Although countries have gone through bloody civil wars and violent dictatorships in the recent past, the military has a particularly constant and high approval rate (I return to the topic in Figure C.1 where I present average means of institutional support). Because I'm interested in the relationships between the military, political institutions, and democracy, I present a brief exercise of bivariate correlation.

Table 2.1. shows the bivariate correlations between trust in the armed forces, the police, and political institutions and support for democracy as the best form of government. The data was extracted from the 2018/19 round of the LAPOP AmericasBarometer, all surveyed countries considered.[35] Pearson

Table 2.1. Bivariate correlation matrix selected variables in the Americas, 2018–19

	Armed forces	Political institutions	Police	Democracy
Armed forces	1			
Political institutions	.33***	1		
Police	.49***	.41***	1	
Democracy as best form of government	.12***	.21***	.18***	1

Note: ***Statistically significant at $p < .001$.
Source: Author's construction with data from LAPOP (2020).

r estimates tell us that all variables are statistically significant among each other, and that the regression is positive.

Table 2.2, on the other hand, present the results of a bivariate linear regression using support for democracy as the dependent variable. We see that the beta coefficient is statistically significant. We would expect that higher the support for the armed forces, political institutions, and the police, the more respondents will agree that democracy is the best form of government. This information is useful to have in mind because it debunks the idea that being a supporter of the armed forces, or the police, equates with authoritarian forms of government. My takeaways are that governments across the region can use this information to their advantage and cede the military broader attributions such as the protection of digital space, as they enjoy a high level of popularity across the population.

Table 2.2. Bivariate regression matrix selected variables in the Americas, 2018–19

	DV: Democracy as best form of government	N
Armed forces	.104***	23,766
Political institutions	.188***	29,759
Police	.154***	28,477

Note: Dependent variable: "Changing the subject again, democracy may have problems, but it is better than any other form of government. To what extent do you agree or disagree with this statement?" ***Statistically significant at $p < .001$.
Source: Author's construction with data from LAPOP (2020).

Digital Pax Latin Americana

For the Latin American states, the current threats in the cyber domain might be damaging for several reasons. First, most of the Americas' sensitive infrastructures (e.g., for national, social, and commercial development) rely on foreign technologies controlled by multinational firms (from both Eastern and Western competing manufacturers). These care little about domestic regulations and are rarely accountable following security breaches. Second, the intraregional competition for power and structural resources makes the possibility of one state using its cyber power against another a supposition that cannot be ruled out completely, especially considering the old rivalries operating within the region, and the fact that the military is steering cybersecurity governance. However, today's most severe challenges in the region are the reiterated attacks from nonstate parties against private and public assets, from Central America to the Southern Cone.

In 2015, OAS and the Inter-American Development Bank (IDB) produced a state-of-the-art report concluding that most nations are only at early levels of creating policies and strategies to counter cyberthreats, equally so in terms of education, awareness, legal frameworks, and adherence to technological standards.[36] Similarly, a group of researchers using the case study of Ecuador found that the most pressing challenges that cybersecurity education faces are cybersecurity skills, structural capabilities, social integration, economic resources, and governance capacity.[37]

Considering such a scenario, the superpowers' technical and policy advice offered to the Americas has focused mostly on how to build up the best strategic cyber defenses in civil and military matters. In showing the Latin American and Caribbean states how to protect themselves, the big players are also advising on whom they should put up such defenses against and how. It is believed that cyber offense and defense advantages are determined not only by cutting-edge technologies but also by skilled organizations that manage the complexities of cybersecurity.[38] As Herbert Lin puts it, "Offensive tools and techniques allow a hostile party to do something undesirable; defensive tools and techniques seek to prevent a hostile party from doing so."[39] This issue is of strategic importance in light of the current foreign influence in the region, an area partially dominated by the United States, but with a growing presence of Chinese and Russian economic interests.

The anti-Western channel RT (previously *Russia Today*) is believed to be the most trusted and watched among foreign news outlets in the region.⁴⁰ Financially, Chinese infrastructure projects in Brazil alone were worth US$53 billion in 2015.⁴¹ In Venezuela, Beijing has lent US$60 billion to various development projects, and Russia has sold (on credit) more then US$11 billion in weapons for the armed forces.⁴² Unable to pay them back, Venezuela has allowed China and Russia to take control over parts of the country's banking and oil industries. Both Beijing and Moscow have proven themselves to be trailblazers in the areas of technology and software products, and consequently, also in cyberwarfare, quickly catching up with Western cyber capacity, and willing to share such knowledge with American partners, something that had seemed quite impossible immediately following the Cold War.⁴³

In the Americas, countries are by default learning the means of cybersecurity governance considering the current confrontations between Eastern and Western powers. It is not a surprise that the old Cold War division of the Western Hemisphere redefines today's digital Pax Latin Americana. While countries such as Chile, Mexico, Panama, and Colombia are fond of the United States' policies in the region, other nations, such as those in the Bolivarian Alliance (the ALBA, which includes Venezuela, Bolivia, Nicaragua, Cuba, and Ecuador, as well as other Caribbean nations), have quite firmly moved away from Washington's sphere of influence during recent decades.

However, for some scholars, Latin American countries have used U.S. hegemony to advance their agendas within the region and the world, sometimes even against Washington's foreign policy.⁴⁴ This said, the ongoing regional effort for cybersecurity governance put forward, for instance, by the OAS, the Police Community of the Americas (Ameripol), and other parties such as the multinational private companies Symantec and Microsoft,⁴⁵ finds itself experiencing a split scenario of power influences. This is evidenced further by states' inherent doubts about collaborating in security issues with old-time rivals, for example, Colombia, Ecuador, and Venezuela; Chile, Peru, and Bolivia; and Nicaragua and Costa Rica, among others.⁴⁶

Nevertheless, and going back to the idea that many of the cyberincidents in the region are instead blamed on nonstate groups, the state-by-state rationale for cyber conflict has been less of a preoccupation in the Americas, although the Western Hemisphere is still a war theatre in which states rely on military defenses. Derek Reveron explained it this way:

> Since the early 1990s, analysts have forecast that cyberwar in the twenty-first century would be a salient feature of warfare.

Nearly twenty years later, individual hackers, intelligence services, and criminal groups pose the greatest danger in cyber-space, not militaries. But the concept of using cyber capabilities in war is slowly emerging.[47]

Most recently, in 2013, the controversial revelations of the worldwide surveillance program set in motion by the U.S. National Security Agency caused governments all over the world to focus on their privacy mechanisms, deterring foreign surveillance by applying the highest security and military means possible to protecting Internet-based networks.[48] In these matters, Brazil leads the way in the Americas with its national cyber defense center created in 2010; Colombia set up a similar task force in 2013.[49] In a powerful move in 2015, the Brazilian political authorities gave the armed forces full responsibility for its cyberspace, both for civilian and military use.[50] In light of the growing trend for soldiers to be placed as digital guardians, the United States, through the U.S. Southern Command, has continuously made efforts to promote multilateral action within the hemisphere, for instance, by sharing OneNet, an information system developed by the Department of Defense, and by extending cybersecurity practices through military branches in partner nations through the coordination of classified data.[51]

Threats and Risks

There has been a rising trend in large-scale cyberattacks across the region. In 2013, the costs of cybercrime in Latin America and the Caribbean were estimated at roughly US$113 billion, "enough to buy an iPad for the population of Mexico, Colombia, Chile, and Peru."[52] By 2015, private cybersecurity firms had estimated nearly four hundred million attempted attacks using malware. Brazil accounted for 27.6 million attacks and ranked eighteenth on a global scale. Other countries were equally vulnerable, including Mexico (15.9 million incidents), Colombia (5.1 million), Peru (4.3 million), Venezuela (2 million), and Chile (1.6 million).[53] A careless attitude from users and the rising trend of mobile technologies has increased the opportunities for cybercriminals, who find little deterrence given the lax enforcement of the states' rule of law.[54]

Data on cybersecurity are hard to project, as there are too many interests at stake, including that of the strategic national security. For example,

Kaspersky Lab, a Moscow-based antivirus and security software firm that produces yearly reports on cybersecurity, has recently fallen out of grace in some of Washington's political and policy circles. This is due to security concerns considering the firm's holding contracts with major government agencies, both in the United States and in other nations in the Americas; these contracts seem inappropriate now that Washington-Moscow cyber bullying is at its highest.[55]

Recent critical information theft has highlighted data security as a significant concern for advanced democracies and a growing worry for less secure countries. For example, in the United States, most recently, a Russian cybercriminal, Roman Valeryevich Seleznev, was charged with racketeering and conspiracy to commit fraud, after the appearance of evidence jointly provided by the Immigration and Customs Enforcement (ICE), Homeland Security Investigations (HSI), and the U.S. Secret Service, which brought more than fifty suspected criminals to justice. The Internet-based organization Carder.su was also charged with conducting fraudulent operations regarding bank transactions in more than 280 cities around the world. Seleznev admitted that Carder.su operated using advanced computerized technologies to protect the enterprise from rival organizations and law agencies.[56] Other cases in the United States have been previously reported.

In Mexico, the number of technology-based incidents affecting the banking sector, including identity theft and fraud, have increased by one-third since 2011, with its top three commercial banks' clients reporting almost a million complaints in 2017's first quarter.[57] In Colombia, the authorities have detected massive malware infections in public services computers sending out fictitious communications from bogus police, criminal justice, and other governmental platforms used by millions of users every day.[58]

The Americas community, led by Ameripol, has set up reactive policing strategies to combat the most advanced cybercriminal organizations. However, stronger multilateral cooperation and further prosecution are still pending, especially in terms of crimes that affect national security.[59] For example, in 2012 a collective of authorities from Argentina, Chile, Colombia, and Spain arrested twenty-five suspects allegedly linked to the hacking group Anonymous, which was believed to have launched a series of coordinated cyberattacks against the Colombian Ministry of Defense and some official websites, as well as Chile's Endesa electricity company.[60] The critical industrial sector seems particularly vulnerable in this region. It is estimated that only Brazil, Colombia, Chile, Mexico, Panama, and Peru have implemented proper ICT infrastructure cybersecurity measures (as mentioned earlier, only at basic levels), leaving the rest of the subcontinent facing severe exposure.

In 2015, more than half of the continent's governments reported that their budgets for cyber securing their critical infrastructure had not risen in the last year, increasing only in Belize, the Dominican Republic, Uruguay, and Chile.[61] Mexico was the fifth-most-disturbed country globally following the ransomware WannaCry attack on Windows-operated IT networks in more than 150 nations. Outside the region, WannaCry hit high-profile enterprises in the public and private sectors such as the Spanish mobile operator Telefónica, the UK's National Health Service (NHS), the French carmaker Renault, Germany's Deutsche Bahn, the Russian Ministry of the Interior, and the U.S. company FedEx.[62] It was calculated that Latin America as a whole does not invest more than US$10 million annually in cybersecurity for its industrial sector, compared to advanced countries such as France, where the budget for this has reached up to US$8 billion.[63] The United Kingdom, on the other hand, has invested amounts similar to France's in order "to make Britain the preeminent safe space for e-commerce and intellectual property online."[64]

State Technologies

In part because the technology used behind massive cyber episodes, such as the WannaCry incident, once belonged to a governmental cyber espionage program that was stolen and later leaked by hackers,[65] no nation can avoid suspecting that other, more advanced states are still developing programs with equally or even more damaging abilities. Take the following example. In 2013, Brazil and Mexico sought answers from the U.S. Department of State after former contractor Edward Snowden leaked data from the NSA revealing that presidents Dilma Rousseff and Enrique Peña-Nieto had had their emails and phones intercepted, the latter even before assuming office.[66] In this affair, the United States only seemed to amass more surveillance power with little discrimination and accountability. The Trump administration recently elevated the U.S. Cyber Command, formerly a division of the NSA, to the status of military authority. Created during the Obama years, the command combines the country's offensive and defensive cyber operations networks with those used for digital intelligence gathering.[67] Its counterpart in the UK, the NCSC, acts as an extension of the intelligence and security agency, the Government Communications Headquarters

(GCHQ). According to one observer, "The idea is to turn the Internet from a worldwide web of information into a global battlefield for war."[68]

For the Americas' states, the above scenario poses at least two strategic challenges. On the one hand, rival states can buy such technologies from the major powers, arguing for the protection of their still vulnerable digital resources. On the other hand, nonstate actors can quickly get hold of these technologies, as has already been evidenced, establishing them as an equally sinister force capable of significant hits to less-prepared nations. Consider the following example. Well before the hacking and leaking of the Democratic Party emails during the 2016 U.S. election,[69] Latin America had experienced its own dose of indirect voting cyber interference. Andrés Sepúlveda, a Colombian serving ten years in prison for hacking his country's 2014 election, told Bloomberg reporters that he had rigged elections in favor of right- and left-wing candidates in Nicaragua, Panama, Honduras, El Salvador, Colombia, Mexico, Costa Rica, Guatemala, and Venezuela. Sepúlveda charged up to US$20,000 a month for using Russian software technologies to create "a full range of digital interception, attack, decryption, and defense" strategies, including the use of fake Twitter and Facebook accounts to influence public opinion and alienate voters.[70]

According to AmericasBarometer, of the respondents included in the 2018/19 round, 60 percent had a Facebook account. The survey also revealed that 78 percent of respondents said they would see political information appearing in their feed. Figure 2.1 unpacks this data and presents a bar graph with the respondents' frequency of receiving political content over the urban and rural divide. The figure reveals that urban respondents are the most exposed to political content in the region. Meanwhile, those living in rural areas do so considerably less frequently. Extracting data from the survey also tells that only 12 percent of the respondents reported having a Twitter account, while 60 percent reported having a WhatsApp account.

Defense and National Security Affairs

Recent cyber incidents have exposed what organizations worldwide can do with such cutting-edge programs, bringing forward the idea that for the Latin American and Caribbean states, what prevails is the cyber protection of their national priorities. This point has been discussed elsewhere by some observers to support the thought that most cyber matters rest within the realms of defense and national security.[71] Two recent examples

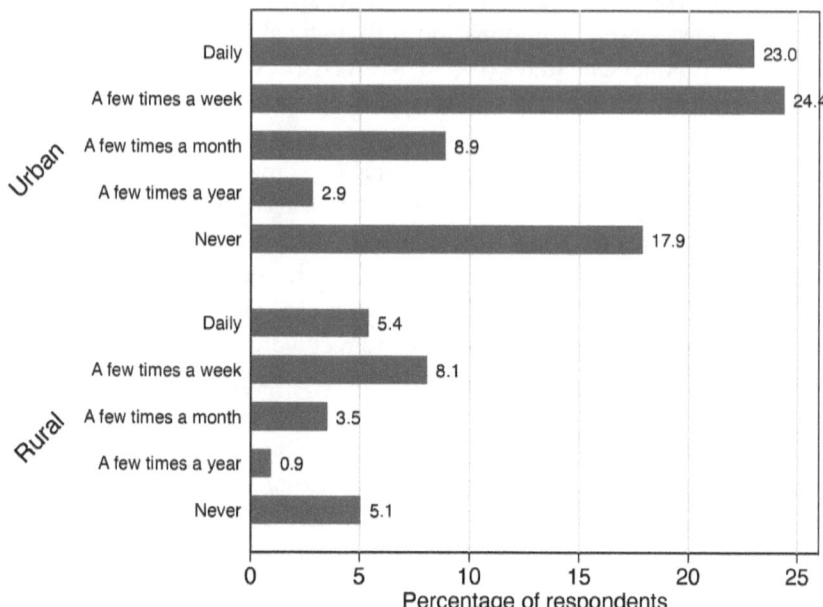

Figure 2.1. How often do you see political information on Facebook? Author's construction with data from LAPOP (2020). Public domain.

can illuminate this point. During a visit from the Chinese president, Xi Jinping, to the United States in 2013, Barack Obama claimed in a speech that state-funded cyber espionage and a nuclear-armed North Korea were the two most relevant issues dividing both countries.[72] Obama made it clear to his counterpart that cybersecurity was now as dangerous and real as Pyongyang having a weapon that could destroy millions of lives in a matter of minutes.

For the emerging and consolidated democracies, the feeling that the cyber factor is a top priority for national security is also becoming deeply rooted. Denmark, for instance, claimed that Russian cybercrime groups (linked to Moscow's security services) hacked their defense system repeatedly in 2015 and 2016. The move prompted Copenhagen to increase its spending on cybersecurity on top of already significant military efforts in response to Russia's missile deployment in the Baltic.[73] Before that, in 2010, Chinese hackers breached Google servers accessing information about U.S. surveillance targets, triggering a major national security investigation. The Federal Bureau of Investigation (FBI) believed Chinese cyberespionage groups could also be linked to the PLA.[74] Following this, Google decided to move its website search engines from China to Hong Kong in response to Beijing's pressure, harassment, and censorship.[75]

The world has seen a rise in cyber-surveillance technologies susceptible to intrusion software recently. Particular ICTs, such as computers, laptops, mobile phones, and a vast array of telecommunications networks, can be remotely monitored and their data exploited, be this content that is stored, processed, or transferred electronically. The EU member states have pioneered novel controls on transfers of software to cover the trade in so-called cyber-surveillance systems since 2012. However, sometimes showing opposing views on which systems should be subject to export controls based on the level of cryptography that they employ.[76]

The 1981 UN Convention on Certain Conventional Weapons (CCW Convention) has created a framework for debate around the possible regulation of Lethal Autonomous Weapons Systems (LAWS). As put by SIPRI researchers, "LAWS present an immense arms control challenge."[77] The problem here is the following:

> Many military technologies tend to automation, through machine learning to artificial intelligence and robotics. They are expected to provide realistic options for offensive weapons that could be deployed in dynamic and complex environments with little or no human input or supervision. An arms control discussion framework has been found for LAWS, but progress is slow. Few governments have defined positions, a minority have decided to address the issue, and discussions so far have been purely informal and mainly for information and definitional purposes. It is not clear if or when negotiations will start or what their objective would be. Options range from a complete ban to regulation of deployment and use. The question must be whether the pace of arms control will match the speed of arms development.[78]

Cyberwarfare and cybersecurity add another layer of complexity to arms control since it is believed that the regulation of the cyber field is more advanced than on LAWS. As mentioned, international cooperation for establishing cyber norms and controlling offensive and defensive capabilities is hard to achieve, more so considering the complex and frequent cyberincidents originated by state, criminal, or terrorist organizations. In Europe, for instance, with the constant disruption caused by Russia-originated incidents, EU member states have not been able to marry their corporate and economic cybersecurity in the private realm with their strategic national security in the political realm.[79] This present vacuum weakens the efficient control of cyberweapons.

In Latin America, I see cyber confrontation occurring as the military tries to deter other state-coined cyber technologies used by organizations acting outside international law. In 2013, for example, a group of hackers based in Peru, calling itself LulzSecPeru,[80] targeted Chile's Air Force (FACh) email servers, accessing information on the acquisition of missiles, radar systems, and aircraft. These included communications held by the FACh with the French companies Astrium, and the Etienne Lacroix Group; the North American firms Cirrus Air Craft, Kaman, and General Dynamics; and the Israeli manufacturer Rafael Advanced Defense System Ltd.[81] LulzSecPeru argued that the hit was revenge over a 2009 incident in which Chilean hackers struck a Peruvian government website posting a painting from the War of the Pacific that had been fought between the two countries between 1879 and 1883 and which has fueled trilateral enmities since Chile ceded the northern territories along the Pacific coast to Peru and Bolivia.[82]

In Brazil, hacktivists exposed personal details of more than fifty thousand Rio de Janeiro military police during the middle of a wave of social protests over police distrust, corruption, and authoritarian violence.[83] Most recently, the website of the Argentinean Army was hacked by an alleged threat from the Islamic State. Although the Triple Frontier area between Argentina, Paraguay, and Brazil is recognized as an organized crime haven with possible links to foreign terrorist groups, authorities have provided no record of members of the Islamic State present in the country.[84] In Colombia, on the other hand, it is said that due to the country's six-decades-long fight against the (FARC), the military and police have acted in coordination over protection and defense issues, setting a precedent for other networked approaches to security. This is reflected in the 2011 national cybersecurity policy, which brought government institutions and the private sector together to coordinate the protection of critical infrastructures in the country.[85] Terrorist networks are gaining more influence in cyberspace, with groups such as al-Qaeda, for instance, finding more "clever" ways to perform the "electronic Jihad." As some observers have put it, just the term *cyberterrorism* is enough to bring together "two of the greatest fears of modern times."[86]

However, cyberterrorism is frequently mistaken for the misdeeds of criminal hackers, and, despite its potential threat, to some cyberterrorism has also been seen as an exaggeration.[87] Scholars Wendy H. Wong and Peter A. Brown came up instead with a potentially more appropriate term, naming those global activists, such as Anonymous and WikiLeaks, who demand political change while employing robust and disruptive Internet technologies, "e-bandits."[88]

Finally, and as an example of the nature of the digital Pax Latin Americana, we can turn to the recent episode wherein governments in Panama and Mexico allegedly used the military spy-hacking software Pegasus to spy on human rights defenders, journalists, political rivals, and anticorruption activists. The scandal meant that Panama's former president Ricardo Martinelli was subject to legal charges.[89] Although the Israeli NSO Group has claimed that it had licensed the use of Pegasus for the sole purpose of fighting terrorists, drug cartels, and criminal groups,[90] the blurred line demarcating what constitutes national security in the Americas, now including cybersecurity, is still tainted by never-ending authoritarian abuse.

Digital Pax: A Double-Edged Sword

The developing states in the Americas are eagerly joining in with the global governance of cybersecurity. This has encouraged cyber readiness at the national level, and cyber responsiveness at the international level, with much overlapping of the two areas.[91] Sound national cybersecurity strategy-making is helping to commit the limited resources of developing countries that might be better allocated to addressing other areas of need, and at a time when social unrest seems to press for better housing, health, jobs, and education. However, a safe Internet is more relevant than ever if the Latin American and Caribbean nations want to build stable institutions and tackle other desperate developmental issues.[92]

Economic growth, as multilateral organizations are trying to represent it to governments, should be pursued in a regulated and protected cyber environment. Yet, incentives for cybersecurity regulation are truncated if countries do not agree on regional practices, especially if they link cybersecurity to public bodies (such as the military) whose job is to suspect and plan for the worst national contingencies. From that aspect, the digital Pax Latin Americana is a double-edged sword. It enhances network governance collaboration for a peaceful cyber environment, but also adds another layer of military and defense planning where states put their national security at the center.

Take two examples that illuminate the arguments presented so far. First, the regional bodies have not only endorsed the militarization of cyberspace as part of the solution but have prompted other states to develop this as a relevant countermeasure. Argentina, Colombia, Chile, Mexico, and Venezuela have emulated Brazil's cyber command and given the armed forces a set of roles and missions in the protection of cyberspace (see Table 2.3). In this vein, Argentina is currently working with the U.S. Department of State on a partnership that addresses the cyber triad: security, defense, and crime. A bilateral working group, set up in early 2017, aims to strengthen Argentina's CSIRT, foster networks of public/private cooperation, and enhance collaboration between both countries' military cyber experts.[93]

In Colombia, cyber capacity building in the defense sector happens at three levels: through the Colombian Ministry of National Defense and its cyber emergency response team; the Police Cybercenter of the National Police; and the Joint Cyber Command (formed by Army, Navy, and Air Force cyber units). The Colombian War College (ESDEGUE) offers a Master's program in which students take classes on "Cyberattack Simulation against Critical Infrastructure for Decision-Making" to learn about decision making during a "cyber crisis."[94]

Cybersecurity in Venezuela is conducted under the authority of Joint Cyberdefense Directorate, (Dirección Conjunta de Ciberdefensa) at the Strategic Operations Command (Comando Estratégico Operacional/CEO), part of the Ministry of Defense and responsible for guiding the operations of the armed forces. The directorate oversees planning, coordinating, and executing cyber operations affecting critical national infrastructure, weapons systems, and the telecommunications networks of the armed forces. Their role is meant to "ensure the use of cyberspace denying it from the enemy."[95] Regarding the development of cybersecurity capabilities for the armed forces, Venezuela recently acknowledged Russia's support through military experts sent to train the local ranks. According to a U.S. government official, a Russian military contingent arrived in Venezuela in late March 2019, believed to be made up of special forces including "cybersecurity personnel."[96] Although most Latin American countries have links to foreign nations for military peer-to-peer knowledge exchange, Venezuela's close relationship with Russia and China is seen with particular suspicion.

Figure 2.2 shows the levels of policy and strategy development in both national cybersecurity and cyber defense above and below the Internet average in the region in 2019 (63 percent). Brazil scored higher in the level of maturity of its cyber defense strategy reaching a level of "established." The figure shows countries' scores, ranked from 1 = very low maturity, or

Table 2.3. Summary of cybersecurity and cyber defense policy developments

	Policy and strategy	Military dedicated agency	Summary of roles
Argentina	National Cybersecurity Strategy Defense White Paper	General Directorate of Cyberdefense	*Responsibilities include the planning, formulation, direction, supervision, and evaluation of cyber defense policies for the jurisdiction of the Ministry of Defence, includes control over the Cyberdefense Joint Command of the Armed Forces.**
Brazil	Information and Communications Security and Cyber Strategy Defense White Paper National Defence Strategy	Cyber Defence Command	*Responsible for planning, coordinating, directing, integrating, and supervising cyber operations in the defense area.**
Chile	National Cybersecurity Policy Defense White Paper Digital Agenda 2020	Joint Cyber Defence Command	*Responsible for planning and executing joint military operations in cyberdefense.*†
Colombia	National Digital Security Policy Policy Guidelines on Cybersecurity and Cyberdefense	Joint Cyber Command	*Strengthening the technical and operational capabilities of the country to enable it to confront computer threats and cyber-attacks through the implementation of protection measures, as well as the introduction of cyber defense protocols.* *Protect critical infrastructure, reducing the computer risks to the country's strategic information.**

continued on next page

Table 2.3. Continued.

	Policy and strategy	Military dedicated agency	Summary of roles
Mexico	National Cybersecurity Strategy National Digital Strategy	Cybersecurity Unit	*Plan, conduct, and execute information security activities, cybersecurity, and cyberdefense.* *Help in the national effort to keep the integrity and stability of the Mexican state.*‡
Venezuela	National Plan for Cybersecurity and Cyberdefense	Joint Cyberdefense Directorate	*Plan, protect, neutralize, coordinate, and conduct operations for cyber defense to ensure integrity in the information systems networks and the Strategic Operational Command's telecommunications, as well as respond to possible cyberattacks, threats, and aggression that could affect the critical infrastructure, weapons systems, and the security of the armed forces and other agencies of strategic national interest, ensuring the use of cyberspace and denying it to the enemy.*§

Sources: * UNIDIR (2020); † Diario Oficial (2018); ‡ Velediaz (2017); § Comando Estratégico Operacional Fuerza Armada Nacional Bolivariana (2019).

"startup" according to the OAS nomenclature, to 5 = very high maturity, or "dynamic." Since most countries scored at a value of 2, the "formative" stage (except for Venezuela), to continue improving national cyberdefense strategies they need to step up their game and comply with international law and with national and international rules of engagement in cyberspace.[97]

Let us consider what happened, for instance, after the terrorist attacks on September 11, 2001. Countries all over the region adopted financial intelligence units (FIU) to counter money-laundering practices; they were supported by professional bodies such as the Financial Action Task Force

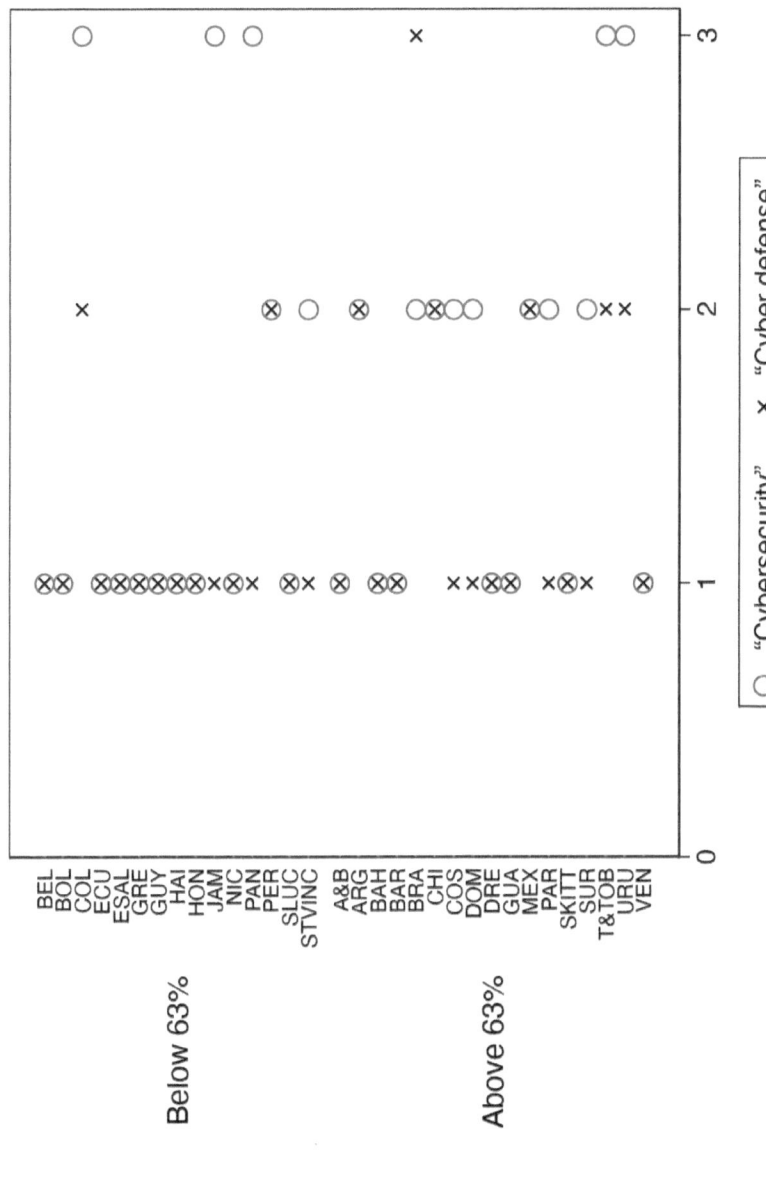

Figure 2.2. Policy and strategy levels of maturity for cybersecurity and cyber defense using the regional Internet average (percentage) as cutting point. Author's construction with data from Observatory of Cybersecurity in Latin America and the Caribbean (2019). Public domain.

(FATF) and backed up by the United States and the European Union.[98] The challenge by then was to counter the finances behind global terrorism. Today, defense against cyberincidents requires, in a similar manner, a supply of knowledge-based responses that use computerized intelligence. Steering and assessing a rapidly changing economic and social environment requires governments to adopt policies that are of both local and global dimensions. In this task, policymakers can find "themselves always playing 'catch-up'" to the empowering, but also threatening, use given to ICTs as measures of deterrence to cyber threats.[99]

As a second example, Internet governance is currently undergoing consultations by the states most eager to advance a freer Internet through meaningful participation by stakeholders and accountability mechanisms.[100] Although Brazil is the flagbearer in such an effort in the Americas, its militarized approach to cybersecurity puts it in doubt whether Internet regulation can bloom in the hands of such a traditionally state-centered actor as the armed forces.[101] Moreover, for some observers, state-centered conflict over power within cyberspace mostly affects human rights and democracy activists. As Brantly puts it, "We posit that the security dilemma in cyberspace is reflected in an arms race between states, resulting in a category of actors being left behind, unarmed (or least significantly underequipped) and vulnerable in a domain that crosses traditional state boundaries."[102]

For the Americas and the Caribbean states, once-strategic continental location has undergone shortened distances and abbreviated time due to the hyperglobalized nature of cyberspace.[103] Once-authoritarian governments in Brazil, Argentina, Uruguay, or Chile now deal with empowered societies where growing populations access and create Internet information copiously. For them, a reverse motion to embrace the Chinese style of Internet control seems almost impossible now.[104] However, the cybersecuritization of the Americas is not so different from that of the more advanced nations. Through greater state-centered policymaking, leading to networks for cybersecurity governance led by the military, countries are increasingly expanding their control over information security through state modernization reforms. Eric Schmidt, former chair and CEO of Google, and Jared Cohen, president of Jigsaw (previously Google Ideas), put it this way:

> Democratic government will most likely be tempted to further their national interests through the same combination of defense, diplomacy, and development on which they relied during the

Cold War and the decades after. But these traditional tools will not be enough: although it remains uncertain exactly how the spread of technology will change governance, it is clear that old solutions will not work in this new era. Government will have to build new alliances that reflect the rise in citizen power and the changing nature of the state.[105]

The Americas' developing nations need to upgrade their game to deal with state-based or state-funded cyber threats.[106] So far, they are preparing indiscriminately for cybersecurity and warfare through their offensive and defensive capabilities. The region has been highly influenced by the political and military postures of China, Russia, the United States, and other advanced nations, whose cyber ethos rationalizes the need to adopt more vigilant positions toward adversaries. As Saltzman argues, "Cyber capabilities do not cause armed conflict, but make decisions to escalate easier and cheaper."[107] In that sense, what can be extrapolated from the Americas is that by improving security over information systems and critical infrastructures, states are often, and indirectly, preparing for the inevitable: upgrading cyber defenses without fully seizing the messy world of cyberwarfare.[108]

Conclusion

In this chapter, I have argued that the "new policy world of cybersecurity" is characterized by networked forums where government agencies, private sector corporations, the military, and users, meet halfway to arrange nonconventional approaches to security.[109] Here, they distribute responsibilities along a continuum of self-interested parties who will initially struggle before reaching consensus. I argued that with the leading military efforts in the region, whatever sort of cybersecurity era Latin American and Caribbean states are entering into seems predisposed by what the men in uniform will be entitled to uphold.

So far, the world seems more experienced in prosecuting the wave of current cyberincidents than in developing militarized responses. This is the topic of the next section of this book. As Joseph Nye Jr. argues, "Preventing harm in cyberspace involves complex mechanisms such as threats of punishment, denial, entanglement, and norms."[110] In chapters 3 and 4, I question whether the military can levy such deterrence. To some scholars, cyberspace favors the offenders because the attacker can hide behind anonymity, and size doesn't seem to matter (weaker actors can have offensive advantages with a low cost and a high payoff).[111]

Since the release of hundreds of thousands of the United States' classified cables by WikiLeaks, and later Snowden's exposure of the NSA's spying programs, governments cannot act hypocritically about their affairs in cyberspace. In this vein, the transparency of cyberactivities has led to little but more profound distrust among nations who promote soft power, on the one hand, but seek to strengthen their hard power, on the other, that is, to gain or maintain strategic advantage.[112] According to Bradshaw and Howard, computational propaganda initiatives led by ministries, the military, and political parties have expanded since 2017, taking place in more than seventy countries, including Argentina, Brazil, Colombia, Guatemala, Mexico, and Venezuela in the Americas. Sophisticated actors from Caracas have also used computational propaganda in foreign influence operations.[113]

The advanced states have proven resilient to the ongoing cyber confrontations. For them, what came after was a rapid process of learning the power game in the national and transnational interactions of the information age.[114] A little after the 2008 global recession, the United States escalated cyberattacks to first place on the list of threats against the country, even before other pressing issues such as the shattered economy.[115] Similarly, in the United Kingdom, the government placed cyberattacks and crimes at the topmost strategic level of international terrorism and state-to-state military conflict, despite their presence in the costly and bloody wars of Iraq and Afghanistan.[116]

From my empirical analysis, I will argue that it is not clear if this is the path for the Americas. However, and given the common ongoing developments and rationales taken for cybersecurity policymaking, such as its pronounced militarization, it is a scenario that cannot be dismissed. In the current digital Pax Latin Americana, the inescapable tradeoffs between cybersecurity and military warfare will shape governance in its own form. Nevertheless, we are already familiar with past and present strategic constraints that might help illuminate the way.

Three

The Military Industry

"The objective is not to have unrestricted cyber warfare, the object is to create structures of deterrence by making our adversaries understand that when they engage in offensive cyber activities themselves, they will bear a disproportionate cost—so they think about it a lot harder before they launch a cyber operation to begin with."[1] In the words of former National Security Advisor John Bolton, the United States is en route to reduced red tape and procedural restrictions to inflict retaliatory attacks on those who lead cyber operations against them. To the most cyber-capable nations, responding offensively to threats is a way of deterring future incidents and attacks. Deterrence, the effort to influence enemy intentions, needs proper muscle to make it a credible option.[2] In the Western Hemisphere, countries are building up their cyber defenses to stop possible threats. However, cyber retaliation seems now a privilege granted only to those that can keep pace with sophisticated cyber technologies.

Highly technical cyberweapons are being deployed more readily than before. One reason is that they do not necessarily need to be high-potential weaponry, but instead, can be generic and low-potential tools.[3] Cyberweapons have destroyed property but not yet killed anyone. This distinguishes them from guns, bombs, or weapons of mass destruction in legal and practical terms. Jeffrey Carr, of Taia Global Inc., a critical data security firm, argues that at least twenty-eight countries have cyberweapons programs. The United States, for example, has used its offensive capability both in combat and in covert operations.[4] State and nonstate actors can use cyberweapons that are targeted (e.g., Trump's order to cyberattack Iran's military computers or Putin's intense cyberattacks against Ukrainian infrastructure) or nondiscriminating (nonstate Russia-backed cyber agents disrupting Georgia's computers with cyberweapons in 2008). What distinguishes cyberweapons the most are their virtual methods, argues Lucas Kello. Malware does not obey the laws of geography. Instead, it uses communications protocols to travel until it reaches an object that can be manipulated. There is no "charge" to cause any damage.[5]

To make sense of this debate in Latin America, this chapter brings into discussion the traditional arms complex and cyberweapons. A few caveats are worth discussing. In contrast to the global arms industry, cyberweapons are usually manufactured away from plain sight of the formal economy. A high degree of anonymity in cyberspace allows creators and attackers to conceal their identities and their coding. Cyberweapons do not carry a "made in" etiquette. The business model of malware seems highly fragmented across jurisdictions and among cooperative developers. As well, the creation of simple malware can be quick and its deployment even faster. Cyberweapons are not subject to long-term political discussion, as some traditional weapons programs are. States do not promote their cyberweapons as they would their defense and arms industries. Also, while the global arms industry adheres to regulation and penalties, cyberweapons technology is very much unaccountable until its mother code has been uncovered, usually after they have been used and rendered obsolete in computer life. Treaties on cyber arms verification and control do not exist. Countries do not publish white papers on their cyberweapons, and so far there is no formal limitation on what states can accumulate in their e-arsenals. The cyber arms race could go on indefinitely. Finally, because cyberweapons are not overtly violent, they expand the range of possible harm and, thus, the outcomes of what we know in the fields of war and peace.[6] Scholars have identified that cyberincidents propagate different forms of "harm," including physical or digital harm; economic harm; psychological harm; reputational harm; and social and societal harm.[7]

This chapter, therefore, does not intend to equate the traditional weapons industry to cyberweapons. The core discussion is to explore how Latin America and the changing character of warfare are creating responses from states and their relationships to old and new technologies to counter traditional and nontraditional threats. Especially, the chapter reviews the weapons industry as one of the drivers of military, diplomatic, and foreign relations between allies and perceived foes. I explore the tools that states have at their disposal, assuming that these can shed light on the capabilities they are harnessing toward current and future scenarios of conflict, whether in the physical or digital domains or a combination of both. At the end of the day, "The domains of modern warfare—land, sea, air, space, cyberspace—call for and indeed enable very different types of operations."[8]

The chapter adds to the contents discussed so far in the book in the following manner. It presents succinctly the regional, economic, and political conditions affecting the geostrategic environment, changes in security threats, and perceptions of risks affecting states' weaponization or disarmament. The evidence presented aims to provide a baseline on which to analyze what states are doing to technologically confront the perceived threats to their security. These might be transnational crime, terrorism, humanitarian and environmental disasters, nationalist rivalries, or the weaponization of competing states (I return to these themes in chapter 4). The chapter will purposefully focus on the defense industry as a unit of analysis. It aims to explore what and if states are considering the use of cyber technologies, nanotechnologies, and other remotely controlled weapons to potentially enhance their position over adversaries and in other military missions of human concern.[9]

To some observers, fears of being attacked with extreme cyber malice are usually exaggerated and based on sporadic evidence. States' authorities, on the other hand, are reorganizing their approach to considering such fears as they restructure their military missions and rewrite doctrines for the use of cyber and traditional weapons. Cyber conflict, understood as "the use of computational technologies for malevolent and destructive purposes to impact, change, or modify diplomatic and military interactions between states,"[10] seems at this level an essential concept to understand in the evolving security and military relations between Latin American states. Of course, relationships between states, especially military-to-military diplomacy, are one domain among a myriad of interactions at the international level. Cyber conflict is therefore understood as though states are belligerent and always mindful of their own power, as well as perceived threats from external actors. In the ever-mutating warfare scenario across the globe, cyber conflict adds another layer of analysis for our consideration.

Weapons in the Western Hemisphere

It is a widely shared opinion that the current peace in the Western Hemisphere diminishes the chances of interstate conflict in the future. Nations, instead, tend to allocate their limited fiscal resources to addressing nontraditional and human security threats that, subsequently, have called for an adequate military industry to meet the new challenges ahead. States adopt new missions and roles and thus the capacity to confront the threats posed by drug cartels, paramilitaries, organized crime groups, gangs, and terrorists.[11] Countries can opt to industrialize the means of war considering this changing character of war. Yet, political and economic factors influence how any given nation will recapitalize its traditional (waging war) and nontraditional (fighting organized crime) military and security capacities. As well, the hyperconnectivity of today's military industrial sectors demands high costs due to its physical and digital operations management. Governments need to provide enough security measures to safeguard their control systems and sensitive information. Generally, industrial control systems can be attacked by "foreign rivals seeking proprietary information, hackers exacting revenge or looking for lucrative loop-holes, or even terrorists hoping to wreak economic havoc."[12]

Since the Cold War ended, nations in the Western Hemisphere have acquired manufacturing capabilities to best suit their strategic concerns. National security policies have led many states to develop in-house skills to produce different types of weaponry and technologies. These include small and light weapons, bulk production of antipersonnel land mines, and other more substantial investments in planes and war vessels. Although some of the equipment is still in production, industrial catalogs have been refreshed to meet the needs of the technologically globalized twenty-first century. What remains understudied is what changing aspects of warfare (including cybersecurity) are considered when recapitalizing the military industry in response to current threats to security.

In the Americas, countries show different security and military realities; they are threatened unevenly by both traditional and nontraditional security risks; they have developed military industries on different stages and up to varying degrees. The political economy of arms industrialization in emerging economies has led to various policy choices adopted to develop the sector (export liberalization, protectionism, and wealth creation). The region offers different realities regarding the indigenous military industry. A majoritarian subset of nations shows no military industrialization at all (Central America) and depends highly on military imports. In other places such as the Caribbean, the old colonial powers (France, the Netherlands, and the United Kingdom) are the main defense providers for the small states. Achieving greater defense self-sufficiency seems more of an issue mainly happening in South America, where the post-U.S. hegemony has reshuffled the overall strategic scenario for arms trading.

Although China, Russia, and Iran are making waves as they gain clients and alliances, the United States, West European countries, and a handful of other states (for example, Israel), remain the primary providers and sellers in off-the-shelf purchases and the co-production of offset defense acquisitions.[13] The replacement of obsolete equipment through domestic industrialization and business ventures with global companies has invited the revival of the defense sector over the last decade, with some cases recently overspending more than others at different points in time (e.g., Chile, Colombia, and Brazil).

Military spending is a useful key to understanding the ups and downs of militarist policies worldwide.[14] A few trends on the topic are worth introducing at this point. First, militarism at the global scale is decreasing if understood as a fraction of GDP destined toward military expenditure (others have shown the contrary if measuring military spending in billions of dollars).[15] Figure 3.1 shows a steadily decreasing trend in military budgeting during and after the Cold War. Since the turn of the century, the global average proportion has not surpassed the 3 percent cutoff point. Meanwhile, the world has gotten dramatically richer. The evolution of GDP shows an increasing trend, most evident from the early 1990s onward.

In Latin America and the Caribbean, the military-expenditure tendency is similar. Figure 3.2 shows that the average proportion of GDP directed toward military spending fluctuated around 2 percent until the late 1990s. After that, we see a decrease in expenditure below the 1.6 percent ceiling. Nonetheless, some world regions remain big spenders. Military spending in North America, the Middle East, and North Africa are well above the rest. The former depicts a steady rise from 2000 until 2010; meanwhile, the latter showed a significant decrease in the late 1990s, only to pick up again in the 2010s with an average above the 5 percent barrier. Most recently, Latin America shows a trend similar to Sub-Saharan Africa's.[16] Global military expenditure rose to US$1,917 billion in 2019, an increase of 3.6 percent from the previous year and the largest annual growth in spending since 2010. The five largest spenders accounting for more than 60 percent of the world's total were the United States, China, India, Russia, and Saudi Arabia. Military expenditure in South America showed little difference from the previous year, reaching US$52.8 billion. According to SIPRI, the average military spending was 1.4 percent of GDP for countries in the Americas,

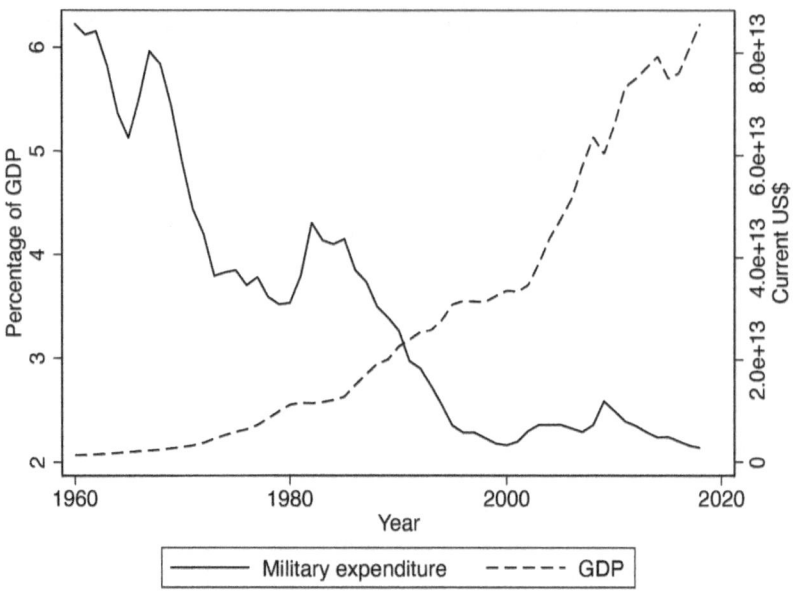

Figure 3.1. World military expenditure and GDP. Data from World Bank (2020). Public domain.

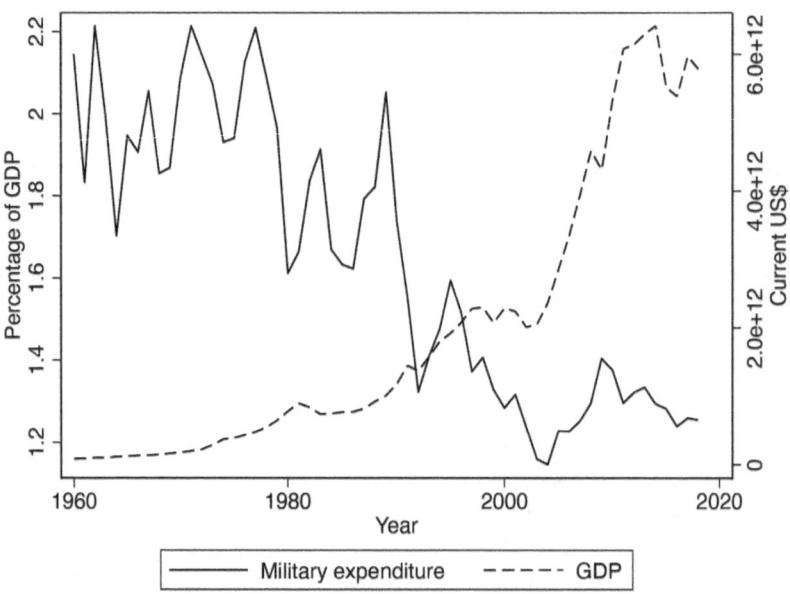

Figure 3.2. Latin American military expenditure and GDP. Data from World Bank (2020). Public domain.

1.6 percent for Africa, 1.7 percent for Asia, Oceania, and Europe, and 4.5 percent for the Middle East.[17]

Economic, social, and political conditions favorable to military industrialization, however, remain tenuous in Latin American and the Caribbean. Security governance is the Achilles heel in the Spanish and Portuguese-speaking countries where generalized corruption, weak institutions, and unfinished democratic transitions undermine the rule of law and the consolidation of peace and security.[18] In this fragile context, military missions are, nevertheless, expanding as a response to the myriad of social, economic, and humanitarian threats that exceed the traditional purpose of sea, land, air, and space security.

As evidenced in other emerging economies that are not fully industrialized (such as Turkey's, Israel's, and others in the Asia Pacific), different policy choices regarding cyber and traditional military roles influence defense industrialization and emerging technologies.

Together with the most cost-efficient defense industries, a handful of states tend to have the highest overall cyber capabilities (conceptualized as the sum of cyber offense, dependence, and defense capabilities). Scholars group the most proficient cyber-capable countries as including Russia, China, the United States, Israel, and Germany among the top in the world.[19] As Valeriano and Mannes argue, "Just who has cyber power often relates back to questions of capabilities and resources."[20] Cyber power can expand a state's ambition to make the most out of technology.

Recent advancements in emerging technologies for strategic use are said to potentially challenge the status quo of great powers military dominance, as weaker powers can now access new technologies. The technological military advantages long enjoyed by the United States, for example, are constantly challenged by the diffusion of emerging technologies among its adversaries.[21] Nevertheless, few technologies have reshaped the dynamics of international conflict, and second, emerging technologies can be both destabilizing and stabilizing, and other contextual factors are likely to mediate the effects of emerging technologies in today's globalized international system.[22] For instance, it is understood that for many years in the developing world the effectiveness of advanced weapon technology was potentially dependent on civil-military relations. Civil-military relations influenced national security policy, which eventually impacted armed conflict in general.[23]

Today more than ever, ensuring a state's strategic sovereignty involves a sum of governance factors that can hinder or enhance cyber military adaptiveness.[24] For local military businesses, this presents a critical juncture that can well be taken to advance digital security in spite of minimal fiscal investment.²⁵ While some emerging economies have abandoned significant spending in developing domestic defense industries, others have arrived at new justifications for the provision of domestic armaments, in part driven by national armed forces entering the global fight against "non-state enemies" in "civil war" battlefield conditions.²⁶ Robert Muggah and John P. Sullivan advocate rethinking military adaptiveness and the rules of engagement in light of the coming crime wars. They argue that

> [t]his cocktail of criminality, extremism, and insurrection is sowing havoc in parts of Central and South America, sub-Saharan and North Africa, the Middle East, and Central Asia. Not surprisingly, these conflicts are defying conventional international responses, such as formal cease-fire negotiations, peace agreements, and peacekeeping operations. And diplomats, military planners, and relief workers are unsure how best to respond. The problem, it seems, is that while the insecurity generated by these new wars is real, there is still no common lexicon or legal framework for dealing with them.²⁷

The absence of a common lexicon constitutes a pervasive problem in cybersecurity governance. States usually define, frame, and operationalize concepts such as cybersecurity in different manners. What concerns us here is that the current reinterpretation of warfare and the evolution of the military industry have inspired the momentum toward the "digitalization" of the security domain.²⁸ Internet-based resources are common in today's conflict interactions. However, the balance between superpowers and emerging states is uneven due to massive distances between their military technologies. Also, academics have taken incongruent positions, with opposite sides arguing about the usefulness of terms such as *cyberwar* and *cyber conflict* (wherein lives are lost) within the traditional war framework.²⁹ What can be said, however, is that cyber interactions call for cyber technologies to be developed and implemented. Whether Latin American states have reached that point of awareness is what I will discuss next.

Defense Industry and Cyberspace

In June 2018, the *Economist* printed a story claiming that demand for cyberproducts grew at twice the rate of that for military hardware. The article stated the the size of the military and civil cybersecurity market had increased from US$3.5 billion in 2004 to US$120 billion in 2017, accounting for an annual expansion of 12 to 15 percent predicted for the next three years, or twice as fast as the global defense equipment budgets. As the magazine put it: "In America, Congress is emphasizing the importance of cybersecurity; Britain's government reportedly plans to shift some of the Ministry of Defense's budget towards repelling cyber-threats. Private-sector companies routinely put cyber-security among their top worries."[30] Defense and tech firms can now compete for clients around the world. Many times, conventional arms producers have a commercial cyber spinoff to feed the new demand for cybersecurity services. Demand has grown in the advanced economies in part due to long-standing advantages in knowledge and expertise in the strategic and intelligence fields.[31]

Innovative defense technical programs require a specific model of societal development, one that promotes independent thinking, aspires to global reach, and reflects on the intellectual capabilities of a nation's people. The challenges to developing a defense industry to meet the challenges surrounding the new "battlefield" of cybersecurity consist mostly of finding the right kind of experts with the "soft skills" to adapt already-existing technologies to new entrepreneurial industrial cycles. Heinz Schulte argues that the likely trend in the future defense industry is clear:

> Metal-bashing in terms of protected vehicles or naval vessels is no longer the prerogative of the West. Sophisticated "systems-of-systems" architecture—which turns an armored vehicle or a naval vessel into a complex software-based fighting system—is the new dimension.[32]

The new technologies dominating the virtual battlefield include "software to dominate cyberspace, laser weapons, autonomous unmanned vehicles (land, air, on water or underwater), and the development of reliable and ample non-fossil fuel for military use."[33] For the emerging nations, the question is how they can promote such a plan for new military forces bringing civilian experts from within and outside of the traditional security agencies. Given the dynamics of the defense industry's redefining what will constitute the core military arsenal of the future, Schulte added:

> Why destroy a bridge physically if you can disrupt the entire flow of traffic by a click of a computer mouse? Cyberwar will be the new and growing dimension of future conflicts. Writing the

necessary software codes on a computer screen does not require the wearing of a military uniform. As long as there is conflict, there will be an industry to serve the protagonists. And as the face of war changes, so do the dynamics of the defense industry.[34]

The worldwide defense industry is reshaping the canon in its rush to address the issue of cyberspace. As Nye Jr. defines it, "The cyber domain includes the Internet of networked computers but also intranets, cellular technologies, fiber-optic cables, and space-based communications. Cyberspace has a physical infrastructure layer that follows the economic laws of rival resources and the political laws of sovereign justification and control."[35] On this, Valeriano and Maness argued even further, adding that "all computer, network, digital, cellular, fiber, and space-based forms of communications, interactions, and interconnections are what make up cyberspace."[36]

The defense industry plays a vital role in defining the cyberspace when it challenges the idea that what happens in the cyber sphere might be chaotic and random. Cyber technologies usually have a strictly defined domain governed via institutions, countries, governments, and networks. There is not much anarchy in the defense industry per se, but it shows up later, when technologies are opened to distributors, buyers, and clients worldwide. Further, the use of cyberspace tends to overlap what are usually divided defense fields, such as air, land, and sea technologies. Strengths in one arena can be used in multiple avenues and at different scales, encouraging cyber technology to become a conduit for possibly common mechanisms and operations. The impact on the weapons industry is massive, as one suitable technology can fit more than one purpose and activate various physical and digital components. The more extensive the development of the defense industry in cyber affairs, the more cyber power a nation can attain. Cyber power is considered

> the ability to control and apply typical forms of control and domination of cyberspace. Cyber politics and power blur, as strong states typically have the most cyber power. What is unique about cyberspace is the ability of small states to compete on the same playing field as strong states.[37]

The focus in the remainder of this chapter and later in the book is on the material power of Latin American states as they take steps to rearrange their defense industries and weapons purchases. It might be premature to talk straightforwardly about cyber power and cyberweapons, as these programs are still in their infancy in the region. The main point to extrapolate across the chapter is how states manage their weaponry expectation and what has this to do with the perceived threat of cyber insecurity.

In this vein, a few questions arise: Is cyber power the next step that some Latin American nations might be looking at to enhance their security? What forces are driving technological innovation, and what degree of readiness do local industries require in order to develop any cyber power? Education, expertise, and technological advancement in the conventional battlefield should leave us with some critical lines of inquiry for studying the evolution of security governance and warfare in the region, the relationship between the state and industry, and finally, the relationship between peace and conflict.

The Post–National Security Era

Conventional deterrence and dissuasion paradigms explained much of the previous warfare thinking and military industrialization programs in the Western Hemisphere. Deterrence, as Richard K. Betts put it, fits two purposes: countering an enemy and avoiding war. "An enemy will not strike if it knows the defender can defeat the attack or can inflict unacceptable damage in retaliation."[38] Globally, most deterrence mechanisms need rules and norms to be defined, refined, and enforced in a variety of formal and informal settings.[39]

In the recent military dictatorships, maintaining strategic stability fueled national security doctrines that saw safeguarding state sovereignty at the core of military capabilities. With democratization, the inclusion of alternative dimensions of security and redefinition of the military mission, however, kicked off a novel approach to risks and threats. Military warfare turned to the logic of collective security, peace-building measures, and common frameworks to enhance transparency in military expenditure and defense industrialization. The post–national security era brought military budgets down (e.g., in Argentina). At the same time, in other countries, they were sustained thanks to natural resources revenues (in Chile, for example, this happened thanks to the now-defunct Copper Law).[40]

Although efforts to recognize the shared burden of waging security over collective threats such as transnational drug trafficking and the illegal activities of nonstate violent groups, unproductive diplomacy and the lack of effective regional security institutions have hindered pragmatic peace approaches to security in the Western Hemisphere. Through the Defense Council of the Union of South American Nations (UNASUR), for example, the region's defense ministers can discuss and agree on subjects of common strategic culture and develop further norms to govern conflict resolution. However, this type of continental institution faces severe constrictions, as security dilemmas are "responsive to the historical suspicions and current political divisions that exist" in the Americas. "The development of effective regional security institutions has been wracked by regional disputes and conflicting attitudes towards the legitimate role of outside powers," some observers have argued.[41]

On top of such suspicions, the region has a recent history of arms and military races that prompt mistrust and hinder confidence building when it comes to military-to-military diplomacy. Although the idea of an arms race is contested at the moment, states seem to be pursuing spending tracks on different levels based on consideration of military investment in their "security web" (i.e., neighbors and other influencing powers) and the likelihood of "potential enemies" posing a hostile threat (i.e., other states or nonstate armed parties).[42]

The uncertainty of cyberspace represents a significant source of preoccupation for some nations. "Weaknesses often are exploited by those who seek to gain, especially in a situation of historical enmity," argue Valeriano and Maness.[43] Nonetheless, many of today's international security relations are based on assuming dependency on Internet technologies and the securitization of the World Wide Web. Even though some believe that cyberspace is quite stable, and Internet interactions among states have proven to be a link of cooperation, perception of fear and cyber threats can dominate strategic thinking.[44]

For example, the U.S. military is eager to advance combined joint operations across multiple theaters and domains in Latin America. Some authors argue that the most likely strategic and operational approaches are planned based on gray zone warfare and asymmetric battlefield. One possible catalyst for military operations in the region is, no doubt, the persistent disorder, both at the political and societal levels. This is not only characteristic of the Western Hemisphere, but it is understood that, globally,

> the U.S. military response to this strategic context is still forming, but there are common themes among the services: the recognition that future operations will entail higher risk; they

need to disperse forces to survive on a more lethal battlefield; a desire to create networked forces attacking with a combination of physical and nonphysical (cyber and electronic warfare); and a rebalancing of force structure in terms of both weapon sophistication and mission type.[45]

Gauging the new role and mission of the U.S. military in Latin America is more complicated than usually thought. Military transformation in the region, as an outcome of imitating the superpowers, opens a conversation where both external risks and domestic policy variables (defense expenditure, fiscal responsibility, security doctrines, and political interests) come into the picture.[46]

In 2016, the U.S. military opted for a national strategy aimed to direct the military response to transregional, multidomain, and multifunctional threats from conflicts that are spilling into several regions, including U.S. territory. To some observers, these arenas of struggle will definitely feature elements of cyber, space, and information warfare, often including a blend of state and nonstate actors.[47] Theoretically, military transformation can be understood as

> [a] virtuous circle of constant iteration between stakeholders with different interests and authority in and outside the state. Rather than a linear process that is limited to the armed forces, congress, and the executive, modern defense governance responds to an ongoing iteration of action and learning from the roles and functions performed by the military and the control over the efficiency of missions executed by the civil-military communities in charge of defense policymaking.[48]

Countries can transform their missions to look either inward (insurgent guerrillas, drug trafficking organizations, social tension, climate change, and natural disasters), outward (interstate conflict and peacekeeping), or a mix of both, in consideration of domestic and foreign threats that drive their present and future military capabilities, doctrines, expenditures, and industrial policies.

Modern armies can dedicate their resources to missions the include: (1) preparing for territorial defense, (2) preparing for or engaging in fighting insurgencies, (3) preparing for or participating in combating terrorism, (4) engaging in peace support operations, (5) engaging in fighting organized crime and street gangs, and finally, (6) providing humanitarian assistance in the case of natural disasters.[49] These categories do not only serve the purpose of revising the nature of civil-military relations and the control and efficiency of the missions of the armed forces, but they are also useful to understanding what purpose military industries fulfill in a country's search for security.

For example, if given country X commits its military forces to undertake a mission in all six categories, hypothetically this means that that state possesses some degree of capability and control over the means necessary to perform efficiently such a set of roles and missions. Countries tend to focus on a select number of areas that better match their security stances and which can be equally sustained through always-limited fiscal resources. Even the rich democracies constrain their missions within military budgets that circumscribe their remit of action. The point to advance here is that any nation develops a defense industry and levels of expenditure that mirror the ongoing military campaigns to which their land, air, space, naval, and cyber domains are committed. One main implication of such thinking is timing. Defense industries produce and procure both short- and long-term goods. However, the manufacturing and delivery process in the defense market, from initial designing or purchasing to final delivery, usually takes months, if not years, to conclude, and then not until numerous political and economic hurdles have been cleared.

Cyberweapons and Strategic Planning

Military industrialization and defense spending priorities are carefully balanced against other fiscal priorities that a state might be faced with in a given time window. Nations invest in their local industries when they see a specific technological capability being maximized through a medium- or long-term contract, which usually might involve taking into consideration the estimated lifespan of a given weapon or technology. Yet, because security concerns can change rapidly, nations tend to counterbalance needs with available resources that can allow them to be more flexible to a set of missions, for instance, to provide for territorial defense while also engaging in fast responses to humanitarian emergencies. Another nation might acquire the means and capabilities to fight insurgencies and combat terrorism while being less worried about peace-support operations or humanitarian assistance.

The tradeoff between traditional and nontraditional security concerns will thus be unique to a nation's strategic threats.

What is relevant to discuss here is that the nature and usage of cyberweapons tend to be short-lived. After each cyber incident, the victim's systems tend to be corrected, and some sort of response is put in place to avoid the same problem happening again. This makes cyberweapons' effectiveness obsolete without the surprise factor, as the victim is now tipped off to their vulnerability. Cyberweapons can be used only once or twice, thus diminishing the returns on the investment.[50] This goes strictly against what the developing countries look for when planning their defenses. The emerging democracies, due to financial limitations, will want arsenals that can alleviate the initial investment over the course of a mid- to long-term scenario. In short, less-expensive traditional weapons systems have a natural cost-effective advantage over some cyberweapons.

The context of the twenty-first century establishes yet another paradigm. Defense industries can acquire or develop cyber defenses, as the information revolution has decreased the costs of computing and telecommunications systems. The Internet is free, and entry barriers for citizens and governments have been lowered to a minimal expression. The power to access and dominate cyberspace is, therefore, increasing. The cyber age in which we live not only promotes the creation of new cyber technologies, but it also empowers cyber interactions. Any program that can potentially be used as a cyberweapon against another actor can be acquired or directly built by developing countries. The difficulty is in pointing out what deserves financing in the overly populated world of cyber information. Countries tend to grasp what is useful for them to develop sources of cyber power. Nye Jr. comments on this in light of "soft power" objectives, including how governments can wield policies that bind together public and private actors behind one common goal.[51] Cyber technologies are crucial to arriving at networked outcomes between government and civil society institutions. Cybersecurity governance, as presented in this book, is networked too. Making a case for public expenditures on seemingly short-lived technologies becomes harder in a scenario where governments are stressed from all flanks to spend wisely.

Add to that the complexity of coordinating trade controls on goods and technologies for both conventional and nonconventional weapons. Governments have to sign politically binding agreements and annually report to expert groups their cross-developments on military technology acquisitions.

More recently, the debate in international forums, such as the multilateral export control regimes—the Australia Group (AG), the Missile Technology Control Regime (MTCR), the Nuclear Suppliers Group (NSG), and the Wassenaar Arrangement on Export Controls for Conventional Arms and Dual-use Goods and Technologies (Wassenaar Arrangement, WA)—has circled the question of how to adjust to technological developments, including the potential exploitation of cyberspace. A significant obstacle along the way has included addressing the task of reaching "political consensus on admitting new applicants, combined with concerns about the viability of large membership and about sharing potentially sensitive information" across these regimes.[52] Moreover, as new software essential to cybersecurity is continuously created, governments and their local IT sectors haven't reached agreement with their counterparts on a common counter-intrusion framework or on vulnerability disclosure standards.[53]

Sustainable Industries

How does a state go through warfare changes while creating a defense industry that will be sustainable in the future? Foreign defense conglomerates have sought worthy investments in the Western Hemisphere mostly to expand offset-induced capacities (understood as a mutual investment in military equipment in which defense contractors agree to transfer technologies to increase the buyer nation's industrial capabilities). Monica Herrera and Ron Matthews argue that Latin American governments have sought offset opportunities and addressed the problem of long-term viable military industries:

> Latin American states are aware of this challenge and also of the fact that they operate at different levels of defense-industrial development; thus, their country strategies are nuanced to match technological imperatives. The strategies have focused on, for instance, revamping capacity (Argentina), creating capacity (Chile, Colombia, and Peru), and consolidating and deepening capacity (Brazil).[54]

The emerging economies want to jump on board the sophisticated military global value chain while also maintaining a certain degree of autonomy. States fall into a "defense trilemma" when choosing between taking part in the globalization of military technology and production, enlarging their economic sustainability, or enhancing security autonomy. Dependency on major-power arms suppliers is seen as reducing the sovereign interests of nonproducing states.[55] The "make-versus-buy tradeoff" is not surprisingly a security paradigm for many middle-ranked developing democracies.[56] Costly military systems (especially the more technologically advanced) oblige states to weigh the option of undertaking autonomous military production by entering the defense production market or advancing secure offset acquisitions.

Raul Gouvea, of the University of New Mexico, coined the term *quadruple helix approach* to assess the internal and external factors affecting the recapitalization of the defense industry. The approach seeks to balance actions by stakeholders, including the defense R&D community, the defense industry, the internal market, the export market, the government and public apparatus in charge of the governance of defense, and, finally, the multinational defense companies that act as global partners. In other words, a networked policy community centered around the defense industry.

Of course, many factors influence how each stakeholder might behave in consideration of uncompromised policymakers, more competitive global defense markets, low R&D expenditure, less competitive domestic industrialization, "make-versus-buy" decisions, and foreign dependency on major defense suppliers.[57] What still needs to be included in the analysis is what the state wants to accomplish with military industrialization considering national and international warfare capabilities. This assessment goes beyond managerial approaches and is weighted more heavily with factors of a strategic and military nature.

Karl DeRouen, on the other hand, fills this void, stating that the military might be expected to push for stronger overall strategic modernization and industrialization because they might be capable of affording and building modern weaponry to operate at maximum efficiency in times of conflict. Another thesis might suppose that the military could be less engaged in the fate of the economy and thus focus solely on their corporate interests. However, this idea constrains the military to current budgets and operational capabilities, which are exposed to the unknown outcomes of a changing security environment. The third strand of analysis builds upon contextual factors, such as regime type, resource constraints, and arms production capabilities, that explain the relationship between defense spending and the country's growth.[58] This track seems more oriented to explain the relationships between autocratic regimes and their militaries, in comparison to democratic ones and theirs. Nevertheless, neither of the above trends necessarily applies to the Western Hemisphere as a whole; instead, they call for a more in-depth exploration of comparative political processes, taking into consideration some aspects of each of these theoretical underpinnings.

Industrial development since the third wave of democratization has occurred at different speeds and in different ways, adapting means of warfare in response to national realities. Trade liberalization, privatization, and market deregulation impacted military industries differently in the region.[59] For example, the debt crisis that hit the Americas in the 1980s forced countries to take into consideration how much impact military spending was having on the evolution of debt.[60] Military industrialization and the missions of the armed forces suffered the same fate, as some countries focused differently on peace support operations, territorial conflict, or countering guerrillas and insurgencies, highly influenced by their transitions to democracy. The industrialized nations that supplied defense contracts to the region during and after the third wave had already been involved directly or indirectly in military conflicts of traditional and nontraditional nature (for example, the United States and European countries with security concerns in the Middle East, Asia/Pacific, Africa, and Eastern Europe). The United States, United Kingdom, France, and Germany had since the post–World War II years created the necessary conditions for public and private military for-profit conglomerates and contractors to flourish in comparison to second-tier actors whose extensive involvement had been more limited (namely, Russia and China).[61] Although the European countries seem to have retreated from participating extensively in global hotspots of conflict, their arms trade balance remains positive, as they have developed domestic markets for land, air, sea, and now cyber capabilities, while also turning into avid sellers to the emerging democracies.

Building Hybrid Capabilities

In the decades since World War I, modern military operations have combined land, sea, air, special operations, intelligence, space, and now cyber forces. Joint military operations depend critically on the technologically advanced uses of military might. More recently, military humanitarian interventions have increased in quantity and scope, forcing both Western powers and emerging democracies to redefine the purpose, role, and scope of the operational areas of the armed forces. For example, in the early post

9/11 era, scholars focused on terrorism and the proliferation of nuclear, chemical, and biological weapons of mass destruction. In that vein, Joseph Cerami and John Young write: "The influence of technology on warfare remains an important focus of study, as does the longstanding significance of arms and weapons-systems research and development, budgets and resource management, manpower, and logistics."[62] Bruce Berkowitz, for instance, argues that twenty-first-century asymmetric warfare and the importance of electronics and cyberspace in the Information Age will determine the future of conflict.[63] Cyberincidents and information warfare campaigns are considered part of what is often called the "grey zone" of military responses. The application of conventional military power can be one way to exploit superiority over adversarial or revisionist states. Grey zone warfare is a tool prone to be used by the more powerful states, rather than the lesser actors.[64] Developing nations instead observe more limits on grey zone activities, as their military arsenals prohibit them from launching combinations of different methods to achieve tactical or strategic effects. Hybrid warfare, as academics call it, implies the opposite of conventional warfare: a combination of "conventional, irregular, terrorist or criminal activity, political subversion, and information operations targeted at multiple audiences."[65] Cyberincidents have become commonplace topics of conversation whenever U.S. and European authorities talk about hybrid warfare. For author and U.S. Army lieutenant colonel J. P. Clark:

> While hybrid warfare is not necessarily a strategy of the weak, the weaker power in a conflict has a greater incentive to resort to all available means. Thus, in an essay specifically concerned with American military operations, it might be assumed that all adversaries will be driven to employ hybrid warfare; the only difference will be what means are available and how expertly they are used. Once again, this creates the paradox that those with the greatest strength in the traditional sense might also be the most potent non-traditional adversaries.[66]

The point to make here is that although some states have created the necessary military means to defeat possibly revisionist competitors, the recently democratized countries can vigorously contest one another's military actions, including a combination of physical and nonphysical (cyber and electronic warfare) means that are cost-accessible to them. I discuss this idea next.

Roles and Missions

In the more cyber advanced nations, cyberwarfare is said to have precedence ove traditional kinetic warfare.[67] The prospect for the armed forces is thus to advance cyber capabilities continuously and aggressively. Consequently, states blur the boundaries between war and peace in a field where conflict seems unregulated. Some form of international control begs the idea that states need to agree on certain preconditions for international cooperation in the political and strategic realms. Whether collaboration happens at all, it is quite clear that the prospects for the armed forces are to change and adapt to the new international reality.

Recently, military roles and missions in the emerging democracies worldwide have unfolded under uncertain conditions, together with the clear danger of not having enough political purpose and institutional control. The armed forces and state security agencies have continued to support autocratic corporate interests that, in some cases, have shaped corrupt policies in the industrial defense system. Consequently, skewed national security interests have impacted military effectiveness and hampered their interactions in the international security system.[68]

Nations have increasingly turned to producing, assembling, exporting, and re-exporting conventional weapons under less efficient arms control parameters. For example, Ken Sharpe argues that the badly politically oriented militarization of the drug war in Central and South America during the early 1990s not only enabled some local defense industries' programs and co-production endeavors with global companies to profit from the new role of the military but has since then sent the wrong signal, namely, that such escalation of resources depends on the decline of drug-producing and trafficking, which up to this day, has turned violently grim.[69] As well, the production of small and light weapons (SALW), particularly in Brazil, which were once thought to increase national security capacities, has fueled the activities of violent nonstate actors that rule urban spaces in cities such as Rio de Janeiro. At least twelve nations in the Western Hemisphere currently produce SALW ammunition, with Argentina, Mexico, Brazil, and Peru having declared that they have exported these occasionally since 2013.[70] SALW are used by the armed forces in carrying out smaller combat missions, as opposed to all-out war scenarios.[71]

The Caribbean states have turned their security forces to enforcing maritime sovereignty and interdicting trafficking routes from South to North

America. Imported naval resources and aerospace technologies work best in their mission to halt narcotics operations. In Central America, the arms trade, legal and illegal, has developed as a continuation of the civil wars experienced in the late twentieth century. Here, narcotics trafficking is of most concern, as violent nonstate domestic groups move through the poorer countries squeezed between Mexico and South America. This has encouraged nations to provide land, sea, and maritime resources to their militaries and police forces with yet mixed results in the fight against criminal groups.

Among the Andean countries, Colombia and Venezuela share a frontier argued to be one of the potential hotspots of conflict in the Americas. Both countries have armed themselves to pursue territorial defense and fight armed insurgent organizations, in the case of Bogota since the 1960s, similar to the Peruvian case in the Southern Andean Cone. Chile, Peru, and Bolivia have engaged in interstate disputes that have led Santiago and Lima to restock their arsenals in the 1990s and then again in the 2000s. Bolivia, on the other hand, remains focused on countering drug trafficking activities and providing their limited security forces with resources also to counter civil tensions that have turned violent in the past (this is also the case of Ecuador).

In the South Atlantic, Argentina, Uruguay, Brazil, and Paraguay have also developed different military capabilities. Buenos Aires and Brasilia have shown the most efforts in terms of advancing their in-house military industries. For the past decade, Argentina has remained in a fragile macroeconomic state, thus diminishing its arms industry's chances of flourishing when compared to a more resourceful Brazil. Recently, the authorities in Planalto Palace in Brasilia have spent considerably on protection of Brazil's national territory, particularly its large deposits of in- and offshore natural resources and its porous borders, which are shared with at least ten other countries.

Because they perceive more urgent security threats, some governments tend to speed up in mastering the most critical areas from which possible real or imagined threats might emerge, whether from land, sea, air, space, or cyberspace. To maintain security over different scenarios, states mobilize resources to control for the known and unknown risks. Since today's insecurities include threats to the environment as much as threats to cybersecurity, we would expect a myriad of capabilities and missions to be industrialized to enhance state defenses. But this is not always the case in the global South.[72]

Technology and Infrastructure

The broad missions performed by the armed forces in the Western Hemisphere, for example, make security outcomes hard to govern because of both political and economic reasons. On the one hand, countries tend to industrialize the equipment of war and security as contributing to macroeconomic conditions, and political consensus favors a positive manufacturing environment. In other words, countries that present the most advanced military industries are those where political stability and business conditions provide essential conditions for operationalizing an arms industry (i.e., where technological and human resources are considered, including labor, knowledge, machinery, and know-how). Estimates of high-technology exports, R&D expenditure, and researchers in R&D show that Latin America and the Caribbean have lower averages than regions such as East Asia and the Pacific, the European Union, North America, and the world average.[73]

Cyberspace, however, demands its own political economy. Costs of information technologies are usually low, and technologies travel rapidly from one manufacturer to another. The physical part of the equation, that related to politics, tends to be slower and very intricate. Cyberspace relies on physical capacities and some political guidance. Valeriano and Mannes argue that the political domain of cyberspace demands not only technologies and infrastructures but also multiple layers of governance where industries, governments, jurisdictions, and the ensuing interactions meet halfway. They argue that this is the "layering" aspect of cyberspace.[74] It is in this multilayered scenario that cybersecurity endeavors take place and are integrated into broader industrialization projects, such as the global weapons complex. States on the periphery can incorporate into the worldwide cybersecurity system by pursuing commercial agreements with transnational companies.

In a thorough review of the literature, Çağlar Kurç and Stephanie G. Neuman conclude that there is a threefold debate regarding the political economy responses that developing states can take when gauging defense industrial policies, including sensitive information and online systems for security and development. The authors argue that the emerging countries that favor export liberalization want their defense industries to be integrated into the global production chains through foreign investment. Governments can withdraw from the production of traditional and nontraditional weaponry and allow local private companies to attract foreign investors to create a competitive market. "There is a need to abandon protectionism and promote competitiveness in domestic defense markets, and local companies can focus on the manufacture of niche defense products," they note. In countries that favor market protection policies, central governments reject an export-oriented, integrative defense industrialization policy. These states seek to decrease dependency on supplier states for access to high technology to gain autonomy and power. Emerging states perceive national defense industries as a tool for "decreasing the influence of supplier states." Finally, countries that seek a wealth production policy believe that the national defense industry serves as a locomotive for the whole economy and enables growth. Human skills gained in military service can be "transferred" to the civilian sector by personnel trained in the military, such as technicians, mechanics, pilots, and health care personnel. Resources such as infrastructure and communication networks constructed for the military also benefit the civilian economy, Kurç and Neuman argue, adding that "in emerging states, the national defense industry is a tool to create an image of a weighty state. It is the symbolic key to attain international stature."[75]

What is standard across the typologies is that direct and indirect Internet insecurity and cyber incidents put all industrial efforts in jeopardy. Not only are unprotected systems more prone to breaches of security but the sense of producing cyberweapons and tech, which are at their point of origin susceptible violation, is profoundly worrying. Information systems within the defense industry are equally liable to fail when a person or a group are irresponsibly unprepared to stop and prevent breaches of personal information, as well as other incidents triggered by malware, botnets, and fraudulent flaws.[76]

The economic challenges faced by the defense industry in the emerging countries are usually taken into consideration by politicians and defense authorities considering the cost-effectiveness of each new asset, whether locally or internationally purchased. They not only want to enhance their grip on the mechanics of today's battlefield but also to effectively integrate both the military and nonmilitary technologies available to them. From an Anglo perspective, retired British senior army general Richard Shirreff has argued that hybrid conflict demands precision-based new capabilities, including what the traditional military tactics might do in combination with cyber capabilities. The imperative in deriving the precision of war strategies is based on acquiring those state-of-the-art technologies best suited to protect and defend the air, sea, land, space, and cyberspace environment.[77]

Conclusion

This chapter has brought into discussion the traditional arms complex and cybersecurity. I argued that the underlying power of conventional weapons in the Western Hemisphere is currently being challenged by the technological and operational differences presented by new cyber technologies. Military deterrence in the region through increased military weaponry was long understood, given traditional threats of force, which are quickly changing. Coercion by tanks, airplanes, and missiles is giving place to coercion via cyber weapons. Today, deterring a cyberthreat is a difficult proposition. As Nye Jr. puts it: "In the cyber realm the effectiveness of deterrence also depends on who (state or nonstate) one is trying to deter and which of their behaviors. Ironically, deterring major states from acts of force may be easier than deterring nonstate actors from actions that do not rise to the level of force."[78] At this point, in the developing Americas the deterrence of cyberincidents might not lack credibility as much as it suffers from a lack of capability. Another strategy of directing deterrence against hostile actors would be through "personalized deterrence," or the act of a government targeted by cyberincidents "communicating directly to individual cyber attackers their intent to hold them personally responsible through denial of benefits and use of criminal law."[79] Although it won't be easy to attribute and identify individuals, it is said to be much riskier and more costly to wrongly put the blame against a state.

The military in the global South must rethink how to best deter potential threats with limited resources. In some countries, the army is there simply to provide uncertainty to an aggressor, though it might not be capable of effectively mounting a defense. In other cases, the military forces are most certainly capable of building a defense that could make things go badly for any potential aggressor. Finally, some militaries pose a threat of retaliation against assets belonging to an aggressor, who might not care to risk the additional costs. But that requires armies capable of reprisals and of making such a threat credible.[80] In cyberspace, some believe deterrence is not useful due to hackers being unidentifiable and invulnerable to counterattacks; thus, a potential waste of resources that might even provoke conflict with another state rather than prevent it.[81]

What I have theorized in the chapter is that preventing a cyberattack, minimizing its effects, and catching its perpetrators requires reinforced public and private capabilities from military and civilian technologies. Their cooperation is vital if the deterrence element is to be enforced. As put by Rebecca Slayton, the costs of cyber operations "will depend not on the features of technology alone, but instead on the skills and competence of the actors and organizations that continually create, use, and modify information technology."[82]

Richer countries can enhance and develop initiatives at all state levels to boost cybersecurity capacity. The gap between them and poor nations, however, leaves them exposed to the harms of cyberattacks, considering all the nations currently active in a connected world.[83] In the Western Hemisphere, low levels of political risk, a healthy economy, and appropriate budgets for the defense sector should, in theory, indicate the capacity of civilians to govern and enforce any technological policy for cybersecurity. In practice, however, Latin American and Caribbean industries do not always reflect the three conditions. The current political and economic crises in Brazil, for example, have not restrained the authorities from creating a potentially self-sufficient military industry focused on aerospace, small arms, and naval production, and increasingly toward cybersecurity too. Brazil has developed relationships and commercial ventures with other international defense companies to co-produce technologies and to transfer knowhow from abroad.[84] Chile, and Argentina to a lesser extent, have grown remarkably since democratization in 1990. However, it is outpaced by Brazil's far more advanced military industry.

As chapter 4 will discuss, countries tend to emulate Brazil's cybersecurity capacity and its success in developing its defense industry, although military austerity haunts the Southern Cone. Countries such as Argentina, Colombia, Chile, Mexico, and Venezuela put political resources and a share of the state's purse to secure certain services that lower their political risk for investment, namely, interstate conflict, terrorism, and physical safety for workers.[85] We thus need to further understand the governance behind cyber operations and the processes of creating and using information technology.[86]

For the second subset of nations, political, social, and business conditions do not predict a bright future, and the chances of relying more enthusiastically on military imports, thanks to economic booms and off-budget purchases, present the most plausible solution (here I'm thinking of Paraguay, Venezuela, Uruguay, and Ecuador). These nations have fallen prone to political instability and populist policies that diminish investments in their economies. Their governments have maintained weak democracies (except for Uruguay's) that allow for gaps in the enforcement of the rule of law and market conditions. Nevertheless, these countries represent a profitable market for regional and extraregional arms sellers whose lower-cost products might seem highly attractive despite a domestic military industry and cybersecurity capabilities. For this second subset of nations, I will discuss in more detail the case of Venezuela.

In chapter 4, I will continue analyzing the changing cybersecurity governance, the perceived strategic threats in the region (interstate conflict, transnational crime, terrorism, paramilitaries and insurgencies, social and ethnic tensions, natural disasters, and climate change) and how these relate to current military-industrial progress. The chapter presents a methodology and six comparative case studies to review my arguments. The case studies selected are Argentina, Brazil, Chile, Colombia, Venezuela, and Mexico.

Four

Strategic Threats

This chapter will assess Latin America's current strategic threats. It surveys the nature of insecurity in the region, defense technological innovation, and cybersecurity governance. The manifold security threats in the region, where death and murder statistics in recent decades are higher than in former civil wars, have moved some observers to call for an "irregular warfare" type of scenario. Under current conditions, armed nonstate groups are able to infiltrate state agencies, assassinate officials, seize and govern pieces of territory, and erode democratic legitimacy.[1] Nations are left with limited fiscal resources to produce the necessary means, including cybersecurity, to counter these phenomena.

This chapter argues that at least seven evident risks to security in the Western Hemisphere can force states to rely on novel military and nonmilitary measures supplied by the defense industries: territorial conflict, insurgencies and paramilitaries, transnational criminality, terrorism, social and ethnic tension, natural disasters and climate change, and cyberthreats. Most of these threats to peace are today considered to bear cross-continental relevance, forcing certain states to care more greatly than others.

Countries most worried about their territorial sovereignty in the Americas have tended to expand their domestic military industries. For example, Colombia, Chile, and Brazil use sizeable portions of their national budgets to secure a flow of monetary resources to their defense sectors, as they feel territorial disputes might rekindle old enmities with their closest neighbors. Countries also depend on military imports to sustain some of the most advanced forces in the developing Americas. These nations have augmented their military capacities by entering into industry partnerships.

For instance, they have created security alliances with the United States, which the anti-U.S. countries in the region, and the less militarily industrialized, have not (e.g., Ecuador, Bolivia, Nicaragua, among others). Because territorial conflict is less than probable in the near future, military industries are turning to manufacturing nontraditional security technologies replacing in stock those destined for traditional military means.

Armed insurgencies and paramilitaries remain in poor and underdeveloped territories in the Western Hemisphere, gaining social control and challenging state forces. Recently, in Paraguay the rebel group Ejército del Pueblo Paraguayo (EPP) has usurped remote areas where state presence is limited. The military industry in Paraguay is nonexistent, and carrying out counter-criminal operations has spread corruption among the security forces. Paraguay is a significant route for drug-traffickers traveling between Argentina and Brazil. Lack of control of the Tri-Border area has allowed illegal operations by nonstate armed groups, including the Asociación Campesina Armada (ACA), to flourish. Paraguay's high political risk scenario and low levels of economic development have put the armed forces at the center of an illegal trafficking ring that diminishes the chances of establishing a military industry and fuels black markets in other hotspots of violence, including, El Salvador, Honduras, Colombia, Peru, and Venezuela.[2]

Countries with more developed military industries, such as Colombia and Brazil, have converted their arms industries to produce anti-insurgency weapons and all-purpose vehicles, to be deployed in both rural and urban areas. These types of military operations in a highly corrupt environment have enhanced the risks of spreading abuses in those areas of defense related to policy, finance, procurement, operations, and personnel issues.[3]

The most pressing security concern in the region is nonmilitary, but it has called for the armed forces to intervene, with poor results so far. Transnational crime is driving the military industry to pursue technological changes, as new means are needed to seize criminal networks (e.g., in Mexico, Argentina, Colombia, Peru, among others). Brazil set an example in the region. Discouraged by the cost of installing sensors along its 17,000-kilometer border, the government decided to focus on the centers of arms and narcotics trafficking, namely, urbanized areas. The industry has been affected nevertheless, as legislation passed in 2016 capped the federal budget, thereby preventing increased investment in border security.[4] Brazil has developed a black market industry that is ready to provide arms and other means to criminal groups, such as the First Capital Command (Primeiro Comando da Capital, or PCC), which operates on both sides of the border with Paraguay. It's no wonder that scholars have suggested that Latin America's unique ecology of crime and violence has fueled domestic cybercrime originating in and targeting local economies. The cybercrime specter surrounding the Americas has been called interesting, complex, and even fascinating given "the unique interaction of the continent's economic structure, organized crime groups' operations, the geographic position and legislative and policy framework," argues Kshetri.[5] Williams and Quincoses have exposed how transnational organized crime groups have strategically diversified their trafficking routes, activities, and markets. Their research findings claim that criminal groups have conducted cyberattacks against Brazilian government agencies in order to obtain false documentation that permits them to conduct illegal logging activities in protected areas of the Amazon rainforest. "Such activities illustrate the growing relationship between technology, illicit behavior, and criminal groups' diverse capabilities," they note.[6] More recently, amid the COVID-19 pandemic, cybercriminals replaced physical markets and opened new online avenues to keep flooding the local population with drugs and counterfeit goods taking advantage of the strict quarantine measures imposed by governments.[7]

Countries affected by terrorism, on the other hand, tend to rely on domestic military industries to create ad hoc means to counter this modus operandi. For example, Colombian industry supplies the armed forces and national police with special antiterrorist equipment, vehicles, and technologies to continue the fight against narcoterrorism. Where lesser military industrialization has occurred, the United States has significantly provided countries with technologies and platforms for their policing, intelligence, and military agencies (e.g., Peru, Colombia, and Central America). In Peru, the United States has agreed to provide military cooperation in counter-narcotics and counterterrorism operations in consideration of Peru's less-favorable military capacity and diminished indigenous industry. During previous dictatorships and civil wars, state-sponsored terrorism pushed the indigenous military industry to produce means for national security. This was mostly a countereffect to offset the weapons embargo put in place by the advanced nations due to gross abuses to human rights.

Marginalization, inequality, and poverty throughout the Western Hemisphere have created the conditions for social and ethnic tension to persist. Political instability has led to clashes between groups and state security forces in most of South America, including countries such as Brazil, Chile, and Argentina, where rural poverty, indigenous rights, and natural resource extraction by private companies has created hotspots of violence and insubordination against the rule of law. Militarized police forces have been sent to control protestors, protect private property, and safeguard roads and critical infrastructure (e.g., Araucania in Chile, northern Amazonia in Brazil, and Cajamarca in Peru, among others). The military industries have provided the militarized security agencies with modified all-terrain vehicles, bulletproof gear, nonlethal weapons, and surveillance technologies.

Recent natural disasters have called for humanitarian responses from the military across the Western Hemisphere, for example, after the 2010 earthquake in Haiti. The United Nations Stabilization Mission in Haiti (or MINUSTAH, 2004–2017) converted its core mission from security to reconstruction.[8] In other countries, such as Brazil, Chile, Peru, and Bolivia, the armed forces are central to post-disaster responses, as they possess the organizational and logistic means to react resourcefully to unpredictable and disastrous events. The military now considers budgets for an array of humanitarian missions, ranging from mobile hospitals to vehicle-launch bridges once intended for use in war scenarios. Climate change has also attracted the attention of the armed forces. There is a growing preoccupation with the scarcity of natural resources, such as water and farming land. Control of water supplies could exacerbate tensions between states, especially those with blurred borders that sit on top of natural reservoirs. Governments have redirected their military to protect basic services infrastructure and natural resources including, minerals, oil, and natural gas.[9]

After recent cyber insecurity incidents, governments have proven not to have the monopoly on violence; just as it happens with some of the above-described threats to security. National control over Internet infrastructures has increased in the region as the government tends to build walls in the digital space. Although the region is exposed to global incidents, most of the cyberthreats affecting Latin America are localized through the in-house CSERTs. The term *cyber conflict* is still a premature concept to be imposed on the region, where state-to-state aggression has not been evidenced as much as nonstate actors attacking economic, political, and social structures. Valeriano and Mannes define cyber conflict as "the use of computational technologies, defined as the use of microprocessors and other associated technologies, in cyberspace for malevolent and/or destructive purposes to impact, change, or modify diplomatic and military interactions between entities."[10] Countries such as Brazil, Mexico, Argentina, and Chile have increased their efforts to assimilate many U.S.-led priority actions around the areas of identifying and prioritizing risks to key national areas (defense and public security, energy, banking and finance, commerce, critical infrastructure, communications, and transportation); build protected government networks; deter and disrupt malicious cyber actors; and improve information sharing, education, and privacy.[11] In general, and as presented in the earlier chapters, Latin America confronts a scenario of increasing cyberincidents, including cybercrime, commercial theft, and foreign operators causing rogue actions against individuals, organizations, and governments.

Analysis and Case Studies

This chapter offers a small N inductive study on the changing security missions in Latin America, military industrialization, and cybersecurity responses. The analysis is focused on nations that have active military industries; thus, the case selection involved choosing countries according to the recent developments in their defense political economy.

The chosen countries were Argentina, Brazil, Chile, Colombia, Mexico, and Venezuela. The World Bank considers Argentina and Chile as high-income countries, while Brazil, Colombia, and Mexico are upper middle incomes. Together the six case studies make up for almost 84 percent of Latin American and Caribbean GDP. Mexico, Argentina, Chile, and Colombia have had a steady defense expenditure since 1990. Brazil and Venezuela make a difference. The latter expanded its military budget massively since the early 2000s, which climaxed in 2011, and after two years of recession it peaked again in 2017, reaching nearly 30 billion dollars. Venezuela has shown highs and lows, however, plummeting since 2013 and recording 464 million dollars in 2017.

Most countries showed a lower percentage of central government expenditure in the military in 2020 than in 2007, with only Mexico showing an increase (see Table 4.1). However, when considered in terms of GDP, Mexico and also Colombia increased their military expenditure (see Table 4.2). The selected cases have undergone an increase in military personnel over the last decade, with only Argentina receding. The most notorious jump is in the case of Venezuela, which went from 115,000 members of the armed forces in 2007 to 343,000 in 2019, an increase probably due to the use of paramilitary forces in support of the regular military forces (see Table 4.3).

Cases that were not included in the sample (Paraguay, Uruguay, Ecuador, Peru, and Bolivia, among others) show smaller defense budgets and lesser or no military industrialization. These nations remain importers

Table 4.1. Military expenditure (% of central government expenditure)

	2007	2018
Argentina	2.7	2.2
Brazil	3.9	3.9
Chile	12.1	7.4
Colombia	11.7	11.6
Mexico	2.0	2.1
Venezuela	—	—

Note: No figures were provided in the case of Venezuela.
Source: Stockholm International Peace Research Institute (2020).

Table 4.2. Military expenditure (% of GDP)

	2007	2018
Argentina	0.8	0.9
Brazil	1.5	1.4
Chile	2.3	1.9
Colombia	3.3	3.1
Mexico	0.5	0.5
Venezuela	1.9	1.2*

Note: *Venezuela's figure corresponds to 2014.
Source: Stockholm International Peace Research Institute (2020).

Table 4.3. Armed forces personnel

	2007	2017
Argentina	107,000	105,000
Brazil	721,000	730,000
Chile	103,000	122,000
Colombia	411,000	481,100
Mexico	286,000	336,050
Venezuela	115,000	343,000

Note: Armed forces personnel are active-duty military personnel, including paramilitary forces if the training, organization, equipment, and control suggest they may be used to support or replace regular military forces.
Source: World Bank (2019).

of weapons and military technologies. In a different categorization, Sanchez Nieto argued that the purchase and production of weaponry in the continent revealed a "soft" arms race between contrasting levels of intensity in expanding defense equipment with "major spenders" (Brazil, Chile, Venezuela), "medium spenders with simmering armed conflicts" (Colombia and Peru), "medium spenders" (Bolivia and Ecuador), "low spenders" (Argentina, Paraguay, Uruguay), and "minor spenders" (Guyana, Suriname).[12]

For the research objective of this book, Argentina, Brazil, Chile, Colombia, Mexico, and Venezuela work as comparative case studies as they present the background conditions that follow the argument of the chapter: due to a changing strategic security scenario their means to assess warfare have changed, affecting in different degrees the defense industry and hi-tech capabilities used for cybersecurity and cyber defense. Nevertheless, this has not caused the domestic military industries to grow equally. By focusing on the selected countries, the chapter provides attention to details that tend to be overlooked when thinking, either of the state of cyber capacities in the Western Hemisphere as a whole or through a case-by-case analysis. The chapter's comparative case selection provides some partial generalizations, which are useful as a first step, but that might benefit considerably if they were replicated in different settings across the developing and emerging economies.[13] The chapter aims for theory development to create arguments and hypotheses useful to, as James Mahoney put it, those "comparativists who seek to move from research topics to specific proposition."[14]

Choosing a different set of case studies might influence the chapter's conclusion in the sense that defense capacity has risen over the last decade, creating initial opportunities for the local defense sector to build partnership and, potentially, offset procurement deals. Peru and Ecuador went to war against each other in 1995 and revamped their armed forces before and after the conflict through foreign purchases. Canada was also excluded from the sample as its defense industrialization is more similar to the United States'. However, Canada's defense spending is considerably lower and comparable to the chapter's case studies and their military expenditure (US$ 20 billion in 2017). Canada's government has been keen lately to introduce foreign participants to its domestic industry and develop further resources to protect passages in the maritime area of the Arctic Ocean.

The seven sources of insecurity (territorial conflict, insurgencies and paramilitaries, transnational criminality, terrorism, social and ethnic tensions, natural disasters and climate change, and cybersecurity) facilitate the gathering of data concerning the current missions of the military. These categories serve the purpose of contextualizing comparison and the scope of the conditions in which to present each case study.

Methodology followed the argument that every case selected can "show sufficient similarity to be meaningfully compared."[15] We can extrapolate that the cases have different security patterns while still presenting common factors relevant to this study: the presence of defense industrialization and evidence of their armed forces changing roles and missions toward new threats other than the military. Countries with the export liberalization of military industrialization are the ones affected by the greater threats to security, be these of traditional or nontraditional nature. Mexico and Venezuela are havens for transnational criminality and paramilitaries holding weapons outside the rule of law. As well, they show high levels of violence and social tension. Both countries have been affected by natural disasters recently (Mexico is prone to earthquakes, and Venezuela to water droughts) that one way or another have forced the military to engage in humanitarian assistance missions.

The countries with midlevels of military industry development (Colombia, Chile and Argentina) have low social and ethnic tension (high when examined in particular regions, but low when the entire country is considered); while natural disasters and climate change have affected Chile most recently (wildfires in 2017, earthquake and tsunami in 2010). Regarding insurgencies and paramilitaries, transitional criminality, and terrorism, Colombia and Argentina constitute a subgroup apart. Chile has by and large kept nonstate armed groups from entering the country; while it only dealt with armed remnants of leftist groups that fought the military dictatorship (1973–1990) during the early 1990s.

Brazil, the only country with a higher level of development of military industrialization, shows a particular security scenario, with low levels of territorial conflict but medium and high levels of the indicators regarding armed paramilitaries, transitional criminality, terrorism, social and ethnic tension, and natural disaster and climate change.

Next, the chapter briefly discusses some recent events in the realm of military-industrial progress and the type of defense industrialization policy adopted in each case study.

Argentina

Argentina is keen to take a more active part as a defense exporter, although it has not produced an output comparable to Brazil's. Its in-house production capacity can make its industry relatively competitive in price in the Latin American context. The government has focused some resources on industrializing ways to complement the territorial mission of the armed forces in light of the dispute over the sovereignty of the Falkland/Malvinas Islands. Lower consideration given to fighting insurgency and paramilitaries has not created opportunities for industry as much as they have for transnational criminality and terrorism. Low social and ethnic tensions and the use of military equipment in response to cases of natural disasters and the effects of climate change have not stimulated the growth of necessary industrial capacities. the exigencies presented by these last three categories have required that the Argentinean armed forces rethink their current military missions and roles. For example, the armed forces are taking a more serious approach to countering secret drug laboratories. When the city of Rosario, among others, became a hot spot for drug production, the government launched a militarized response in 2014, together with an expanded military presence in the remote northwest zones. Criminal syndicates from East Asia and the Middle East, with some alleged connections to militant groups, have been suspected of using the borderlands with Brazil and Paraguay for fundraising through illegal economic activities.

Overall, the armed forces in Argentina remain poorly resourced in the face of pressing concerns to recapitalize their air controls, provide land equipment to access remote areas, and acquire sea technologies to stop the flow of drugs crossing the Atlantic to Europe. Due to a decrease in territorial conflict with their neighbors, the recent government has not been moved to increase defense budgets and is more concerned with cleansing public finance. However, the ministry of defense is expected to invest in submarines, multi-role aircraft, and transport aircraft. This includes a contract with France for the delivery of five Super Etendards to the Argentine Navy.[16]

Although Argentina continues to produce, domestic defense capabilities are restricted to the manufacture of trainer aircraft, transport aircraft, submarines, and small arms through a manifold of local companies, such as the Fábrica Argentina de Aviones SA (FAdeA), Astillero Río Santiago, The Institute of Scientific and Technological Research for the Defense (CITIDEF), Investigación Aplicada (INVAP), Dirección General de Fabricaciones Militares (DGFM), and Tandanor, and it cannot compete with more cost-effective world producers. Although Argentina is a frequent member of multilateral bodies, such as the UN Security Council, and takes part in the G-20, observers argue that its "lack of operationally effective armed forces makes it an anomaly in the international system."[17]

The small size of the armed forces keeps the domestic market from growing. Recent defense industrial policies have been oriented to protect the local market, with relatively low economic openness and limited incentives for defense cooperation and foreign manufacturers of equipment. Government intervention in the economy does not provide investors and companies with a clear legal framework for business. Corruption and limited technological capability also limit market entry for foreign companies.[18]

Argentina has sought collaboration from foreign partners in affairs of defense and cybersecurity. For the G-20 summit in Buenos Aires in 2018, the U.S. military provided security, reportedly using supersonic planes, an aircraft carrier equipped with missiles and radars, and cyber defense equipment.[19] Argentina has advanced the most of its cybersecurity preparedness thanks in part to U.S. bilateral diplomatic and defense cooperation. In April 2017, both governments signed a partnership on cyber policy in the areas of cybersecurity, cyber defense, international security in cyberspace, and law enforcement responses to cybercrime. A bilateral working group was set up to "developing national cyber policy frameworks; the critical role of Computer Emergency Response Teams in protecting networks and managing cyber incidents, information sharing and the protection of critical infrastructure, with an emphasis on public/private cooperation; strengthening the cooperative relationship between our military cyber experts; ensuring effective and independent investigation and prosecution of cybercrime through cooperative law enforcement efforts such as the Budapest Convention." Washington's side was led by the office of the cyber issues coordinator at the State Department, including representatives from the Department of Homeland Security, the Department of Justice, the National Security Council, and the Department of Defense. The Argentinean counterpart was led by the undersecretary for technology and cybersecurity at the Argentine Ministry of Modernization partnering with representatives from the Ministry of Foreign Affairs, Ministry of Interior, Ministry of Security, Ministry of Defense, and Ministry of Justice and Human Rights.[20]

The move differed quite radically from previous years when Argentina sought to scale down U.S. technological and defense knowhow in the region.[21] In September 2013, Defense Minister Agustin Rossi and his Brazilian counterpart Celso Amorim pledged to cooperate closely to improve cyber defense capabilities following revelations of the NSA spying on Latin American countries. "We have established that we will hold a meeting in Brasilia before the end of the year to intensify our complementarity in the matter of cyber defense," Minister Rossi said. Amorim announced that

Brazil would provide cyberwarfare training to Argentine officers, adding that their nations' software industries had "great capacity" that could support any initiatives in the cyber defense area.[22]

Argentina's government seeks investments from military and technology firms annually through a security and surveillance technology conference called Segurinfo. Among the companies that usually attend the conference are Dreamlabs, Widefence, Blue Coat, Cisco, Alcatel-Lucent, and 3M. A report by Privacy International also revealed that the German government had sold surveillance technologies to Argentina for a decade. In 2015, it was established that the Italian spyware vendor Hacking Team negotiated with several intermediaries in Argentina that were said to have ties to law enforcement agencies and state intelligence bodies. This information came after a major leak of internal information from the Italian firm, but it is not clear that any transactions took place. The exchange of information between Argentina's cyber intelligence agencies and foreign partners, including Germany, the UK, the United States, France, Australia, China, and Israel, has increased over the last few years. For example, the Argentine Ministry of Defense signed a deal with its counterpart in Israel for the total sum of US$5.2 million to provide cyber defense and cybersecurity services aimed at strengthening their CSIRT.[23]

In conclusion, the defense industry in Argentina does provide equipment and hardware to the local and regional armed and security forces, however, with modest growth to allow major modernizing programs for new cyber technologies. While the defense budget goes to cover personnel costs, the government has relied on foreign partners to fill the vacuum of cyberspace capabilities. The nation's recent economic crisis and fiscal austerity brought about by President Macri's reforms have led some to forecast a gradual decline in military spending driven by a cost-cutting model to cover the barely functional capabilities of the armed forces. Modernization for the cyber era will depend mostly on purchasing secondhand equipment or through preferential agreements such as the one with the United States. A smaller and better-prepared military might be ready to handle cyber technologies if the knowhow keeps flowing and the government decides to acquire relatively cheap technological and cyber capabilities. Argentina's niche technologies supplying military aircraft and aerospace components are expanding in the defense electronics domains through new endeavors led by Fabricaciones Militares and INVAP, two of the main military companies in the country. It is expected that foreign partnerships and technology transfer agreements might boost the country's defense suppliers, offering them an exclusive opportunity to acquire cybersecurity and warfare components, which could be replicated domestically soon. The lack of concern for cybersecurity-related technological buildup makes Argentina dependent on defense imports.

Brazil

Recent defense expenditure in Brazil has shown an upward trend as the armed forces assume new roles in domestic security, border surveillance, and cybersecurity. The procurement of military equipment has been favored by an export-oriented and wealth production combined industrial policy. Challenges remain after Brazil's economy stagnated after the commodities boom of the early 2000s. Brazil is a regional power and seeks world status through its military and security capabilities. Military modernization combines home-made equipment and purchases to defense conglomerates in military aircraft, armored vehicles, warships, and submarines. Among other projects, the native company Equitron Automacao Eletronico Mecanica and the Brazilian armed forces produced the Engesa EE-9 Cascavel, a 6x6 reconnaissance vehicle.[24] Also, in its attempt to install cutting-edge border security, the government planned the Sistema Integrado de Monitoramento de Fronteiras, a border control system that includes radar and detection equipment and sensors. Another example of Brazil's burgeoning industry includes the REMAX lightweight remote-controlled weapon stations from ARES Aeroespacial e Defesa ordered by the Brazilian Army Logistics Command.[25] A changing scenario for the military provides the necessary ground for the defense industry to respond to a demand for alternative air, sea, and land capabilities. The country's efforts to counter transnational criminality plus domestic sources of insecurity and social tensions in urban areas ignite defense spending and steady growth in defense exports. As well, a large armed force represents a favorable market and R&D environment for the defense industry to grow in the medium and long term.

Brazil is using its in-house defense industry to jump into the flourishing cybersecurity market. Brazil's aerospace design and technological developments, for example, have pushed military manufactures, spending, and procurement. Many of the cybersecurity platforms remain reliant on foreign suppliers who can offer products of better quality and lower prices. Brazil's willingness to overcome insecurity threats (criminality in the Amazon, international peacekeeping and, at home, humanitarian roles) have convinced the authorities to continue funding capabilities in surveillance and cyber defense procurement contracts with domestic enterprises dedicated to R&D.

Very much like Argentina, Brazil has set up a bilateral cooperation working group on cyber defense and Internet policy with the United States. The U.S.-Brazil Internet and Information and Communication Technology (ICT) Working Group assembles senior representatives and other nongovernmental stakeholders working together "to strengthen our joint commitment to an open, interoperable, reliable, and secure cyberspace." This group is an example of networked cyber governance, as theorized earlier in the book. Although less emphasis is put into accelerating the defense industry to grow local capabilities, the motley group of senior officials cares for many areas, including cyber defense and cybercrime, data protection, and data privacy, ICT procurement, and military and law enforcement cooperation

Apart from the evident hands-on role of the U.S. Department of State, the interagency nature of the group is palpable, as many other stakeholders take part in this cyber policy network. For example, the United States participates with representatives from the Department of Homeland Security, the Department of Justice, the National Security Council, the Department of Defense, and the Department of Commerce. Experts also come from the U.S. Trade Representative's Office, the Federal Communications Commission, and the Federal Trade Commission. The Brazilian side is made up of senior representatives from the Department of Science, Technology, and Innovation at the Ministry of External Relations, and the Secretariat for Digital Policy at the Ministry of Science, Technology, Innovation, and Communications. The Brazilian delegation also includes officials from the Office of the Presidency, Institutional Security Cabinet, Cyber Defense Command, the telecommunications regulatory body Anatel, as well as Brazilian Embassy attachés to Washington from the Federal Policy, Ministry of Defense, and Brazilian Intelligence Agency.[26]

Would this translate into a boost for science, R&D, and the military industry? Defense budgets in advanced military technology are now a priority in Brazil with 1.25 percent of GDP allocated to military science and technology, by far the largest commitment in the region. According to Mora and Fonseca, the increased focus of some countries in increasing science and technology spending reflects their greatest perceived threats and "tracks with Washington's threat perceptions in the region." The authors cite Ricardo Arias, program manager at the United States Southern Command's Science, Technology and Experimentation Division, who argues that Washington's science and technology efforts in the region have similarly been oriented toward improving Latin America's capacity to counter illicit trafficking and criminal networks and monitor land, air, and maritime spaces threatened by non-state actors. U.S. agencies have invested millions of dollars in research efforts that support the region's capacity to combat these asymmetrical threats. The U.S. Departments of Defense, State, Homeland Security, Transportation, and Treasury have all supported science and technology in the region.[27]

Brazil's Sistema Integrado de Monitoramento de Fronteiras (Integrated Border Monitoring System—SISFRON) is perhaps the country's most refined attempt to enhance overall interagency coordination on a large scale, using a mix of surveillance radars, remote sensing networks, complex communications systems, cyber technologies, and unmanned aerial vehicles (UAVs). SISFRON is a large-scale electronic barrier across Brazil's 10,500 miles of borders with Argentina, Bolivia, Colombia, French Guiana, Guyana, Paraguay, Peru, Suriname, Uruguay, and Venezuela. The fence was developed by the Embraer-run Tempro Consortium and some foreign technology companies.[28]

Although one might say that cybersecurity operates entirely in the digital space, most defense and homeland security endeavors are a bridge between IT and operational technologies. Brazil's aerospace and defense industry have followed this rationale to provide an extended security ecosystem aimed at rendering impenetrable the government, private businesses, citizens, and infrastructure. Aerospace and defense companies have adapted quickly to cyber-specific issues, such as securing highly sensitive data and controlling critical national systems.[29]

In April 2017, President Michael Temer inaugurated the biennial version of LAAD Defense and Security, Latin America's biggest and most important military industry exhibition with more than six hundred exhibitors and forty thousand visitors from across the globe. Temer stressed the importance of the national defense industry in the recovery of the Brazilian economy and announced the creation of a new division in the Chamber of Foreign Trade, CAMEX, to promote the export of defense products and services.[30] In 2016, the Brazilian government set up the Cyber Defense Command (CDCiber) within the Army (EB, per its Portuguese acronym), to play a fundamental part in the National Defense Strategy. CDCiber serves as the planning, coordination, direction, integration, and supervision body of cyber operations in the defense area and is the shining star of the

Cyber Defense Program in national defense. The entire defense project was valued at about US$104 million. It included the development of hardware and software solutions, and the procurement of supercomputers and digital research materials to foster the growth of this field in Brazil. ComDCiber projects include the National Cyber Defense Academy to train civilian and military human resources; and, the Cyber Defense Products Standardization and Certification System, which will certify hardware and software for use in the cyber sector. ComDCiber also set up a Cyber Defense Observatory and Cyber Defense Talent Management to promote research and development in domestic technology.[31]

CDCiber also pushed for more resources. In 2013, Brazil bid to buy a strategic communications satellite responding to the community of experts' call for more protection in Brazilian systems. Coordination and training have also taken an extra step, for example, in the months before and after the Football World Cup and the Summer Olympic Games. Cyber officials and other personnel are trained at the Institute of Cyber Defense at the Cyber Operations Simulator (SIMOC). In 2018, military and civil organizations ran Cyber Guardian, a cyber operations exercise that utilized the SIMOC.[32]

Chile

Chile is a defense importer, mainly from the United States, Israel, and Western Europe. Chile's military procurements were financed by a 10 percent tax on the revenues of the state-owned copper miner Codelco, under a program known as the Copper Law. Lack of transparency and corruption has encouraged some politicians to change the financing system of the defense sector. Chile's military is well provisioned, mostly for the purpose of deterring interstate conflict, and, most recently, has turned heavily to humanitarian and natural disaster relief tasks. Because the armed forces do not deal with insurgencies, criminality, terrorism, or social conflicts, it is puzzling to understand whether these highly common threats across the hemisphere will have an effect over changing military missions and the defense industry. Different factors would seem to demand a change in Chile's military missions that will probably require a substantive renovation of its defense sector, from doctrine to equipment.

Governments have been willing to upgrade military equipment and maintain a technological edge over neighboring armies, which drives local industrialization and offsets purchases. The focus has changed lately to air and ground transport equipment, which can be used for both combat and nontraditional operations, as well as high-tech defense systems, with a strong focus on the navy. For example, in September 2016, the Chilean Air Force announced the purchase of six S70i International Black Hawk helicopters and is also seeking to purchase amphibious aircraft.

Chilean FAMAE is an independent state company producing arms and ammunition, weapons systems, and R&D for military and civilian products. Like Brazil's, Chile's industrial policies are driven by wealth creation and export consolidation. FAMAE's strategic alliances with local and foreign companies make the co-production a priority. In June 2016, FAMAE and Israel Weapon Industries (IWI) presented their new assault rifle Galil Ace 22N-C which will be used by the Chilean army. Increasing participation in international operations makes Chile a stable business environment for foreign investors. However, FAMAE has reported a decline in sales, and the government recently withdrew funding for ASMAR shipyard expansion.[33] Defense expenditure is determined by commodity prices that sources overall government spending. Low prices of copper exports can affect the outlook for defense imports and local investments.

Cyber technologies tied to the defense industry are mostly absent in Chile. Although major efforts have been made in creating a governance framework for cybersecurity, most endeavors are occurring at the political and policy level. Over the last years, the Ministry of Defense proposed a series of civilian-led policies on military education, cyber defense, the military-industrial complex, transparency, and weapons acquisitions, among others. However, no projects for the local industry to grow cyber hardware, software, or any other technologies were in the schedule.[34] In 2015, Chile set up an interministry committee on cybersecurity, whose main objective was to devise and propose to President Michelle Bachelet a national policy on cybersecurity, published officially two years later. Since then, Chile has moved closer to the United States in terms of cyber consultation. In September 2018, senior representatives from Chile traveled to Washington, D.C., "to facilitate stronger bilateral cooperation on cyber issues, including government capacity to address emerging challenges and shared threats in cyberspace." In the two-days visit, the officials discussed ways to improve cybersecurity, including the protection of critical infrastructure, combatting cybercrime, and enhancing cyber defense. Although it was highlighted in the media briefing that "additional discussions with the private sector emphasized the benefits of public-private partnership on cyber policy," no details on this were made publicly available.[35] What is clear from these types of meetings is that cybersecurity governance occurs among various state stakeholders within the security and military branches of government. In the United States, the lead on interagency activity seems to come from the secretary for cyber and international communications and information policy at the Department of State, and involves senior representatives from the National Security Council, the Department of Homeland Security, the Department of Justice, the Department of Defense, the Department of the Treasury, and the Department of Commerce. On the Chileans' side, the undersecretary of defense led the delegation that also included officials from the General Secretariat of the Presidency, Ministry of Foreign Affairs, the Public Prosecutor's Office, the National Intelligence Agency, and the Ministry of Interior. It is unclear if any of these delegates represented the interests of the defense industrial complex.

As it did with Brazil and Argentina, the United States has sought in Chile a partner to promote consensus on voluntary and nonbinding norms of state behavior in cyberspace during peacetime and international cyber stability. Although these terms are contested among scholars, for countries they seem to take on value as the "advancement of efforts within the Americas to build trusting partnerships among like-minded nations," to "deter malicious cyber activity contrary to this framework." This means for the United States to work closely in collaboration with nations to address matters related to "critical infrastructure protection, incident response, data protection, information, and communication technology procurement, international security in cyberspace, and military and law enforcement cooperation."[36]

The pressure to respond with a better governance framework made President Sebastián Piñera sign a bill on computer crimes and issue a directive to public bodies to take urgent measures to implement the new legislation.[37] The move came closely linked to the appointment of a national cybersecurity officer below cabinet level, to oversee all government action on cybersecurity. The presidential order mandates the appointment of a high-level cybersecurity officer in each service; the application and updating of technical regulations on cybersecurity; identification of internal cybersecurity risks; revision of networks, systems, and digital platforms in public operation; surveillance and analysis of the operations of the technological infrastructure of state administrative bodies; compulsory reporting of incidents; and response to cybersecurity incidents. At the time of writing this book, the model was still in the works, awaiting the implementation of the new model of national cybersecurity policy. The lead agency was specified as the Ministry of the Interior, and not Defense. No details regarding the creation of domestic cyber technologies were detailed, either.

Finally, in Chile, responsibility for domestic cybersecurity seems to have been allocated to the Ministry of the Interior, while internationally the agenda resides in the hands of Defense. When Jim Mattis, U.S. Secretary of Defense, went on tour in Santiago, President Piñera and Defense Minister Alberto Espina, joined him in signing a new bilateral accord regarding cybersecurity. Mattis agreed in the same terms to other bilateral alliances with Argentina, Brazil, and Colombia, the remaining stops in his subcontinental tour. Espina argued that "the United States has the technology and the knowledge, and for us, it is fundamental to have an agreement with them to set up training courses and exchange information."[38] Mattis also noted that the United States would collaborate in other military fields such as medical research and the development of remotely piloted aircraft systems (RPAs, or drones). In his speech, Mattis, who shortly afterward parted ways as Secretary of Defense with Donald Trump, thanked Chile for helping to maintain democratic stability in a "sea of uncertainty." Beside him when he spoke was Vice-Admiral Craig Faller, who then became chief of the U.S. Southern Command, operating in Latin American and the Caribbean. [39]

Colombia

Colombia's armed forces are among the most experienced in the Americas at performing both traditional and nontraditional military roles. Active warfare operations include territorial conflict deterrence, combating counterinsurgencies and criminal activity, and antiterrorism operations. Due to recent political stability, social tension has decreased, thus not requiring armed forces intervention in public security. Natural disasters and climate change, on the other hand, are demanding changes in military missions, with the Army developing in 2009 a battalion to attend to emergencies and provide relief operations.[40] Although the armed forces and national police require equipment and technologies to assume these manifold conflict scenarios, securing the post-peace accord with the FARC and the pursuing ongoing negotiation with the insurgent group Ejército de Liberación Nacional (ELN) take up most of the armed forces budget especially since the dissolution of nonstate armed groups has gone hand in hand with the fight against drug trafficking and production.[41] Power vacuums in previously FARC-dominated territories can strain the armed forces' resources as nonstate actors (right-wing paramilitaries, organized crime, and FARC dissidents) have rushed in to fill the void left behind by the departure of the guerrillas. Unrest over border issues with Venezuela also requires military attention, although lately diplomatic conversations have been held periodically.

After Brazil, Colombia has the second-largest armed forces in Latin America and is expected to keep it that way. Its defense industry, on the other hand, does not reflect the size of its army. The state-owned naval company COTECMAR builds ships for the navy (offshore patrol vessels, coastal patrol vessels, amphibious assault ships, and fluvial support patrol vessels).[42] INDUMIL, on the other hand, manufactures weapons (the Galil assault rifle, produced under Israeli license), ammunition, explosives, and other equipment for the armed forces and the police.

Colombia's defense industry aims to calm domestic demand with some small-scale exports to Israel and other countries.[43] The armed forces continuously replace aging equipment in order to continue confronting ongoing security threats, which demands high costs. The government struggles to approve acquisition and investment budgets in light of more pressing social needs. Through export liberalization policies, the advanced equipment needed for counterterrorism and insurgency operations is purchased abroad as Colombian industry lacks modern design and development capabilities. To promote favorable investment openness, COTECMAR and INDUMIL want to establish cooperative ties with international partners in offset and joint programs. Corruption and the high costs of initiating business in Colombia, however, remain a deterrent to foreign direct investment.[44]

In 2018, the Corporación de Alta Tecnología para la Defensa (High-Tech Defense Corporation, CODALTEC) of Colombia and Indra, the Spanish technology and defense systems company, agreed on the delivery of a cutting-edge tactical air defense radar system for the Colombian Air Force, specially designed to detect aircraft flying at low altitude. This is the first military air surveillance radar that has been fully manufactured in Colombia, and it will be marketed by CODALTEC in the region.[45] Colombia has sought technical collaborations in order to establish more autonomy in its defense systems, among other initiatives led by the Grupo Social y Empresarial de la Defensa (Social and Business Group for Defense, GSED). Indra, on the other hand, is a global manufacturer of command-and-control systems providing data control for military scenarios. Its systems are usually interoperable and integrate equipment used by different military branches of other countries. Indra has formal relationships with 287 companies all over the world, and an ecosystem that includes many IT companies providing cybersecurity solutions, such as Kaspersky, Alcatel, McAfee, Apple, Microsoft, Oracle, Dell, Cisco, FireEye, and Defensoft. Developing strategic capabilities for Colombia opens up the scalability of its defense establishment and the flexibility of its military systems, encouraging them to operate in and secure new areas, such as IT security and cyber defense. Indra is already operating a portion of NATO's radar systems on the southwest flank of Europe. In Colombia, Indra established a Cybersecurity Operations Center (i-CSOC) in the capital city, Bogota, plus a network of offices in Barranquilla and Medellin from where its technology hubs drive projects devoted to technological development in infrastructure, public administration, energy, and smart cities. The array of mobility projects includes a tunnel, subway, streetcars, and highway. Indra has also developed an e-government platform in the Andean city of Tunja, as well as organizing Colombia's 2018 presidential elections.

The CODALTEC startup caters ad hoc solutions to the public forces, including the militarized police, allowing the country to become relatively self-sufficient in some of its technological capacities. The corporation is highly integrated into the Industrial Defense Base, which is what they call the public and private sector companies that meet the technical requirements of both the armed forces and industry. Most of its efforts have been put into developing and designing locally made unmanned remotely operated aircraft simulators, armored vehicles, and radar systems. Not-for-war systems include SALUD.SIS, the vertical health information system for the military forces.[46] Since 2013, Microsoft and the Colombian government have collaborated in the areas of cybersecurity, education, and innovation. The agreement has benefited the National Training Service (SENA) and led to exchanging information on areas in which Colombia sought better training, including how to counter organized crime, child pornography, and identity theft. Microsoft technology has also been used by the NextCity Lab, an endeavor to drive infrastructure growth and the application of ICT throughout Colombia.[47]

In late August 2017, the Ministry of Information Technology and Communications, the Chamber of Computer Science and Telecommunications, the Ministry of Defense, and the OAS all met in Bogotá for the Fourth Digital Security Forum. Government representatives, academics, and entrepreneurs from the private sector discussed the state of the art regarding Colombia's defenses against cybersecurity and cybercrime. Since the 2011 publication of the "Cyberdefense and Security Guidelines," a policy document laying out the country's strategy for countering cyberthreats, Colombian authorities have given resources for the creation of a cybersecurity network across the public and private institutions. Cyber capacity building is said to be performed within the defense sector at three levels: the Colombian Ministry of National Defense and its cyber emergency response team; the Police Cybercenter of the National Police; and the Joint Cyber Command (made up of Army, Navy, and Air Force cyberunits).[48] The Colombian War College (ESDEGUE) teaches a Master's program in which students take classes in cyberattack simulation against critical infrastructure for decision making during a "cyber crisis."[49]

Colombia's Defense Ministry has also taken an active voice in claiming responsibilities for cyberincidents affecting public infrastructure. In the run-up to the March 2018 parliamentary elections, more than fifty thousand attacks on the national voter registry (a digital platform gathering the identification data of more than thirty-five million voters) were detected, according to military intelligence officials. According to Defense Minister Luis Villegas, some of the repeated hacks were linked to Internet addresses in Colombia. Others came from Venezuela. "Russia is increasingly using Venezuela as a base for covert operations in its growing rivalry with the U.S. for international influence that is starting to affect Latin America to a point not seen since the Cold War," said an intelligence analyst and former senior intelligence officer in the Colombian army. The same source added, "The attack on the voter registry was conducted by Cuban computer experts trained over decades in electronic intelligence by Russia, which manages Venezuela's voter registration and national identification systems."[50] In May 2018, Colombia joined NATO as a global partner allowing the Southern nation to participate in initiatives in cyber and maritime security.

Mexico

Mexico's security crisis has called for a fresh wave of defense modernization, especially when it comes to dealing with organized crime violence. The country has moved away from territorial hypotheses of war and concentrates the armed forces' efforts in military hardware for its navy and surveillance material. Recently, foreign suppliers have sold Mexico aircraft and helicopters, naval vessels, and special operations helicopters. It is observed that the Mexican defense industry is small and employs the low-tech subsidiaries of foreign companies benefiting from free trade agreements and lower costs of labor.[51] The lack of international threats to Mexican sovereignty since the nineteenth-century war with the United States has meant that the authorities have never fulfilled the promise to build an army capable of protecting its large geography. As well, and similarly to the case of Colombia, the lack of a central government presence and the existence of smaller militias has meant that the rule of law never spread across the territory in an even manner. Since the revolution in the early twentieth century, Mexico has organized a local industry to provide the local forces with small arms and ammunition (through the Fábrica Nacional de Armas), while falling behind other regional players such as Chile, Brazil, Argentina, and Venezuela. More recently, Mexico has invested considerably in narrowing the gap.[52] Terrorism seems to fade away in Mexico, with some elements of the guerrilla group Ejército Zapatista de Liberación Nacional (Zapatista Army of National Liberation) receding to join mainstream politics. At the same time, other factions remain in strongholds in the mountains.[53]

What remains dangerous is the alleged corruption and collusion between the state's security agencies and some of the most powerful criminal and drug cartels. Police and other law-enforcement agencies are suspected of turning a blind eye to many wrongdoings so that they don't get killed later. As well, due to low salaries, many in the armed forces have decided to cooperate with the organized crime groups by providing information and doing the dirty work of the cartels, although constantly in fear of being arbitrarily killed anyway. This is what they call *plata ó plomo*.

The country's most relevant concerns are, therefore, how to restrain organized criminality and the violence of the recent years of fracturing cartels and vacuums of power from imprisoned drug lords. As well, social tension and natural disasters have put the armed forces together with the police to spend resources on emergency duties due to Mexico's seismic-prone geography. In 2017, the country was almost paralyzed due to social unrest inspired by a tax on gasoline. Citizen uproar followed later, with more protests regarding forced disappearances. The armed forces and the police have prepared for an explosive scenario in which migrants, militias, protesters, and drug cartels put to the test the state's handling of security. This mashup of violence and vulnerability also worries the United States, which has doubled its effort to militarize the border under direct orders from President Trump.[54]

Some key new developments in Mexico's military arsenal have included the Navy's offshore patrol vessel *Oaxaca*, entirely locally designed and built, likewise the Tenochtitlan-class coastal patrol vessel, and the DN-XI armored vehicle.[55] The United States sold fourteen Black Hawk helicopters.⁵⁶ Mexico's Secretariat of Defense (Sedena) planned to move into aeronautics development and manufacturing, to locally manufacture a twin-seat fixed-wing aircraft and a pair of experimental planes for basic flight training in 2018. Training by the United States in counter-narcotics tactics has featured a particular focus on helicopter repair and the maintenance of surveillance and patrol aircraft. European suppliers are also eyeing Mexico's profitable market, taking advantage of the Mexico-EU free trade agreement, with Germany, Holland, Poland, and Spain providing weapons systems and maintenance services. Other businesses that have landed in Mexico include Brazil's EMBRAER and South Korean manufacturers. It is believed that the lack of know-how in producing modern technologies such as drones inhibits Mexico's domestic defense industry's ability to expand and modernize in the short term, deepening its dependency on foreign actors. As well, current high levels of corruption while conducting business in Mexico make investment expensive and risky.⁵⁷

Under President Enrique Peña-Nieto (2012–18), the Sedena created the Center for Cyberspace Operations, the leading state apparatus in charge of cybersecurity and cyber defense. Together with the U.S. North Command, Mexico has participated in cyber exercises that put its military personnel in close relationship with U.S. military experts. As part of launching a cybersecurity National Strategy in 2017, Mexico created the Subcomision de Ciberseguridad, in which the Sedena is a permanent participant, together with various other state entities dedicated to public security and scientific and technological endeavors.⁵⁸ That same year Microsoft landed a contract that helped launch the country's Centro Regional de Ciberseguridad (Cybersecurity Engagement Center) and reached an agreement allowing Microsoft to share services on risk analysis and prevention with state bureaucracies. The agreement was led by Mexico's Policía Federal and not the armed forces. The military in Mexico has been less favored in terms of building up cyber defense mechanisms.

Although still in the making, Mexico's use of cyber tools was heavy criticized after a report disclosed evidence that the Israeli Pegasus spyware technology had been aimed at international officials investigating the high-profile 2014 disappearance of forty-three students from the Ayotzinapa teaching school, in the state of Guerrero. According to a media report in *Foreign Policy*, "The surveillance technology used appears to be Pegasus, a software created by the NSO Group, an Israeli arms manufacturer. It uses phishing links to hijack smartphones—once a device is infected, the software can monitor a person's every move through access to phone calls, emails, and texts, rendering encryption useless. It also can take over the phone's microphone and camera to record conversations and surroundings."[59] It is estimated that Mexico purchased $80 million worth of spyware destined initially to track down terrorist activity and organized criminal groups. Spyware use has grown in Mexico. Joaquín "El Chapo" Guzmán used it heavily on his closest circle as a surveillance tool. El Chapo would spy on lovers and lieutenants, helped by an IT specialist working for the drug lord. Later, at Guzmán's trial in New York, it was revealed that the FBI had flipped the IT specialist, giving to the law enforcement agencies access to El Chapo's encrypted phone systems.[60]

Other hi-tech systems such as unmanned aerial vehicles (UAVs) have been used by both the armed forces and the Mexican cartels along the U.S.-Mexico border as intelligence-gathering tools.[61] The "narco drones" discovered by authorities have included commercially available Chinese models. Commenting on these commercial-off-the-shelf (COTS) and do-it-yourself (DIY) capabilities, Sullivan and Bunker argue that "[t]he *plazas, colonias, favelas,* and megacities of the drug war zone will become the operational space of future state vs. non-state conflict (embracing local high-intensity crime through criminal insurgency). The use of drones is no longer merely a tactical issue; it has strategic and operational potentials for states and their competitors."[62]

Mexico's cyber situation is still reactive to everyday incidents. In 2011, the Sedena received a DDoS attack that left the server offline. In 2013 Sedena's website was defaced for various hours. In 2015, the "Mexicanhackers" group conducted a hackathon against the Sedena using the catchphrase "#ADIOSNARCOGOBIERNO" and exposing twelve objectives in the Sedena systems. In June that year, a policy document entitled "*Análisis costo-eficiencia: Construcción de un centro de operaciones del ciberespacio*," obtained by the magazine *Proceso*, exposed a series of vulnerabilities in the Sedena and lack of ICT infrastructure to prevent, detect, and solve cyberincidents. The document also highlighted Sedena's limitations in equipment, software, and operational infrastructure to visualize and interpret data. Until the creation of Mexico's Cyber Cyberspace Operations, cyber defense and cyber intelligence were poorly developed and highly vulnerable to exploitation.[63]

Venezuela

In the last three decades, Venezuela has experienced and a growing security threat from transnational organized crime that has found operating space in the broken social fabric left by the administrations of Hugo Chavez and, later, Nicolas Maduro. Territorial conflict remains a midlevel threat as continuous impasses with Colombia in the past have led the central government to reinforce its borders and send troops to specific locations that connect with Colombia and Ecuador. A growth in military expenditure occurred during the rise of oil prices that allowed Chavez to deal with Chinese and Russian suppliers.[64] Threats from insurgencies and paramilitaries and terrorism by nonstate actors remain at low levels. However, social and ethnic tensions have been on the rise due to the polarization of the country. The armed forces and the militias Bolivarianas have spread across society, and the state uses repressive machinery toward both the country's opposition and its citizens. To contain mass protests, the country needs to ensure sufficient resources for the armed forces and police.

Venezuela's military industry remains small and insufficient to cover the necessities of the armed forces, thus, forcing the country to rely on modernization programs supplied by partners from foreign defense companies. Fiscal vulnerability is high in Venezuela, particularly given the presence of low oil prices. However, the armed forces have secured a consistent influx of public money as scope of humanitarian tasks has expanded, especially in distributing goods and performing other humanitarian roles following natural emergencies. Recently flooding and earthquakes have hit the country. The UN put Venezuela on its list of countries with the most deaths caused by natural disasters during the last twenty years, a ranking topped only by Haiti, and followed by other nations in the global South such as Indonesia, China, and India.[65]

Recent expenditures have included the allocation of resources to the Compañía Anónima Venezolana de Industrias Militares (CAVIM) the state-owned company created in 1975 to manufacture assault weapons, explosives, ammunition, and military machinery. CAVIM has worked with Iran on the production of unmanned aerial vehicles, and with China for terrestrial vehicles. The PSUV-led government of Maduro discussed with Russia's Vladimir Putin the purchase of submarines. Various military exercises have occurred in Venezuela's Caribbean Sea with the presence of Russian forces, something observers have called a "21st-century throwback to the Cold War Soviet-Cuban alliance."[66] The link with Moscow is strong. CAVIM announced a partnership with the Russian consortium Kalashnikov to produce ammunition and assault rifles in two plants in Venezuelan territory, giving proof of the country's dependency on anti-Western governments and little connection with the rest of the Western Hemisphere's arms industry.[67]

Regarding the development of cybersecurity capabilities for the armed forces, Venezuela recently acknowledged Russia's support in the form of military experts sent to train local personnel. According to a U.S. government source, speaking on condition of anonymity to *Reuters*, a Russian military contingent arrived in Venezuela in late March 2019, believed to be made up of special forces including cybersecurity personnel."[68] Although most Latin American countries have links to foreign nations for military peer-to-peer-knowledge exchange, Venezuela's close relationships with Russia and China are viewed with suspicion. For many observers in the Washington establishment, the presence of the United States' confirmed rivals isa threat to the Americas' peace and stability in the post–Cold War era. As explained in earlier chapters, Venezuela has actively sought military alliances with Eastern powerhouses in exchange for oil and other commodities sold at bargain prices. President Nicolás Maduro has few states in the world accepting his legitimacy as the elected authority. Since early 2019, when the opposition leader Juan Guaidó was recognized as the country's legitimate president, Maduro has sought refuge in countries such as Cuba, China, and Russia. Moscow has described this as a U.S.-backed coup against the socialist government. The *Reuters* report went on to argue that "the U.S. determination that the Russian contingent includes cybersecurity specialists suggests that part of their mission could be helping Maduro's loyalists with surveillance as well as protection of the government's cyber infrastructure."[69] Both Moscow and Beijing have put a foot in Latin America, helping shield more than one Venezuelan regime from criticism of authoritarian practices. Both countries promised military, financial, and humanitarian aid to Hugo Chávez and later his successor Maduro, in return pocketing huge investments, mostly in energy. The Russian foreign ministry has said that the presence of Russian military specialists in Caracas and other Venezuela's cities is an aspect of military-technical cooperation between the two countries.[70]

Cybersecurity in Venezuela is under the authority of Dirección Conjunta de Ciberdefensa (Joint Cyberdefense Directorate) at the Comando Estratégico Operacional (CEO, Strategic Command Operations of Venezuela) part of the Ministry of Defense and responsible guiding the operations of the armed forces. The directorate is in charge of planning, coordinating, and executing cyber operations affecting critical national infrastructure, weapons systems, and the telecommunications networks of the armed forces. Their role has been expanded to "ensure the use of cyberspace, denying it from the enemy.[71] In contrast to other less-prepared nations in the region, Venezuela created the directorate in late 2014. Since then, the military has claimed that attacks have constantly "intervened" in state's services and websites. General Vladimiro

Lopez, former chief of the CEO, said in an interview that cyber defense tools were not supposed to "attack anybody, on the contrary, to defend the integrity of the nation," and to "preserve the secrecy of our operation and the state's."[72]

In August 2018, the hacker group The Binary Guardians shut down dozens of Venezuelan websites, including the portals for the government, the supreme court, the legislature, and several private companies such as the subscription TV service DirecTV and telephone provider Digital, in support of an armed rebel group responsible for staging a raid on an army base in the city of Valencia. "Our struggle is digital. You close the streets, we do so to networks," the hacker group claimed during "Operation David" urging antigovernment protesters to demonstrate "and support our valiant soldiers."[73] Later, during 2019, independent security researchers in Venezuela and abroad exposed a hacking attempt to steal activists' usernames and passwords for email and social media platforms Gmail, Facebook, Microsoft Live, and Twitter, allegedly performed by pro-government computer experts.[74] According to Oliver Marguleas at the University of Washington, "Venezuela's Internet provides an important role for Venezuelans as it allows for space the government cannot effectively monopolize." The constant weakening of vital information available online "is hampering the quality of life for all Venezuelans and reducing the voice of civil society. Examples of a weak Internet can be observed by Venezuela's abysmal Internet speed, as well as Internet content itself, which is evidenced by degrading of freedom scores by Freedom House," Marguleas added.[75]

Under this scenario, Venezuela's military cyber defense has yet to be tested. Most recently, Maduro's government set up a militarized response to protect basic services such as electricity and water after a massive blackout hit the country in March 2019. Maduro claimed responsibility for the cyber hit on the United States and insisted that the "successive attacks" were directed by John Bolton, National Security Advisor to President Trump (I return to this episode in the Conclusion).[76] The armed forces responded almost immediately with "Ana Karina Rote 2019," an exercise in which troops were summoned nationally to public utility stations to ensure the physical and virtual state of water, telecommunications, electricity, and oil supplies. Military authorities also appeared at the metro lines in Caracas.[77]

Conclusion

In this final section, I shall briefly discuss a few issues brought forward in the chapter: Why do countries industrialize the means of war, considering changing military missions? What political and economic factors help or constrain the buildup of domestic military capabilities? And, finally, what strategic threats are considered when recognizing the importance of cyberspace for defense and security?

Latin American states can now access transnational production and development networks that can lead to indigenous virtual and physical arms production. As shown above, in the Western Hemisphere there exist many examples of the roles and missions of the armed forces being altered to confront a manifold of military and nonmilitary challenges, both in and out of cyberspace. Countries seem to industrialize the means of war, considering changing military missions when new technologies are sufficiently cost effective to allow them to quickly address a security issue. The use of military equipment in roles such as hybrid missions in response to both military necessity and criminal threats seems to be the trend among those able to purchase new material. The relative cheapness of technology, and foreign alliances with the military-industrial complexes in the advanced democracies, has allowed civilian authorities, the industry, and military officials to access means of warfare, not so much for waging war against each other as to address immediate threats. Cybersecurity readiness and militarization have responded under a similar paradigm. Current military buildups (either substantial or more discrete) do not increase the chances of countries entering into interstate wars but rather might leverage new situations that can affect their immediate perceptions of threats and risk. There is still a lot open to empirical investigation. Cyber tactics seem to be at infant levels in the region. This does not mean that, as with air, land, sea, and space threats, countries are not preparing for cyber threats. In the six countries analyzed in this chapter, we see that, despite the priority given to gaining the industrial means (both abroad and locally manufactured) to militarily address territorial conflict, insurgencies and paramilitaries, transnational criminality, terrorism, social and ethnic tension, natural disasters and climate change, authorities have allocated resources to accumulate cyber governance expertise and learn cyber defense mechanisms within the frame of their military buildup. It is also still too early to be able to identify whether major power conflict within the middle states can set a course toward inserting cyber technologies into combat. The states with traditional arsenals can use them to control the risks to national security. Not so much with cybersecurity measures, as the countries explored seem more eager to dedicate cyber technologies to ensure critical services interconnectedness, first, and then to espionage, surveillance, and intelligence-gathering purposes.

What are the political and economic factors that encourage or constrain the buildup of domestic military capabilities? The response here has been to cite economic reasons affecting policy priorities in favor of and against military buildups. Countries that have the means to invest in military arsenals seem to depend more on digital communications and hi-tech weaponry that are both dependent on and vulnerable to cyber malice. Countries that remain analog in their military acquisitions might be offline. However, those that have had access to cutting-edge technologies know more about the revolutionary effects of escalating military arsenals and growing cyber-dependent resources. Aerospace technologies used by the emerging Latin American states seem the most prone to emphasize the importance of the cyber realm. As Valeriano and Maness argue, "Traditional powers such as the United States, China, and Russia seem to be the most dominant cyber actors because they have the resources, manpower, and money to support massive cyber operations."[78] The emerging states that follow behind can, potentially, expand their influence in the fifth domain by enhancing the other four through the cheaper acquisition of military hardware and software that extends traditional capabilities in cyberspace. Others have measured the scale of a state's cyber capability, quantifying its offensive capabilities (i.e., how sophisticated a state's cyberweapons are, as well as how well trained and computer savvy its citizenry is); its cyber defenses (how protective a state is over its cyberspace, what control it has over its Internet service providers, and how resilient it is against potential aggressors); and its cyber dependence (to what extent do a state's daily activities and infrastructure on the Internet make it vulnerable to unwelcome connection).[79] It is still too early to say how proficient Latin American states are by these types of measures. Still, certainly it is a start to identify their willingness to create cyber defenses, to harness cyber offense capabilities designed either at home or abroad, and to assess their cyber dependence, especially when it comes to revealing how connected to cyberspace their critical infrastructures, banking, defense, and power grids seem to be. To quote another extract from Valeriano and Maness, "Cyber offense gives powerful states another option in their arsenal when employing action against their weaker adversaries. An offensive strike such as Stuxnet could deter states from launching cyber operations against the United States, out of fear of retaliation in kind by similar sophisticated offensive incidents. The weaker state might be better off investing in defensive measures that protect against the offensive strike, rather than trying to match their stronger adversaries in cyberspace."[80] We still need to see whether Latin American states will take the offensive or defensive path.

Finally, what strategic threats are considered when recognizing the importance of cyberspace for defense and security? Take the case of Plan Colombia, targeted at the insurgency of the FARC. When a country assesses the military and nonmilitary means to combat such a complex threat, many strands of the state's resources are pulled together. Military, economic, humanitarian, and political tools are combined to deal with modern, transversal threats. Recognizing the role of the cyber domain has forced Latin American countries to join resources, although more drastically under the armed forces command, as the case studies above showed. The interdependence of security phenomena added to the unintended consequences of globalization makes former military objectives blurry and operations theatres mixed. Not only Colombia suffers from previous and current threats to security that are occurring simultaneously at the local, national, and international levels, as well as attacking individual and collective social organizations. Cyberspace seems to unite the many threats that bedevil a more widely connected world, such as "terrorism, illicit arms traffic (including of potentially very dangerous chemical, biological, or nuclear material or information), drug trafficking, cybercrime, health security threats, intellectual property theft, and other smuggling (including of human beings, often trafficked as modern-day slaves)" that have affected the United States and other countries in the Americas.[81]

In this vein, Washington has recognized such multilevel threats in cyberspace. It has made it clear in its National Security Strategy that there is little room for cyber actors to threaten its national security, whether they are cybercriminals, either individually or in groups, or rogue states. "We will strengthen America's capabilities—including in space and cyberspace—and revitalize others that have been neglected. Allies and partners magnify our power. We expect them to shoulder a fair share of the burden of responsibility to protect against common threats," the document argues.[82]

The next chapter will continue to address the crossing-over of threats, more specifically the concern over Latin America's Western and Eastern influences in today's strategic scenario. The chapter takes into consideration the peripheral conflict in cyberspace and the hybrid and proxy warfare between countries with historical disputes. As Lora Saalman aptly puts it, "While hybrid warfare may be a well-worn concept, a new key element in this 'soft war' and the future of hybrid warfare is cyberspace."[83]

Five

U.S.-China Rivalry

In today's cyber connected world, we live under a completely new geopolitical paradigm. In this chapter, I shift gears to theorize on the eventual rejection of American hegemony and shed light on the states that see cyber operations as a way to advance their own influence at the expense of the United States and its allies across the globe.[1] I argue that the ongoing rearrangement of U.S. interests in the Western Hemisphere has produced dispersed outcomes, most notably leaving the doors open to the arrival of foreign players, among them China. The Sino-U.S. rivalry evidenced globally posits a conundrum of choices, privileges, alliances, and power moves. In this sense, the debate over who is Latin America's superpower of choice is far from reaching a consensus.[2] The thirty-five independent states in OAS, for example, hold simultaneous trade and security deals with both the United States and China. This scenario mirrors that in East Asia where Washington and Beijing exert the most influence. South Korea, Japan and the states in the Association of Southeast Nations (ASEAN) are witnessing firsthand the dynamics of mistrust and zero-sum balance created by the eagle and the dragon as they vie for opportunities for both trade and security.[3]

With Barack Obama in the White House, U.S. policymakers tried to reset stagnated interest in Latin America and the Caribbean. However, Obama's policy record against the arrival of foreign emerging powers resulted in a mere "low-key success."[4] Under Donald Trump, scant effort has been made at the top tier of decision making, that is, besides revising trade tariffs with NAFTA members and building a wall on the Mexican border (I revisit these and other policy events in more depth in chapters 6 and 7). Policy toward the rest of the subcontinent has been relegated to midlevel governance via the State Department's Bureau of Western Hemisphere Affairs.[5] This despite the fact that, to some observers, the Americas is the newest battle zone for East and West superpowers.[6]

President Xi Jinping wants China to enjoy South-South cooperation, and the Belt and Road Initiative (BRI) provides a steady influx of Sino presence in the region, something that is likely to keep growing.[7] Launched by Xi in 2013, the BRI seeks to integrate Asia throughout the rest of the world through land roads and maritime routes. To some scholars, the BRI "signals a new emerging trend in China's foreign economic policy and has been described by Beijing as the third round of China's opening after the development of Special Economic Zones in the 1980s and accession to the WTO (in 2001)."[8] For China, Latin America is "eager" to join the Belt and Road.[9]

But what are the tangible effects on the ground? For some time, the United States has enjoyed unrivaled security and economic "jurisdiction" on the South American continent, providing access to security, diplomatic status, and trade with the main emerging economies. However, now that the Washington Consensus is passé a rather different game is being played.[10] States gathered in forums such as the ALBA, CAN (Comunidad Andina), CARICOM (Caribbean Community), Mercosur (Common Market of the South), CELAC (Community of Latin American and Caribbean States), and the UNASUR have reshuffled their dependency on the United States in many ways.

In 2016, and among the region's "pink-tide" countries, Ecuador signed an agreement to boost its national defense industry with goods and equipment from Chinese companies. The move came together with a massive influx of Beijing capital to the oil sector. Other states, such as Chile, Colombia, and Brazil, now have China as their top trading partner. However, policymakers in Santiago and Bogotá remain close to the U.S. security community. In Brasilia, the subregional powerhouse has sought its own path to diplomacy, although eager for Sino investment. A third group remains dependent on U.S. commerce but over time has fallen for Chinese military bargains, for example, Peru and, most notably, Bolivia under President Evo Morales and Venezuela since the late Hugo Chávez came to power. In some of these cases, the trend has been marked by ad hoc political affinities. In Argentina, for example, China almost ties the United States in value of export destinations.[11] The late president Néstor Kirchner and his wife and successor Cristina Fernández sought economic and military assistance from Beijing and Moscow, when they were heavily chastised by the International Monetary Fund (IMF) during several years of economic crisis.[12] Right-winger Mauricio Macri managed to hang onto Sino investments following his election in 2015, although he remained reliant on U.S. security institutions to fight transnational crime.[13]

Observers argue that cyber operations among great powers "are about collecting and protecting information," making states compete in cyberspace for "strategic advantage, not to score a decisive victory."[14] Given such a scenario, how do Latin American states navigate Sino-U.S. rivalry? Why do they seek partial alliances with either Washington or Beijing in instances of security and trade? What can they achieve using this two-track policy? How do cybersecurity developments reflect this duality? How do we convey cybersecurity in today's international security environment, heavily marked as it is by the U.S.-China rivalry?

These questions are of timely importance and geostrategic relevance. There are at least two ways to move the debate forward. First, the United States and China seek to play utilizing their own grand strategies, without necessarily minding the cascading outputs in the Americas.[15] Put briefly, the emerging democracies of the continent are the unintended overseers of U.S. geopolitics, while regional governance does not offer explicit rules for such global competition. Second, if, in the Asian-Pacific, the American and Chinese search for primacy responds to strategies of mistrust and counteractions to each other's influence,[16] in the Americas the outcome of Sino-U.S. rivalry is in many ways equally provocative and worrisome. States are thus left to gamble on their futures as the tectonic plates of international politics shift.

The chapter surveys international relations theories borrowing concepts from the literature I introduced in chapters 1 and 2 and discusses why the Americas is of importance for U.S. strategic policymaking. In doing that, it seems relevant to present the case of the telecommunications company Huawei and its presence and plans for Latin America. The idea here is to discuss what rewards and opportunities states face in the presence of the U.S.-Sino rivalry, including security and economic tradeoffs, considering current U.S. hegemony and the arrival of Chinese wealth on the continent. Are cyber issues influencing the Americas' emerging democracies when partnering with the United States or China? To answer that, I explore public opinion toward China and the United States and shed light on practical answers to U.S.-Sino rivalry and the puzzles posed for those dealing with cybersecurity policy in the Americas.

States and Global Politics

The academic debate has come to consider that the emerging countries in the world "are a new arena of rivalry for Sino-U.S. relations."[17] Scholars have addressed the issue by collecting ideas from a subset of international relations theories that reflect both an institutionalist and realist cut. Realism argues that states care about their own wealth and security over everything else, while institutionalists would argue for more liberal interpretations of international relations. Middle state theory, on the other hand, borrows from traditional approaches to offer an analytical perspective for understanding the complex nature of the strategic environment and its public policy implications.

Not all emerging states, however, behave according to the predictions of middle states theory. Rather, I find that this theory helps us identify where ideas accord with the case of the Americas, and why this is a relevant paradigm for the U.S.-China dealings in the region.

The so-called middle states hold an intermediate position in terms of their relative power in the international order. This role tends to be particularly prominent in terms of its regional aspects, positioned below great powers, but above minor states.[18] To start with an example, in Europe, middle powers play a diplomatic and security role in advancing collective security through NATO. Sweden, Denmark, Norway, Belgium, and the Czech Republic, among others, participate in intergovernmental military alliances, using a deterrence mindset to protect their sovereignty, along with the major powers, for example, the United States, Germany, and the United Kingdom. Other examples around the globe include India and South Africa, plus regional actors such as Iran, Saudi Arabia, Egypt, Indonesia, Thailand, and Turkey. While some countries, such as Japan, might behave like middle powers, their relative power puts them closer to the superpowers.[19]

The middle powers do not usually have nuclear weapons, nor do they hold a permanent seat on the UN Security Council. Nevertheless, their security contribution to world affairs is real and constant, with their participation in major armed conflicts, humanitarian interventions, and peacekeeping operations, in pursuit of their strategic concerns. They maintain capable armed forces and have a record of using force against intra- or extracontinental state and nonstate actors. Middle powers are also said to be able to provide human capabilities, knowledge, and technologies that the superpowers do not always possess or are reluctant to utilize. The role played by middle powers in global politics is limited by their own strengths and weaknesses, which usually keep them away from escalating armed conflict, and, with some exceptions, focused mostly on the hot spots proximal to their own geographical locations.

Linked to their desire to promote global democracy, on commerce and trade issues middle states play a role in legitimizing international organizations and regional cooperation forums. They have a need to maximize their economies, which, although not always, are often moderately developed. While seeking higher levels of industrialization, they compete with similar states that offer products and services of comparable value. They usually have unique comparative advantages in terms of natural resources that make them attractive to advanced nations. Additionally, their ability to offer products and services allows them to dedicate a sizeable portion of their GDP to social policies that alleviate poverty and underdevelopment.

In short, middle powers exist in a security dimension, moving toward peace initiatives and conflict-mediation roles. In the systemic dimension, they are counterhegemonic, pro-multipolarity, and consensus seeking. Regarding the rules of the international order, they tend to promote human rights, support international institutions, and favor region-building processes.[20]

Beijing wants to exert influence on other states considering the roles they play globally and given their regional scenarios. China's eager policies can, for example, be seen manifested in the 2016 White Paper for Latin America and the Caribbean Countries and the China–Latin America Cooperation Plan (2015–19).[21] China's 2016 Defense White Paper also assumed that the growing size of its military and economic influence abroad would help shape the strategic outcome of other nations.[22] These power moves contrast with the reluctance of U.S. policymakers to engage more substantially with the Americas.

Regarding multilateral institutions, the most important political forum, the OAS, provides a security charter for regional peace through mechanisms aimed to avoid or end attempts at breaking the rule of law, signed under the democratic charter. Surprisingly, OAS has rarely been a comfort zone for Washington. Rather, it has become a place where aspiring liberal democracies engage in peer-to-peer dialogue to address continental issues. Its most recent secretaries general, Chilean Jose Miguel Insulza, and Uruguayan Luis Almagro, have advocated for democratic consolidation, as opposed to autocratic scuffles in the region (e.g., the 2009 constitutional breakdown in Honduras and the ongoing crisis in Venezuela).

For example, during the 70th OAS assembly in 2018, some *cancilleres* (foreign affairs ministers) engaged in a tense dialogue with their Venezuelan counterparts, after discussing a multicountry and U.S.-backed proposal to impose sanctions on Caracas. Chile and Brazil, for example, wielded their foreign policies to put pressure upon, and investigate, possible state-sponsored crimes against humanity during Venezuela's recent wave of national protests. Following the legacy of the late Hugo Chávez, the country is now ruled by a ruthless executive and a pro-government constituent assembly. The socialist party has almost no competition and almost all opposition leaders are in jail, banned, or exiled.[23] The Obama administration backed a mediation to help solve the ongoing humanitarian, economic, and political crisis, although alignment with the OAS has not been enough to make a difference. President Trump canceled his participation in the Americas Summit in 2018 at the last minute.[24] The continent had expected him to embrace the cause of democracy more strongly.[25]

Political antagonism aside, the Western Hemisphere has benefited from a long-standing Latin American Pax. States have refrained from direct conflict despite the presence of conflict hot spots. Instead, they have turned their energies to extending alliances between their militaries.[26] Of course, they have not yet achieved the likes of NATO's or ASEAN's more advanced collective security networks. Through forums such as the Inter-American Defense Board and the South American Defense Council, the Americas' states have opened a channel of international security and military isomorphism. Nevertheless, some observers are wary of the Americas' overlapping security forums and the value these add to democratic consolidation.[27]

Emerging security alliances have grown between territorial neighbors and former rivals (i.e., Chile and Argentina) and with extracontinental partners (the European Union). It is in the form of alliances that the United States develops its current security policy toward the continent. The Pentagon executes military and strategic interests through the Southern Command.[28] The Fourth Fleet, a war-ready division of the U.S. Navy, plays a manifold role within the command's reach on the continent, including nontraditional missions (humanitarian aid, counter-piracy, and anti-drug-trafficking operations), plus annual military exercises where interstate war scenarios are simulated. In April 2018, for example, the Silent Forces Exercise (SIFOREX) was carried out on the coast of Lima, Peru, along with Colombian forces and observers from both Brazil and Argentina. The biannual exercise has been a routine activity for the last seventeen years.[29] The UNITAS, another show of force, goes even farther back in time and in 2017 it celebrated its fifty-eighth anniversary, with participation from a group of states, such as Australia, Chile, Colombia, Ecuador, El Salvador, Guatemala, Mexico, New Zealand, Panama, and South Africa.[30]

A Case Study: Huawei

In this section, I put theory into practice. I briefly present the case of the Chinese telecommunications company Huawei, and the ramifications it presents for security and foreign policy in the subcontinent. Huawei was blacklisted from selling its telecoms kit in the United States and other Western nations. However, it cushioned the blow by swiftly rolling out in Latin America's emerging economies. Huawei has opened branches in Argentina, Brazil, Colombia, Chile, Ecuador, Mexico, Peru, and Venezuela, and is bidding for the region's next-generation 5G networks. Despite U.S. pressure in the region urging governments not to deal with Huawei, nations of the likes of Brazil, Argentina, Colombia, and Chile have publicly welcomed the Chinese tech group in their bids for current and future state and private-led megaprojects.

In April 2019, Chilean president Sebastián Piñera visited Beijing and met privately with Huawei's chairman of the board Liang Hua to discuss various issues, including the future of 5G networks, data protection, cybersecurity, and the transoceanic fiber-optic cable project between Chile and China. That same month, U.S. Secretary of State Mike Pompeo visited Santiago, where he reprimanded the emerging economies for their favoritism to Chinese lending and their trading with Sino-companies. Pompeo argued that doing business with Huawei puts U.S. cooperation at risk. Chilean Minister of the Interior Andrés Chadwick responded that his government "did not take warnings on foreign policy decisions."[31] A similar attitude has pushed Brazil's president Jair Bolsonaro toward doing business with China, regardless of his being cataloged as an esteemed ally of President Trump.

I present the following subsidiary impacts. First, Huawei is contributing to the Latin American countries' digitalization and the development of ICT infrastructure (I have covered the importance of these topics in chapters 3 and 4). This includes efforts to win telecom and optical network contracts in the public and private sectors. Second, the Chinese company is focusing on 5G, IoT, and cloud computing solutions. And third, Huawei has been welcomed by universities, local and national governments, and private businesses.

The scatter plot in Figure 5.1 shows the countries in the Americas where Huawei has office branches. The axes measure Internet penetration (percentage of population) and how easy it is to do business in these economies. Regardless of how countries rank, the Chinese corporation has entered the ICTs market bidding for contracts in North and South America.

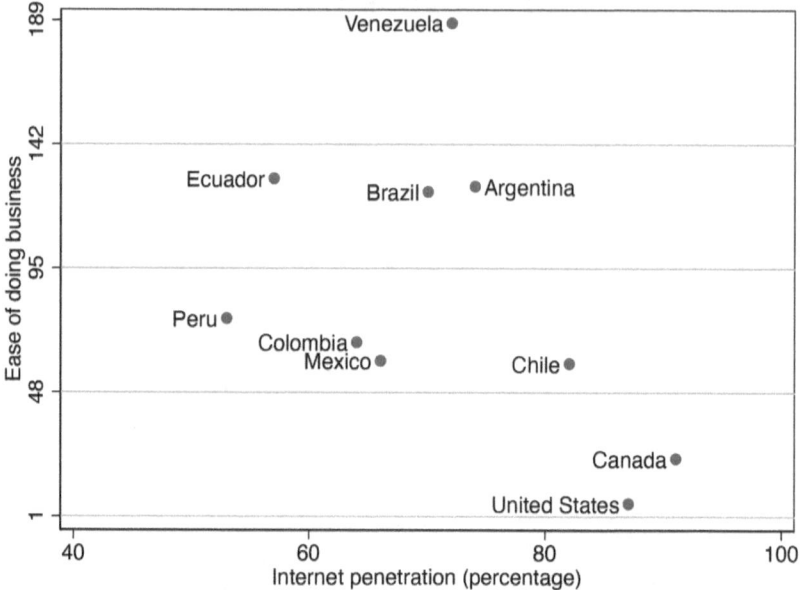

Figure 5.1. Huawei branches in the Americas. Author's construction with data from World Bank's Doing Business (2020), ITU (2019), Huawei (2020). Public domain.

My analysis is that Huawei is eagerly trying to catch up with competitors in the technology and consumer sectors across Latin America despite obstacles to doing business that include corruption, overly regulated markets, and unwieldy bureaucracies. The company's global reach has allowed it to set up a robust market presence in Europe, Asia, and Africa, where they have gained plenty of knowledge to deal with a variety of economies and their ad hoc business cultures.[32]

However, there is concern about Huawei's cybersecurity track record and its suspected link to the Beijing government, which the intelligence agencies from the "Five Eyes" alliance believe might enable spying. For this, Mark Stokes at the Project 2049 institute argues that the United States "appears to be taking the Chinese cyber challenge seriously and dedicating resources to countermeasures," such as "enhanced counterintelligence, greater cooperation with international partners such as Taiwan, and imposing costs through effective deterrence."[33]

Latin American countries, on the other hand, have set up their own cybersecurity governance to manage their increased digital connectivity and the rollout of IoT. They see in Huawei a beneficial partnership to achieve these and other goals. In Brazil, the fourth-largest smartphone market in the world, Huawei aims to revamp sales to a double-digit market share, much like it enjoys in the rest of the region. Huawei launched a smartphone in 2014 with little success; however, it was announced that two premium devices from the new P30 Series would enter the market. There were plans to partner up with the local company Positivo, Brazil's biggest electronics manufacturer. However, negotiations failed in late 2018. It was publicized that local retailers Magazine Luiza, Via Varejo, B2W, Fast Shop, and wireless carrier Telefónica Brazil would be among those selling their phone devices.[34] The Chinese company remains one of the major suppliers of telecoms equipment for Brazil's carriers, where it has had a market presence over the last twenty years. "In Brazil, For Brazil" was the slogan used by chairman Howard Liang during his recent visit to celebrate the "new era" of Huawei in the country.[35]

Huawei's Consumer Business Group, which includes smartphones, is managed in Latin America from Santiago. Huawei's phones have been a strong competitor in the Chilean market, allowing the company to open new retail stores. In December 2018, Yong Dou, Huawei's CEO in Chile, announced his intention to "dominate" the fiber optic and 5G networks market share in the next three years. Huawei is the second-largest mobile kit provider in the Southern Cone country, with 17 percent of the market, close behind Samsung (18 percent). The group also participates in Chile's Proyecto Fibra Óptica Austral, the underwater connection to the country's southernmost territories.

Huawei has benefited from the rapid digitization of the emerging economies and looks forward to capitalizing on the spread of 5G networks. To secure new endeavors, Huawei has targeted local partnerships with public and private labs for innovation and technology growth. In Colombia, the company launched its first channel summit in 2018 to celebrate its partnerships with local businesses, companies, and state and local governments bidding on the country's digital transformation. The Chinese group aims to expand the business of cloud computing and IoT, on top of opening more retail stores in Bogota, Barranquilla, Cali, and Medellin.[36] Huawei also opened a Colombia-based contact center recently, previously centralized in Mexico.

In December 2018, Huawei announced a significant association with the Institute of Technology (FIT) in Sorocaba City in the state of São Paulo to open an IoT center in the same city. The Chinese group has two other IoT initiatives in Brazil, one with telecoms provider Telefonica and another with PUC-Rio Grande do Sul University. Huawei totals almost 1,400 IoT partners worldwide.

The Chilean newspaper *La Tercera* reported in February 2019 that Huawei's offices in that country were planning to open the company's first data center in Latin America. Government-backed InvestChile assisted Huawei in analyzing regulatory, operational, and infrastructure issues.[37] It is believed that Huawei's offices in Argentina and Brazil are also proposing their own data centers. Latin America has also attracted Western-based companies such as Google and Amazon, which announced new investments to expand on data centers.

Huawei knows the market and the potential that Latin American countries present. In the past, Huawei has collaborated with Telefonica to provide cloud services in Chile, Brazil, and Mexico. Current long-term projects include Huawei's high-tech training center for telecoms operations, businesses, and government agencies in Puerto Madero, Buenos Aires. The 1,300 square meters building's inauguration was attended by Argentine ministers and Chinese diplomats in August 2016. Under President Macri's economic recovery plans, foreign companies have been allowed to set up local manufacturing plants, partly by exempting them from taxes.[38] Huawei hired an Argentine assembly line for mobile kits in Ushuaia, the capital of Tierra del Fuego, a faraway city on the Southern Atlantic coast. Huawei also planned to open a smartphone plant in Brazil.[39]

The substantial use of new sensitive IT technologies by both public and private organizations is notorious in Latin America. Privacy International highlighted Argentina's increased tech and intelligence collaboration with advanced and developing countries, including China. Partnerships for this purpose occurred most recently in preparation for the G20 summit celebrated in Buenos Aires in December 2018. The potential danger of such technologies entered the public debate after the recent attempt to introduce aerial surveillance balloons in the city of Buenos Aires. Checks and balances on privacy and surveillance in Argentina and across the subcontinent are a matter of ongoing discussion and possible regulation.

Latin America's push for smart and safe cities has also inspired concern regarding some of Huawei's moves. In Argentina, for example, local civil society organizations protested a visit from Huawei's representative to meet with the governor of Mendoza, a province in the western central part of the country.[40] It was publicized that the local authorities discussed with Huawei acquiring facial recognition and other surveillance and public security programs from Huawei.

The Trump administration harshly criticized Sino-Latin American security, economic, and political growing links. Kimberly Breier, Assistant Secretary of State for Western Hemisphere Affairs at the U.S. State Department, said in April 2019 that Chinese practices in Latin America fail to meet standards for transparency, sovereignty, and anticorruption. Nevertheless, and besides Huawei, Sino-technologies have found ample business reception across the region.[41] These include Chery cars, Lenovo computers, DJI drones, China Southern Airlines, Mobike bike-sharing, Hisense televisions, Didi Chuxing car service, and Alibaba online shopping. In the field of clean energy, Huawei supplies inverters, a critical technology used to turn power from panels into useable current, at South America's largest solar farm located in Jujuy, Argentina.[42]

Finally, recent international developments, for example, the case against senior Huawei executive Meng Wanzhou, who faces charges in the United States and Canada linked to alleged dealings between Huawei and Iran, has created a major "rift" between Canada and China. Countries in the Americas might face a similar fate if the company's doings are somehow challenged at either the diplomatic or judicial level. Beijing has arrested two Canadians since the charges against Meng were brought, something Prime Minister Justin Trudeau has called an arbitrary retaliation.[43] The potential problem for third countries is that Meng might disclose illegitimate and damaging practices elsewhere linked to the Chinese technological expansion overseas. As Nigel Inkster at the IISS think tank in London put it, ZTE and Huawei have "extensive penetration overseas, particularly in the developing world, where they have built and operated core networks, including as part of the Belt and Road Initiative."[44] Due to strategic competition, most cyber projects are kept confidential, especially in areas of security, research, and development, making the high technology sector an obscure battleground, where complex legal and financial strategies are often secretly put in place.

Rewards and Opportunities

As the case of Huawei has shown, Sino-U.S. rivalry has encouraged regional states to seek access to better trade and commerce opportunities from the international order, which goes together with their security status. I will look at some of the most interesting military and security rewards and opportunities for the Latin American states.

First, the more experienced players are willing to share resources and provide much-needed opportunities for long-term partnerships. The United States, through the Southern Command, offers many ways to educate and empower other states' armed forces, including in the field of cybersecurity, although its effectiveness and efficiency have not been adequately measured.[45] Most recently, it has created a rapid response network for continental security threats, which stretches from the Gulf of Mexico to the Magallanes Strait and has the capability to pursue stabilization operations on short notice with partners from Central and South America plus the Caribbean. The military authorities proposed a blueprint entitled the "Theater Strategy, 2017–2027."[46] The document is quite transparent in terms of how the U.S. security establishment sees the arrival of Chinese influence on the Americas:

> Over the past decade, China, Russia, and Iran have established a more significant presence in the region. Their actions and rhetoric require separate and severe consideration. These global actors view the region's economic, political, and security landscape as fertile ground through which to achieve their respective, long-term objectives and to advance interests that may be incompatible with ours and those of our partners. Their vision for an alternative international order poses a challenge to every nation that values non-aggression, the rule of law, and respect for human rights—the very same principles that underlie the Inter-American system of peace and cooperation.[47]

Despite the warnings, however, states have grown immune to U.S. concern regarding the emerging powers. They have long been playing the game whereby their options are tilted to each side in the U.S.-Sino competition. In the opinion of some observers, Latin American states' coming of age in broad regional security matters happened through actions taken on the ground. For example, after the 2004 intervention in Haiti, the U.S. military contingent exited, leaving the regional states to fill in the void with reconstruction and stabilization of the battered island. Under UN Security Council resolution 1542, the MINUSTAH was led by a conjunct command that depended on troop contributions from Brazil, Argentina, Uruguay, and Chile.

For countries such as Brazil, Chile, Peru, and Colombia, the idea of safeguarding sovereignty in a technologically interconnected world tends to be equally crucial in both securing their national economies' prospectuses through foreign investments and sustaining a unique international reputation. Emerging states sometimes seek to elevate their position via peer-to-peer competition in a world characterized by realpolitik, leaving little space for friendly competition. U.S.-China competition highlights this. In the developing Americas, only Mexico and Chile belong to the OECD. This sets them at a higher standard than the rest of the region, echoing the argument over the unfair competition that Washington consistently uses against China. Most recently, in an attempt to rekindle the Trans-Pacific Partnership, the Americas' emerging states have shown a willingness not to depend solely on U.S. commercial policy, as the Huawei case shows. Nevertheless, if clinging to bilateral Chinese or North American support for business and technological interests aids their current accounts, states will almost certainly bypass any other regional or multilateral endeavor.

In contrast to the modern unipolar history of Latin America, the U.S.-China rivalry impacts profoundly on the region's democratic pathway; this is due to the economic and security penchants of a bipolar world. For several reasons, Chinese meddling in Latin America poses more pressing questions for those states siding with U.S. security dogma (e.g., Chile, Colombia, Panama, Argentina) than for those that have turned their backs on Washington (Venezuela, Peru, Brazil). The former group are stuck with U.S. foreign and military partnerships and deprived of Chinese ventures which can appear tantalizing. This is the price they pay in exchange for U.S. commitment. The ALBA countries, on the other hand, have gained security means, thanks to a wealthy trade opening with the Far East. The defense industry in Venezuela, for example, was almost obsolete two decades ago, managed by corrupt corporate controllers and hardly capable of making a profit. Recently, Caracas has been one of the largest weapons buyers in the region, thanks to sweet and novel Chinese and Russian deals, along with a rise in the price of oil.[48] Off-budget military spending allows Venezuela and other countries, such as Peru and Ecuador, to access the profits from state-owned corporations, national banks, and revenues from natural resources (oil, gas, and mining), which remain "subject to very limited levels of transparency or legislative review."[49] In the period between 1990 and 2018, only six countries have acquired weapons from the Asian giant: Argentina, Bolivia, Ecuador, Mexico, and Peru, with Venezuela making the most significant investment starting in 2006.[50]

Countries can access the Eastern technology and weapons market because military budgets tend to be isolated from political crises and shifts in commodity prices worldwide. Brazil, for example, is currently going through one of its worse economic and political disasters since the return to democracy. Still, in April 2017, Russia completed a transaction involving powerful missile defense technology, which had been long awaited by Brazilian officials. Russian and Chinese weapons providers have also completed arms deals with Mexico, Nicaragua, Peru, and Venezuela.[51] Russia sits in second place, behind the United States, as the world's largest arms exporter, with a particularly strong global position in Asia (most notably India, China, and Vietnam), the Middle East, and Africa. Between 2000 and 2016, Russia sold 4.6 percent of its arms exports in Latin America, with more than 80 percent of deliveries going to Venezuela alone, followed by Peru and Brazil.

While others in the region are still buying from U.S. and European traditional and nontraditional (cyber) arms companies,[52] the picture might be ripe for change, due to two factors. First, sour relations with the Trump government are keeping even the closest of the United States' partners wary of what the future might bring. Mexican authorities recently signed a deal with Spain totaling $148.5 million for defense products, including helicopters and armored vehicles. These were to supply the armed forces in their fight against the drug cartels.[53] On the other hand, starting in 2015, old-time U.S. buyers, such as Chile, Colombia, and Uruguay, appear to have stalled their arms imports.[54]

The U.S. partner countries have remained competitive and technically competent through bilateral purchases and other national operations, such as Chile's naval industry and Colombia's weapons and explosives manufacture. However, military budgets are costly, and political support is short. The strongest markets in the Americas face a defense-related austerity wave as fiscal resources might be better spent in education, health, and social security. Now, even with China and Russia building up to the level of U.S. military influence in the Americas, Washington is still an uncontested military power on the subcontinent and across the globe. Nevertheless, falling defense budgets in the United States need to be a matter of consideration since Eastern powers are filling a growing vacuum in the Americas' most unstable areas.[55] This point is further discussed next.

Alliances and Enduring Security Issues

Although Chinese military and market opportunities have been exploited by some Latin American states, the link has been limited to purchases and service contracts, removing any reciprocal security assistance or hands-on policy commitments from the equation. On the other hand, states from the Americas who are fond of the U.S. can gain considerable advantages for themselves, not only in terms of obtaining military and other security equipment and technology transfer but also the privilege of entering an ongoing security alliance with a long-standing precedent.

The United States and China came to an agreement on issues of regional and global governance; nevertheless, Washington does not want to give up the control that it has enjoyed over many issues for a long time, including trade, security, and foreign policy.[56] As well, just as in East Asia, the role of the United States in the Americas' political economy is manifold. During the nineteenth century, the United States set a historical precedent for offering support to independence movements, on the one hand (Chile, Peru, and Argentina), and to colonialism, on the other (Puerto Rico, Cuba, Dominican Republic, and Haiti). The United States has occupied territories (Dominican Republic) and toppled presidents (Guatemala and Chile); sided with parties involved in interstate rivalries (the United Kingdom against Argentina in the Falkland/Malvinas Islands War); sent military interventions abroad (Panama and Haiti); and supported counternarcotics operations (in Colombia, the Andean countries, and Central America).[57] Regardless of the outcomes of such interventions, the bottom line here is that the United States is a historical protagonist in the Americas, while China is not. In sum, this is the significant difference in the East Asia scenario, where Chinese influence is contested by the United States' record of influence over the last two centuries.[58]

However, and in light of recent trade and security policies toward the Americas, an essential part of U.S. influence seems to be diluting. For instance, the U.S. Southern Command is considered to be responsible for security cooperation "in the 45 nations and territories of Central and South America and the Caribbean Sea, an area of 16 million square miles."[59] That means that Washington is providing security coverage to both supporters and opponents. Some of the anti-U.S. governments who have embraced China the most eagerly also get a considerable chunk of U.S. foreign aid.

In August 2015, a bipartisan motion was reached in the U.S. Senate to spend US$675 million on development programs in El Salvador, Honduras, and Guatemala alone. In 2016, President Obama sent his last budget to Congress, requesting $1,784,379,000 in assistance for Latin America and the Caribbean. This was mostly for economic and military initiatives to support the peace process in Colombia and the humanitarian and violence crises unfolding in Central America. The budget showed lesser amounts of foreign aid given to Haiti, Peru, and Mexico.[60] So, while cheaper weapons, technology systems, and trade ventures from China arrive with great fanfare in the region, it is U.S. assistance on the ground that could produce real outcomes for regional security and development.

A few warnings need to be taken into consideration. The expensive assistance packages given through Plan Colombia and Mexico's Merida Initiative (which provided around 80 percent of its budget to judiciary, police, and military assistance) did not reflect well in terms of making these countries any safer.[61] Regional states have done little to pursue common security mechanisms and stop spillovers of violence onto the rest of the continent. Besides this, the heavy hand of the U.S. in nations such as Brazil, Chile, Argentina, or Uruguay has not provided any meaningful aid. Most recently, the Colombian government asked Chile to send seventy-five military officials and act as observers in the ceasefire and peace process with the FARC guerrillas.[62] A favorable post–peace process scenario in Colombia might help secure regional peace and boost infrastructure investments at the local level. Indirectly, it would also benefit Beijing, whose most active initiatives rely on intervening with construction programs in developing countries. Besides that, little else could be considered as a diplomatic triumph for China in Colombia.[63]

Another point to make is that Latin American states place too much value on sovereignty and foreign policies. These are rigorously tailored to fit the dictates of international law. At the same time, state-to-state cooperation is delicately handled, to avoid offending any third parties with possible alliances that might disrupt the balance of power in the region. China, on the other hand, plays by a different set of rules. Because it has no physical presence in the region, China's security influence is not seen as a threat. At the end of the day, what matters is what countries can do with Chinese suppliers, and not the harm that China might directly inflict in the region. For the United States, the history of past and current military and security interventions creates a different puzzle. Many governments reject the idea of a U.S. military presence across the region.[64] While under President George W. Bush some argued that "an unvarnished sense of superiority, displayed proudly on the regional and global stage, has revived the resentment and distrust of Latin Americans toward the United States,"[65] the sentiment today is surprisingly different. The current ongoing security crisis has people clamoring for more U.S. military presence; this is happening most notably in Central America, where there is higher approval of the United States than in South America.[66] In the latter, intra- and interstate security has partially improved; however, transnational crime has been kept at bay only in a few places (e.g., in Chile).[67]

Geostrategic competition in the Asia-Pacific has encouraged Chinese and North American leaders to announce foreign policy measures designed to counterbalance one another. Washington set up its "Pivot to Asia" strategy to contain Beijing's expansion by linking the East China Sea, Taiwan Strait, South China Sea, Malacca Strait, and the Indian Ocean. Washington also set in motion the TPP, although it was later abandoned by President Trump.[68] In this context, the BRI has become somewhat fruitful for Latin America. President Xi intends to portray China as the "new trailblazer of global capitalism," considering Trump's more protectionist policies.[69] Political risk, for example, does not seem to matter too much to the Chinese. Issues in domestic politics appear as windows of opportunity for closing new deals. Recently, Brazil's opening talks about a military coup[70] did not mean that the Chinese authorities avoided signing deals with companies such as Embraer (the conglomerate that produces commercial aerospace and military products and services) and Vale (a logistics company involved in the extraction of minerals and metals).[71] In a visit to Lima in February 2018, former U.S. secretary of state Rex Tillerson warned against excessive Peruvian dependence on China. Only a month later, President Pedro Pablo Kuczynski lost his office after congress ousted him. Later, in June, Peruvian officials announced a new port and a mining project adding up to $ 4 billion in Chinese investments.[72] The worrisome disparity is reflected in the fact that the continent sends commodity-based imports to China. At the same time, the Americas' emerging economies use their current accounts and debt capacity to buy back manufactured goods, partially undermining their own industrialization.[73]

Observers claim that China's trading policy tends to overlook issues of human rights, labor, and environmental affairs while focusing mostly on trade negotiations.[74] Within the various regional projects in the Americas, not a single cooperative framework has emerged to combat the disintegration of foreign investments, even during the golden years of the defunct Washington Consensus.[75] China is optimistic that it will increase Latin American cooperation through smaller forums. On that matter, Chile, Colombia, Mexico, and Peru have recently expressed hopes that a greater Chinese presence might command more telecommunications and information technologies, energy, infrastructure, agriculture, and tourism investment from the Alliance of the Pacific.[76]

Concerning security alliances, including cybersecurity, the scenario is quite similar. China can take advantage of disagreements over regional governance and the constant suspicion that occurs between states. Now that the division between the United States and the rest of the Americas seems more pronounced, it is unlikely to expect that an authoritative voice on topical issues, such as humanitarian and political crises, underdevelopment, impoverishment, and fear and lawlessness, will emerge. As long as this scenario remains active, U.S. policymakers will struggle to find the echo of a "compelling message" and to make sure that states in the Americas want to work in partnership with Washington.[77]

China, on the other hand, is calling for greater cooperation in and out of the BRI umbrella. Beijing is now expanding its purview to include a "digital silk road" as well as a "silk road of satellites,"[78] calling for joint projects in these areas supported by the consortium of Chinese companies that promote military and technological goods, among them China Aerospace Science and Technology Corporation (CASC), China Aerospace Science and Industry Corporation (CASIC), and China Electronics Technology Enterprise (CETC). Chinese defense conglomerates now rank among the top companies worldwide selling sophisticated technologies to the world.[79] China comes second only to the United States in terms of defense budget at US$150.2 billion in 2017, much of it believed to be funding the expansion of its defense industry. China is now capable of building aircraft carriers and quantum-technology communications, as well as other combat-management systems. Chinese arms transfers grew in volume by 211 percent between 1998 and 2017.[80] Beijing now sits among the top six weapons exporting counties in the world in the last three decades, together with Russia, the United States, Germany, France, and the UK.

China's military burden has remained steady since 2010. Despite the recent lowest levels of economic growth, it has continued financing military spending for missions and new equipment. However, we tend to know only the generalities of it since the details regarding many items in its defense budget remain unavailable to third parties. What we do know is that since 2015 the People's Liberation Army's personnel strength had been reduced in numbers from 2.3 million to 2 million. Recent modifications of its forces include creating a joint command that combines the armed infantry with the Air Force, the Navy, the Rocket Force (strategic nuclear forces), and the Strategic Support Force (space, cyber, and electronic warfare). In the words of SIPRI researchers, "The formation of the Strategic Support Force emphasizes the growing importance of space and cyber as areas for military operations and is also an indication of substantive investment in them."[81]

Acquisitions in the security industry have allowed Chinese and Western competitors to enhance their arms-producing companies. The global defense industry sells products and services as diverse as alarm systems, electronic access controls, biometrics, surveillance cybersecurity-related products. Cybersecurity and intelligence have been key sectors for security industry acquisitions since the 2010s. Boeing, DynCorp International, Harris, QinetiQ, Raytheon, SAIC, EADS, BAE Systems, and Ultra Electronics have purchased a significant number of smaller companies that specialize in these areas, many located in the United States, France, and the UK. Arms companies have benefited from the intense actions of governments in buying technologies to confront a myriad of perceived threats posed by international terrorism, organized crime, illegal immigration, and piracy.[82] The diversification into cybersecurity outside of traditional arms production and military services areas has allowed companies to overcome losses in traditional military products while entering the booming markets of dual (military-civilian) cyber technologies. Bridging the civilian and military sectors of government, many cyber products are focused on the protection of critical infrastructure, competing with traditional companies in the IT sector for a slice of the cybersecurity market.[83]

The network of state-owned companies that trade on ICTs has spawned a wave of contractors, together with greater Internet diffusion across Eastern Asia. Besides China, Singapore, Hong Kong SAR, Japan, and South Korea have had Internet diffusion patterns showing high levels of cyber development. Although it is recognized that Asia lags behind North America and Europe in cyber knowledge and practice, it is believed the high Internet penetration plus the growing R&D environment will shrink the gap considerably.[84]

Allegedly, China's industrial advances have been partly supplemented by the systematic hacking of Western technology designs.[85] Today, states, companies, and civil society groups in Asia are exposed to severe incidents originating on the Internet, including information attacks and hacking to damage computer systems, censorship, spam abuses, and the promotion of violent ideologies and the coordination of cyber or physical violence.[86] North Korea, for example, is said to conduct hybrid use of force "by the combined and coordinated use of often deniable or covert acts; including, conventional but limited military action, cyber warfare, criminal behavior, and the use of semi-state actors."[87]

The sophisticated cyber capability of Asian actors has allowed attacks on targets around the globe, as recorded by the Council on Foreign Relations' "Cyber Operations Tracker."[88] Hybrid operations of both military and intelligence nature have persisted in Asia during recent peacetime. Many Asian actors remain inferior players to the Western superpowers. Thus, the sustained use of asymmetric capabilities includes the deployment of cyber technologies aimed to disrupt international computerized networks. For countries such as North Korea, the use of cyberweapons is "value-added as its main adversaries, South Korea, the U.S. and Japan are highly networked states that are reliant on the cyber domain for both military and civil purposes." For China, the same principle applies.[89] In these cases, cyberincidents are used for many purposes, including espionage, disruption of infrastructure, private hacking of institutions, economic crime, and attacks to undermine the political systems of other nations.

To James Andrew Lewis at the Center for Strategic and International Studies (CSIS), formerly the rapporteur for the UN Group of Government Experts on Information Security, China is not engaging in "cyber" Cold War with the Western states; rather, Beijing is spying heavily on the United States, and spying is not a crime under international law. U.S. companies and the intelligence community have reacted ineptly to what Lewis believes is not even China's "A-game" yet. Moreover, Lewis questions whether China is any more capable of attacking domestic infrastructure than it is to launch cyberattacks on military forces deployed in the Pacific if it sees itself as cornered and without any avenues of escape.[90] What is a certainty for Lewis is that "the primary source of risk in cybersecurity comes from conflict between states."[91] Russian and Chinese cyberactions create a sense of vulnerability among the United States and its allies. Lately, official responses from U.S. partner states across the globe, including the Western Hemisphere, as the previous chapters have shown, have been to invest in cyber-resilience plus defense and offense capabilities. In 2016, the United States and South Korea, for example, established a cyber task force to set up dual plans and preemptive action.[92] A year later, the Pentagon requested a record $7 billion to help improve cyberwarfare defenses and develop cyber tools and infrastructure to provide offensive cyber options.[93] To some observers, the Pentagon should take the lead on military systems as rivals Beijing and Moscow have strengthened their ties to develop new technologies.[94]

Public Opinion toward China and the United States

The 2017 edition of the U.S. National Security Strategy (NSS), published under the Trump administration, puts China as a challenging state to American power. It goes on to argue that Beijing is attempting to expand its influence in Europe, Western Hemisphere, and Africa.[95] Since at least

2006, the Pentagon's official publication has gradually warned of China's potential to compete militarily with the United States. Clearer strategic concerns have been expressed since 2010 regarding Beijing's military modernization, increasing cyber and space warfare technologies, and overall attempts to undermine U.S. dominance in critical areas of the globe.[96] This final section puts to the test whether U.S. or Chinese influence is pivotal to the citizenry in the Americas. It describes patterns in public opinion of Latin America's citizens, using two criteria for determining support or trust for the United States and China: (1) citizens' preferences for economic development models considering different alternatives, and (2) trust in foreign government and international organizations. In the Appendix file, I present three-way contingency tables A.2 to A.4 to assess the relationship between views on the Chinese and U.S. governments in the instances of weekly and daily use of Internet among those that favor the Chinese or the U.S. model for future development.

I argue that citizens' views on foreign governments and institutions are essential when talking about cybersecurity. In one sense, we might expect that the superpowers and countries such as Russia and China will continue experimenting with infiltration and spying on other states, as the revelations of Wikileaks and NSA data surveillance have previously shown. Developed and highly cyber-capable states can manipulate data. It is believed that China has the ability to interfere with ICT services provided through enterprises, such as Huawei, and military firms seeking to maximize their national strategies in the international strategic environment, albeit while purposefully avoiding the armed conflict threshold.[97] Since 2010, China has bolstered the installation of a "coordinated development" policy between the national defense and the economy that has hit Latin America as well as other parts of the developing world.[98] In a variety of white papers and government documents, the Department of Arms Control has utilized a concept to inform the world that China is on a path of "peaceful development" and has a "defense policy which is purely defensive in nature." At least on paper, the Sino defense policy builds on assurances of trust, cooperation, stability, and development within the international regime.[99] To pursue its "coordinated development" policy, Beijing's national defense budget uses around 2 percent of its overall fiscal budget, compared to more than 3 percent in the United States, 5 percent in Russia, and 10 percent in Saudi Arabia.

To provide more detailed insights into mass opinion, I briefly provide statistical findings from opinion surveys of citizens in six Latin American countries. Data was taken from the 2016–17 round of the AmericasBarometer.[100]

Table 5.1 summarizes citizens' preferences for economic development models. Respondents were asked, "In your opinion, which of the following countries should be a model for development for the future of our country?" Alternatives included developed and developing countries such as China, Japan, the United States, Brazil, and Russia, among others. One of the choices allowed respondents to select "none and follow our own model," while a final selection was "other." The same question was presented to respondents surveyed in Argentina, Brazil, Chile, Colombia, Mexico, and Venezuela.

At first sight, we see that preferences across the six case studies were varied. Three "country models" came with the most mentions, China, Japan, and the United States. Only respondents from Argentina and Mexico chose China as the first option. Meanwhile, respondents in Brazil, Chile, Colombia, and Venezuela thought the United States was the best economic model for development with the lowest score in Chile (25 percent), and the highest in Brazil (36.7 percent).

Surprisingly, Table 5.1 indicates that respondents in Venezuela, a country that has been under a socialist rule for the last two decades, and whose leader keeps using strong anti-Washington rhetoric, more than one-third

Table 5.1. Best country as model for future development, valid percentages

	Argentina	Brazil	Chile	Colombia	Mexico	Venezuela
China	22	18.6	18.8	12.7	26.6	19
Japan	21.4	29.3	17.9	18.2	24.1	9.9
India	.5	1.2	.7	.7	.7	.5
United States	20.5	36.7	25	34.5	20.2	35.8
Singapore	2.5	1.3	4.9	2.5	1.8	2
Russia	3.9	1.7	5.4	3.0	4.8	4
South Korea	1.7	3.1	3.3	1.7	2.4	2.1
Brazil	8.9	—	3.6	6.	5.6	10.2
Venezuela	1.4	2.1	.8	1.9	3.8	—
Mexico	2.6	4.3	3.5	6.2	—	4.9
None/We need to follow our own model	11.4	1.3	14.6	11	8.3	10.9
Other	2.8	.3	1.5	.9	1.8	.6
Valid N	1320	1435	1470	1428	1314	1419

Source: Author's construction with data from LAPOP (2018).

of the respondents thought the United Stated represented the best model for development. As I have said before in the previous chapters, Venezuela has developed deep economic, social, and military links with nations other than the United States, most notably China and Russia. Nonetheless, the Chinese model for development comes in second place with 19 percent and Moscow with a low 4 percent, well behind Japan, India, Brazil, and those that preferred none or their national models.

What is also telling is that respondents in Mexico, the Latin American country nearest to the United States and historically a bilateral trade and commerce partner, preferred the Chinese model first (26.6 percent) and the Japanese model (24.1 percent) in second place. Brazilians, on the other hand, mentioned the United States model of development the most with a substantial 36.7 percent, while the Chinese model came in third place (18.6 percent), after Japan (29.3 percent).

Based on the two countries that preferred the development model of the United States and China the most (Brazil, and Mexico, respectively), I decided to try to shed more light on the respondents' preferences for governments, and to see if there is a match between economic model and political leadership.

Table 5.2 depicts citizens' trust in foreign governmental and international organizations. Respondents were asked separately how much trust they put in the governments of China and the United States. To grasp a better perspective, I also included citizens' opinions of international organizations, the OAS, and the UN. Originally, respondents were asked to place their level of trust in these governments and institutions, using a four-point scale graded between 1 = "very trustworthy," and 4 = "not trustworthy at all." I recoded the data into a new variable to reflect an ascendant 4-point scale to make the values comparable and following the convention that a higher number would indicate a higher score. The new values in the recoded variable were 1 = "not trustworthy at all," 2 = "little trustworthy," 3 = "some trustworthy," and 4 = "very trustworthy." All other values were coded system missing.

Mean values (average value of a distribution) in Table 5.2 show that Brazilians (who overall had the most preference for the United States' economic model) do not place a great deal of their trust in foreign governments or international organizations. Unlike their economic development model, means indicate a higher level of confidence (still reasonably low) in the government of China than the government of the United States.

On the other hand, Mexican respondents (who mentioned China the most as the developmental model to follow) showed a clear preference

Table 5.2. Trust in foreign governments and multilateral organizations

	Brazil			Mexico		
	N	Mean	Std. Dev.	N	Mean	Std. Dev.
China Gov	663	2.61	.95	704	2.62	.86
U.S. Gov	1020	2.24	.98	1188	1.76	.90
OAS	581	2.56	.92	673	2.46	.81
UN	981	2.89	.89	1061	2.71	.82

Note: Trusts values were recoded as 1= "not trustworthy at all," 2 = "little trustworthy," 3 = "some trustworthy," and 4 = "very trustworthy." All other values were coded system missing.
Source: Author's construction with data from LAPOP (2018).

(although still very low in levels of trust) for China (mean value of 2.6) over the United States government (mean value of 1.76). Regarding respondents' trust in international organizations, respondents in Brazil and Mexico showed similar levels of trust in OAS and the UN (again with low mean levels). In both countries, respondents awarded a slightly higher mean to the United Nations over the Organization of American States. For the case of Mexico, all means were higher than the respondents' trust in the U.S. government.

Table 5.3 shows the matrices for bivariate statistics.[101] By correlating the variables presented in the previous statistical exercises in pairs against each other we can indicate the degree of association. It should be noted that even if two variables are highly correlated, one cannot infer causality between them or, in other words, "assume that change in one variable causes change in the other in the absence of other conditions."[102] In chapter 7, I provide further theoretical explanation in lay English on the purpose of bivariate correlation (some repetition along the way is therefore inevitable). Table 5.3 shows the Pearson r values for the correlation between the scaled variables trust in China's government, trust in the U.S. government, trust in OAS, and trust in the UN. So far, I want to treat the variables as if they were of equal status. It is often the case that variables are asymmetric, one being thought as a dependent variable Y and the other as the independent or predictor X.[103] When the researcher assumes an interest in prediction, regardless of whether causality is implied, the language of "outcomes" and "predictors" is preferred to the language of "dependent" and "independent" variables.[104] I will return to the scientific basis for looking upon one variable as potentially causally dependent on the other in chapter 7 when I perform statistical regression as an analytic procedure.

Table 5.3 shows the statistically significant variables, which means the observed effect is probably not due to chance alone. The value given in each cell uses asterisks to denote how low the probability is that the observed effect is due to chance alone. Here, the values are presented with a probability of being by chance less than 1 percent. The lack of an asterisk means that the corresponding pair of variables do not have and statistically significant effect. Large (positive or negative) coefficients indicate the strongest effects. From reading the tables we see that the Pearson r values are statistically significant and have positive coefficients. The strength of the associations is rather weak, with only the sample from Mexico indicating a moderate strength value between trust in the UN and trust in OAS (r coefficient is .544).

In Table 5.4, I report the results of another bivariate correlation exercise, this time including the scalar variable asking respondents, "How frequently do you use the Internet?" Respondents were originally asked to answer using a 5-point scale where 1 = "daily," and 5 = "Never." As before, I recoded the data into a new variable to make the values comparable (a higher number would indicate a higher score). This way, values indicate 1 = "Never," 2 = "Rarely," 3 = "Sometimes a month," 4 = "Sometimes a week," 5 = "Daily." All other values were coded system missing. Simple scientific hunch suggests there would be a correlation between being exposed to information and contents available online that lead to one person

Table 5.3. Bivariate correlation matrix (a)

Brazil

	China Gov	U.S. Gov	OAS	UN
China Gov	1			
U.S. Gov	.285**	1		
OAS	.371**	.444**	1	
UN	.247**	.178**	.454**	1

Mexico

	China Gov	US Gov	OAS	UN
China Gov	1			
U.S. Gov	.353**	1		
OAS	.319**	.360**	1	
UN	.306**	.238**	.544**	1

Note: **Correlation is significant at the .01 level.
Source: Author's construction with data from LAPOP (2018).

Table 5.4. Bivariate correlation matrix (b)

Brazil

	China Gov	U.S. Gov	Internet Usage
China Gov	1		
U.S. Gov	.29**	1	
Internet Usage	.044	.058	1

Mexico

	China Gov	U.S. Gov	Internet Usage
China Gov	1		
U.S. Gov	.353**	1	
Internet Usage	.184**	.188**	1

Note: **Correlation is significant at the .01 level.
Source: Author's construction with data from LAPOP (2018).

building an informed opinion (or maybe badly informed too!) on foreign governments. The rationale here is to extrapolate whether there is any relationship between using the Internet and trusting foreign governments, the strength and the direction of the association.

Table 5.4 reports that the variable Internet Usage is not statistically significant with trust in either government for the Brazilian sample. On the contrary, for the Mexican sample, Internet usage is statistically significant with trust in the Chinese and the United States' governments. The r coefficient is positive but weak at .184.

To summarize these findings, I propose that the maximum correlation does not imply causation. Causal claims cannot be inferred or defined in terms of statistical associations alone.[105] However, the absence of correlation implies the absence of the existence of a causal relationship.[106] This means that assessing for the effects of a research factor provides statistical insight on the desired effect of other factors. Simply put, correlational measures shed light on the opinions regarding China and the United States considering the previous formal theory and research explained in the chapter.

Discussion and Conclusion

In 2017, the commander of the U.S. Southern Command, Admiral Kurt W. Tidd, argued about the emerging powers' expanding presence in this way:

"If we're concerned about them [Russia, China and Iran] on a global scale, that means we need to make sure that we pay attention to what they are doing in the Latin America region."[107] At first thought, the U.S. security and political establishments seem conscious that their hegemony in the Americas is being disputed and that some novel energy will be necessary to avoid displacement from their own backyard. [108] The arguments of this chapter have aimed to explore such events, first asking how the Americas' states have navigated Sino-U.S. rivalry and then, What impact do cyberaffairs have in the matter? It must be said that despite the distance between the world actors involved in the contentious cyber dispute between East and West, both the eagle's and the dragon's global reach is triggering realignments for security purposes that have been unfolding gradually since the Cold War era. The movement away from U.S. hegemony has offered some of the Americas' states the chance to increase their bargaining power by making new partners in the multipolar world order. Is this a beneficial approach for cybersecurity in the long term? I argue that it will be difficult to surpass the historic containment of U.S. security and defense policy in the Americas, even for the biggest of eastern superpowers. The cyber power of the U.S. and the reach of the Fourth Fleet and the Southern Command pose a digital and physical deterrent to any capable force from outside the region. Even the most committed anti-U.S. leaders in the region depend on the security that only such a powerful force can provide. Equally, the United States plays an indispensable role in on-the-ground minor-scale security policies, as was recently shown during military operations in Central and South America.

I also questioned why the Americas' states seek partial alliances with either Washington or Beijing, either in instances of cybersecurity or in other aspects such as trade? Latin American states can maintain separate tracks for security and trade purposes mostly because China is a distant-supply player, and cares more about a proxy engagement with the United States in East Asia rather than in the Americas. Bilateral relations with the United States seem to be more important than compromising the security balance in the Americas.[109] Under President Xi, China's "peaceful rise" cares greatly about a stable environment that is conducive to economic development. Hongying Wang at the University of Waterloo put it this way:

> For the recent past and the near future, good relations with the United States is China's top foreign policy priority. Its management of relations with other countries and regions is sensitive to the effect on Sino-U.S. relations. Beyond what they consider

to be China's core national interests, Chinese policymakers have shown remarkable flexibility and pragmatism, trying to minimize provocation of the United States.[110]

It is also claimed that Chinese economic prowess still has an underdeveloped grand strategy regarding security in its own region. This seems particularly to be the case now that the United States has decided to rebalance its forces in East Asia.[111] Also, China is far closer to other more important developing states, such as India, Iran, and Pakistan, that have more pressing capabilities in terms of tilting the balance of power in Southeast Asia through both symmetric and asymmetric means of warfare, including cyber capabilities. In these matters, some scholars seem most concerned, as June Treufel Dreyer noted, that the United States assume a clear set of guidelines under which "all Chinese provocations must be answered in some way," yet, "this does not necessarily mean a kinetic response."[112]

My final concerns regarded what Latin American states can achieve under the two-track polity and how public opinion plays a role in the matter. States can accomplish much more in promoting their security, technological advancement, and commerce policies to fulfill their citizenry's desire for developmental advancement by holding simultaneous deals with the United States and China. Ever since the global financial crisis, Chinese penetration of supply and demand in ICTs has clearly favored the developing states' populations. Economic booms have kindled the appetite for Chinese products, and in times of crisis Chinese investment is well received. This might be the point where the United States loses the most. The Chinese double game of selling cheap and capitalizing big is difficult to imitate. Therefore, among other reasons, security alliances with the United States do not translate into commercial security (Argentina and Brazil). Some states gain the most through non-U.S. security and commercial ambitions (Peru, Venezuela); for some other nations (Chile, Uruguay, Mexico, and Colombia), security alliances for mutual strategic interests have so far proven resistant to Chinese persuasion.

Six

Alliances with the United States

On July 19, 2019, U.S. Secretary of State Michael R. Pompeo visited Argentina to attend the Western Hemisphere Counterterrorism Ministerial Plenary. In Buenos Aires, Pompeo delivered what he called President Trump's effort to re-engage with U.S. partners in the hemisphere: "Indeed, we believe this is a new era in relationships between the United States and South America. As I said in Santiago in April, we have an enormous opportunity to help each other continue to have prosperity and security."[1] Since Trump's arrival at the White House in 2017, Washington had offered regional leaders his support in countering transnational crime, terrorism, and cybersecurity.[2]

An alliance is a formal or informal alignment for security cooperation between two or more states, intended to augment each member's power, security, and influence, argues Stephen Walt. The critical element in any alliance, Walt adds, is a commitment to mutual support against some external actor.[3] As I contended in the previous chapters, the notable cyber power resting in the hands of the United States has a profound impact on shaping new alliances with different actors. The effort to seek cyber alliances comes at a time when the United States is reorganizing its cyber forces at home while shielding itself from foreign cyber intrusion. "We are engaged in a long-term strategic competition with China and Russia. These States have expanded that competition to include persistent campaigns in and through cyberspace that pose long-term strategic risk to the nation as well as to our allies and partners," says the Department of Defense's 2018 Cyber Strategy.[4] Its "defending forward" philosophy—getting as close to adversaries as possible to see what they're planning as a means of informing others to prepare or take action themselves[5]—has prompted the U.S. Cyber Command to help defend networks of other nations learning about persistent threats and malware coming from adversary states.

Through its Southern Command, the Pentagon guides military actions "to protect the homeland" by building partnerships and countering threats in Latin America and the Caribbean. "Without action, the United States will continue to cede influence to China and Russia in the hemisphere," argues the command's 2019 strategy.[6] U.S. partnerships are not just military-to-military but also include diplomatic efforts by the State Department and by other organizations, such as the U.S. Agency for International Development and the departments of Justice and Homeland Security. In return, Washington has found willing partners to cooperate and promote dual-use ICTs across the hemisphere. "We are connected, not only in the traditional domains that we think about when we talk about military operations—those domains being land, sea, air, space, and cyber. But we're connected, importantly, by values and democracy," said Admiral Faller.[7]

Perceived threats in the region, digital or physical, are a chief concern to Washington. The sophistication of drug traffickers, human smugglers, terrorist supporters, arms dealers, and money launderers has prompted the Pentagon to rethink new ways to disrupt these threats. "Threat network enablers like facilitators, suppliers, recruiters, and technicians provide criminals and extremists with unprecedented global reach, and the ability to operate stealthily in both the physical and the cyber worlds. Criminal networks leverage all means available to move lethal narcotics, people, weapons, and dirty money into and out of Latin America and the U.S. homeland. Extremist networks like ISIS reach deep into our hemisphere, inspire would-be terrorists to conduct attacks in the region, or to attempt entry into the United States to do our citizens harm," argued Admiral Tidd, Faller's predecessor, before he retired.[8]

In this scenario, what are the prospects of U.S.–Latin American cyber partnerships? Tracing recent U.S. policy toward the region shows Washington's stance in making cybersecurity alliances a reality among regional governments. U.S.-led cybersecurity efforts will continue to be welcomed as long as Latin American states show the political and financial will to adapt their military's role and mission, which has brought forward, among other issues, the creation of cybersecurity forces and cyber commands at the national scale. Being able to count on U.S. support in building up cyber capabilities has proven precious to both old reliable and newly U.S-friendly nations. To stress my argument, later in the chapter I review three positive cases where cyber partnerships have happened (Argentina, Brazil, Chile) and three negative cases (Colombia, Mexico, and Venezuela) where they have not.

Cyber Alliances

International actors can combine their capabilities through alliances in ways that further their interests and security goals when considering an external threat. The expected response of those actors forming a partnership will be to attempt to strike a balance against threats and aggressive intentions. States with similar ideologies can enter an alliance with one another. Sometimes states can bandwagon with existing threats if they believe that resistance is impossible. Nations can ally in exchange for economic or military external assistance to deal with perceived danger.[9] Today, U.S. allies and other like-minded states face two options: to cyber partner or not to cyber partner with Washington's support. The decision requires strategic decision making that intertwines among the economic, military, and diplomatic levels. Latin American countries, on the one hand, are welcoming Chinese ICTs with open arms. Since the election of Jair Bolsonaro, Brazil has engaged in serious talks with the Chinese company Huawei to launch its 5G mobile infrastructure. The same has happened in Chile since the election of right-wing Sebastian Piñera, and likewise in Colombia under President Ivan Duque.[10] Both these countries, on the other hand, have a long record of diplomatic and military-to-military relations with the United States that has prevented them from expanding links with Beijing despite the Sino frenzy in the region.

Alliances are usually formed as a response to a threat. States can enter into an alliance in opposition to a principal source of danger (*balance*), or ally with the state that poses the primary threat (*bandwagon*). Weaker states are usually attracted to more powerful ones when this is demonstrated. When a state's strength and power are in decline, allies might opt to stay neutral or to defect to the other side.[11] Just like with NATO allies (Canada, France, the Netherlands, and the United Kingdom), the United States is trying to strengthen relationships and interoperability in areas of emerging threats, most notably in the space and cyber domains. Improving integration is deemed to enhance regional security structures and reinforce Washington's global network and theaters of operation. By building sustainable partner capacity and increasing collective capacity, the United States and its partners can "respond rapidly to crises and contingencies."[12] In theoretical terms, *balance* and *bandwagoning* can thus be framed in terms of "threatening power." This means aligning with the weaker side or choosing the stronger, in consideration of factors affected by the level of threat that states may pose.[13]

Some scholars theorize on the U.S.-led regional alliance systems as standard models of multilateralism (e.g., NATO), or through a bilateral system of alliances in which different allies build bilateral strategic ties with Washington, but not with each other (the "hub and spokes" model).[14] A third model has emerged depicting U.S.-led alliance systems as a hybrid category that combines overlapping bilateral, minilateral, and multilateral initiatives, or a nodal defense system organized around specific functional roles to tackle different threats.[15]

Cyber affairs between the United States and Latin America occur in what authors have described as Latin America's dual-track policy for international engagement (which I introduced in chapter 5), one that favors the nodal defense approach only to a certain extent. Meanwhile, countries have signed bilateral agreements of cyber cooperation with Washington; they have also sought assistance elsewhere under other overlapping multilateral organizations. For example, Brazil has a mutual cyber agreement with the United States and bilateral ones with Argentina, Turkey, Germany, and the EU. Brazil is also part of various cybersecurity multilateral fora under the auspice of OAS, the UN, and Europol. Through alliance diversification, Latin American countries have had no problems in securing trade and commerce opportunities with U.S. military rivals if both tracks are not mixed. The problem here is that because protecting cyberspace requires leaders to govern between private and public interests at the micro (i.e., citizens) and macro (i.e., critical infrastructure) levels, many times security and technology interests overlap when it comes to strategic national projects. The nodal defense model has, in my view, an obvious downside. Allies are pragmatic, and they will always seek to put their national strategic interests first. Partnerships and alliances are nuanced and not solemn endeavors.[16] Moreover, Latin America's dual-track policy is not new to Washington. It has been in the making since the end of the Cold War and became newly evident most recently with the weakening of U.S. hegemony in the Americas.[17] While many Latin American countries are accelerating their efforts to escape underdevelopment, cheap and ready to use ICTs are always convenient. Having a safer Internet, on the other hand, requires more bureaucratic capabilities, financial resources, and developing long-term human skills.

Among the U.S. military leadership, commanders have started to reevaluate what success means in cyberspace and how partnerships influence results. Brigadier General Timothy Haugh, commander of the Cyber National Mission Force, has said that cybersecurity is not the Pentagon's sole role anymore, but enabling domestic and international partners to defend critical assets to success. By monitoring adversaries' activities, at home and abroad, the United States can alert the most relevant agency, from the Department of Energy to the FBI, to act. This way, new cooperation routes to chastise cyber adversaries include using the Department of State to engage in diplomatic sanction, the Department of Homeland Security to alert and monitor industrial interests, or the Treasury to apply economic sanctions.[18]

From Obama to Trump

Since the time of the Cold War, the United States has had the substantial power to deploy its military prowess almost anywhere in the world. Contemporary alliance dynamics are shaped by this distinctive feature that, along with other conditions, defines today's moment of unipolarity. States are influenced thus by the large diplomatic and economic shadow of the United States, especially those in the Western Hemisphere, who share not only geographic location but also liberal ideals and specific historical features.[19] During Barack Obama's administration, for example, U.S. diplomatic engagement with the Americas was, for many observers, a low point in continental relations (except for policy toward Cuba). Nonetheless, while many issues drew the attention of Washington elsewhere in the world (Al-Qaeda and other groups in Afghanistan, Pakistan, and around the world; a new START Treaty with Russia; the Iranian and North Korean nuclear programs; Middle East peace; and reorienting U.S. military forces to the Pacific, among others), securing cyberspace was a landmark in terms of including international partners willing to pay careful attention to the issue.

In Latin America, Obama's Pathways to Prosperity Initiative sought to bring countries together based on mutual respect and mutual responsibility to confront twenty-first-century challenges, including energy security, climate change, and technological and economic prosperity.[20] Relations turned sour in 2013 when it was revealed that the NSA had spied on world leaders, including Brazil's president Dilma Rousseff, who denounced Washington's espionage and cyber-surveillance at the United Nations. Later, in 2015, at the Summit of the Americas, Obama and Rousseff turned the page as they agreed to resume talks devoted to the creation of the U.S.-Brazil cyber working group.[21] This type of U.S.-led bilateral working group promotes policy and knowledge transfer on cyber and ICT issues at various governmental levels, usually including representatives from the military, diplomatic, and trade policy circles. This way, Washington has expanded its influence in other countries' key areas, including e-government, the digital economy, cybersecurity, cybercrime prevention, capacity-building activities, international security in cyberspace, and research, development, and innovation.[22]

For the United States, securing cyberspace in a multilateral way was ratified as an item of concern in the 2010 NSS under the pretext that the U.S. "will deter, prevent, detect, defend against, and quickly recover from cyber intrusions and attacks," and calling for international partnerships "to ensure an organized and unified response to future cyber incidents," at home and abroad.[23] Obama's administration focused on making it clear that "because cyberspace crosses every international boundary, we must engage with our international partners."[24] National doctrines and policies elsewhere on the globe followed suit, with like-minded countries setting up national CERTs and translating defensive and civilian capabilities into military cyber capabilities.[25] Like the U.S. Cyber Command, at least a dozen other states had military cyberwarfare organizations by 2013, including Argentina, Brazil, Canada, China, the Democratic People's Republic of Korea, Denmark, Germany, India, Iran, the Republic of Korea, Switzerland.[26]

Obama left the White House with a mixed record in national and economic security in cyberspace. While its Cybersecurity National Action Plan (CNAP) encouraged the deterrence and disruption of malicious cyber activity at home, it was also used to amass cyber power through the operational capability of the Pentagon invested in the Cyber Command structure and directed against Russian, Chinese, Iranian, and North Korean cyber forces.[27] By then, the world had witnessed severe cyber incidents, among them, the Sony Pictures attack, and attacks on Georgia, Estonia, Iran, and Ukraine. The lessons to be taken for the lesser international actors were many. One was that in the last decade, those countries that had instituted frameworks designed to raise the level of cybersecurity and cyber defense were able to take better control of their digital security. This included the crucial task of directing national government policy to take new actions in responding to and recovering from cyberincidents.

Washington has used the cyber issue to warn Latin American governments against a possible reconsideration of its defense and economic interests if they don't step up their game. Once Trump came to power, U.S. views on the need to ensure that data partners are trusted became very well known. In the words of Secretary of State Pompeo, "The United States has an obligation to ensure the places where we operate, places where information is, places where we have national security risks, that they operate within trusted networks and that is what we will do."[28] For Latin America, the China problem lies at the crux of the U.S. preoccupation. "The more China becomes intertwined with our Western Hemisphere nations, the more likely it is they can surveil, they can monitor, they can surround the United States," said Republican Senator Cory Gardner, who leads the Foreign Relations subcommittee on the Asia-Pacific region.[29]

Reacting to New Threats

By the end of 2012, Obama's administration rebalanced its strategic priorities away from the Middle East and toward the Asia-Pacific region. The move meant a relocation of a significant portion of its naval, space, and cyber assets, consequently diminishing U.S. resources elsewhere in locations where it had enjoyed long-standing security partnerships since the Cold War (e.g., NATO). The increasing concerns coming from China meant the likely loss of attention in the campaigns in the Middle East, Eastern Europe, and North Africa. Among the allies of the United States, the repositioning of its forces ultimately brought a rethinking on responses to many challenges, including terrorism and counterinsurgency, counter-proliferation, piracy, and hybrid threats and cybersecurity.[30] Developing according to the changing security environment would thus require cost-effective strategies for working with partners that might not only receive U.S. assistance but help conduct trilateral and multinational responses. In a recent example, Estonia, Latvia, and Lithuania worked with the U.S. National Guard in Maryland, Michigan, and Pennsylvania in a U.S. European Command (USEUCO) cyber defense exercise preparing for a cyberincident that required a multinational response.[31] While in the past cyber partnerships were mostly directed by the secret service's cyber intelligence efforts, the U.S. Cyber Command

and its military hierarchy have operationalized cyber relations differently, by focusing not only on traditional partners, such as the Five Eyes intelligence alliance, but also on like-minded nations that can cooperate in the international front.[32] The cyber issue is used as one tool that the United States can rely on when requesting or lending support to friendly nations. General Paul Nakasone, commander of the U.S. Cyber Command, argued in a testimony before the Senate committee on armed services that the Command is actively pursuing five critical areas of development: supporting strategic competition; establishing a war-fighting ethos across the Command; improving the readiness of cyber forces; enhancing partnerships across government, allies, and the private sector; and deploying improved operating infrastructure.[33] On confronting "capable adversaries" through strategic alliances, Nakasone argued: "Our efforts to act against them and to enable our partner combatant commands, government agencies, and allies have helped to defend our nation and its interests. Those efforts, however, must rapidly become more agile, more capable, and more sustainable."[34]

Cyber alliances have allowed the United States to maintain constant contact with allied nations that face common threats from a variety of malicious cyber actors, including nonstate and criminal organizations, states, and their proxies. To Nakasone, "USCYBERCOM ensures two critical capabilities against these threats: it enables partners in whole-of-nation efforts to build resilience, close vulnerabilities, and defend critical infrastructure; and it acts against adversaries who can operate across the full spectrum of cyberspace operations and who possess the capacity and the will to sustain cyber campaigns against the United States and its allies."[35] The 2017 NSS emphasized the emergence of great-power competition in cyberspace and the long-term strategic competition with China and Russia countering their efforts to "weaponiz[e] information, stealing intellectual property, and mounting influence campaigns."[36] The NSS contends: "We will strengthen America's capabilities—including in space and cyberspace—and revitalize others that have been neglected. Allies and partners magnify our power. We expect them to shoulder a fair share of the burden of responsibility to protect against common threats."[37]

The recently announced cybersecurity bureau within the State Department aims at filling a gap in the U.S. government's diplomatic abilities by appointing an ambassador-at-large for cyberspace security and emerging technologies.[38] Once it has been officially rolled out, the bureau is expected to take a leading role in building up the cybersecurity capacity of developing countries' governments. Beijing, on the other hand, has responded actively in advancing efforts to dictate Internet governance and provide allies with high-tech exports and surveillance capabilities, such as security cameras with facial recognition software, which have been sold so far to Argentina, Bolivia, Ecuador, Panama, and Uruguay. China's best bet is for a more energetic presence in multilateral bodies where the United States has decided to retreat."[39]

Punishing hacking nations and working with allies seems to be Trump's new approach to cyber affairs. By jointly naming and shaming Moscow, Beijing, and Pyongyang, the United States hopes to redraw the boundaries of cyberspace action. Robert Strayer, deputy assistant secretary for cyber and international communications and information policy and the State Department's top cybersecurity diplomat, embarked on the mission of getting foreign officials to impose joint criminal charges against hackers or financial sanctions as quickly as possible after officials determine a nation is responsible for a hack. The Justice Department, for example, coordinated with the UK, Canada, Australia, New Zealand, and Japan to condemn APT10 or Stone Panda (said to be supported by the Chinese Ministry of State Security), accused of targeting a global collection of government agencies and companies.[40]

Enabling global cyber ops would require a vast number of networked agencies and resources in the planning and execution of cyberdefense and offense actions. For the pool of like-minded nations to grow, for instance, throughout Latin America will require a common front of partners with similar capabilities, something not all countries are ready to offer. Recently, for example, hackers from North Korea have targeted financial institutions in Latin America, such as major banks in Mexico and Chile, to make money for the regime.[41]

The OAS Cyber Observatory describes Latin American states still in the early stages of cyber maturity development (I review this extensively in chapter 7).[42] While many governments are generally not trained in cybersecurity, countries' primary units dedicated to cyber-defense and cybercrime prosecution have received training from the United States (Grenada, Paraguay, Uruguay, and St Vincent and the Grenadines). In a similar way to the ASEAN-U.S. Cyber Policy Dialogue,[43] Washington has discussed broadening regional cooperation on cyber capacity building to protect critical infrastructure and combat cybercrime, as well as to counter the use of ICTs, including the Internet, for terrorist purposes in the Americas.

The Trump administration rekindled relations with countries such as Argentina, Brazil, Paraguay, based on reciprocal security cooperation. "On this continent, the United States is showing up as never before, reminding our friends of how much we have in common, of how much our interests align, and how much we love you," Pompeo said.[44] Argentina, for instance, refocused efforts on the Triple Border issues in exchange for Washington's support for macroeconomic reforms, an IMF bailout rescue plan, and its candidacy to join the OECD. Normalizing U.S.–Latin American security relations based on economic rewards carries considerable risk. Washington has promised similar financial rewards to El Salvador's linked to the success of their counternarcotic cooperation, the Mexican government and the border question, and Ecuador's revamped security cooperation, which brought USAID back to that country.[45] Many times in the past, Latin American countries have not delivered consistently, or U.S.-led programs have brought unwanted consequences (the two most infamous cases are Plan Colombia and the Merida Initiative in Mexico). For cybersecurity matters, the problem relies on Trump's militarization of cyberspace, which requires a perpetual state of digital weapons carrying strikes.[46] While whole-of-government approaches have been promised before, such as in Obama's 2011 Strategy to Combat Transnational Organized Crime, which hoped to raise international awareness of TOC and build multilateral cooperation against it.[47] Latin American states struggle to employ the full range of government agencies and authorities. Many times, foreign policy interests collide with defense interests, intelligence bodies are understaffed and underfinanced, and, most notably, law enforcement agencies are pervaded by corruption.

Cyber military operations reached a new height when, in August 2017, the Trump administration elevated the Cyber Command to the status of a unified combatant command for offensive and defensive roles and mission. The move also meant that the command could advocate for its own investments and resources and report directly to the secretary of defense. What remains to be done is a total separation from the NSA that could lead to the command's independence from the intelligence community to generate its war-fighting capabilities.[48] The move sent out a clear message for the world and the region: the United States is getting ready to step into a whole new level of cyber capabilities and military-cyber elements within each service branch.

For the Pentagon's cyber allies in the rest of the world, the consequences are by now twofold. One the one hand, cyber strategies cite cyber defense as the primary strategic threat to other perceived menaces (e.g., terrorism) that have been at the top of the list since 9/11. Budgets for cyber missions within the armed forces and the defense establishment are on the rise, as well. The full spectrum of cyber military capabilities is merging with cyber- and electronic warfare while involving a broader number of actors and institutions across governments for the protection of critical national infrastructure. On the other hand, the military is getting involved with civilian agencies in establishing clear and synchronized potential targets of national importance, as well as the risks to governmental functions, such as its ICT networks, coming from cyberspace.

Canada, for instance, launched a defense review in 2017 that declared purely cyber defensive capabilities no longer sufficient to protect its networks from external threats. The next steps include blending military and civilian expertise by recruiting personnel that will revitalize the Canadian Forces Network Operations Centre.[49] The overall militarization of cyberspace, which I have described in this book, has also been felt in other parts of the world. In Russia, for instance, the 2011 "Conceptual Views on the Activity of the Russian Federation Armed Forces in Information Space," considered cyber warfare as an extension of its information warfare capabilities. In 2017, Moscow officially acknowledged the formation of a new information warfare branch of the armed forces. In Ukraine, the National Cyber Security Strategy, which is assigned the purview of the National Security and Defense Council, brings together the Ministry of Defense, the Armed Forces, and the Security Service. Ukraine has had a close relationship with NATO states through the Cyber Defense Trust Fund.[50]

Elsewhere, in the Asia Pacific region for example, the trend is also noticeable. In Taiwan, under threat of Chinese cyber intrusion in 2017, the Ministry of Defense launched its Information and Electronic Warfare Command. In Australia, the government's recent defense white papers aim to bolster the Australian Signals Directorate (ASD) and joint capabilities at the Defense Force Headquarters. China's PLA has been developing a new concept of "information confrontation" in electronic and nonelectronic information warfare under the Strategic Support Force (SSF). The Indian armed forces have their cyber centers, which will establish a unified cyberspace command. In Japan, the Cyber Defense Group, launched in 2014, came to provide an integrated space for the armed forces' cyber-warfare functions monitoring ministerial networks and cyber-attacks. In Singapore, the Ministry of Defense announced the creation of the Defense Cyber Organization responsible for cyber policy, training, and defending military systems with a staff of 2,600 soldiers. In North Korea, foreign governments estimate that Pyongyang has a unit of cyberwarfare specialists, adding up to seven thousand experts and officers. In South Korea, the Defense

Cyber Command sits on the Joint Chiefs of Staff, whose chairman controls authority for cyber operations and cyber capabilities for national security.[51]

The idea that governments rely on the military, considering the latter's positive public image and its perceived capacity for planning, organizing, and coordinating security roles, is supported for instance, in the Military Balance published by the International Institute for Strategic Studies (IISS). This widely read state-of-the-art report on security and defense affairs states that in Latin America:

> Legislative changes reflect the changing security environment in the region. Armed forces are envisaged as taking on an increasingly broad range of roles, moving away from traditional territorial-defense tasks towards tackling non-traditional challenges such as those from transnational organized crime and drug trafficking and now also including cyber threats, deployments overseas on peace-support tasks, and HA/DR missions. Increasingly, multinational regional exercises also focus on these tasks. Common challenges continue to lead more defense ministries to consider the benefits from greater regional cooperation, not only through training but also in areas such as information sharing.[52]

Similarly, the 2017 U.S. Presidential Executive Order on Strengthening the Cybersecurity recommends protecting critical infrastructure and federal networks through international engagement.[53] Through cooperation with foreign partners and allies and engagement with all stakeholders, as appropriate, the United States seeks to pursue five objectives including (1) reduce the risk of conflict stemming from the use of cyberspace (2) identify, detect, disrupt, and deter malicious cyber actors (3) uphold an open and interoperable Internet whereby human rights are protected and cross-border data flows are preserved (4) maintain the essential role of nongovernmental stakeholders in how cyberspace is governed, and (5) respect the global nature of cyberspace.[54] Are all these objectives viable? Can they be enforced via partnerships? In the face of U.S. unipolarity, weaker powers in the international system ultimately have three choices: ally with each other to mitigate the influence of the hegemon, align with the United States and support its actions, or remain neutral.[55] Because I have argued that cyber alliances can be a reaction to U.S. dominance, there are several implications and effects worth emphasizing.

In the next section, I explore some of the features of cyber alliance and international engagement, taking the forms of direct diplomatic action, joint military exercises, and participation in policy and professional bodies. I then conclude by debating how the United States maximizes the returns from its international engagement toward improving the cybersecurity of the nation and the hemisphere.

Group 1: Argentina, Brazil, and Chile

Capacity-building programs seem to be the preferred way that the United States can benefit from establishing partnerships. The move brings two crucial rewards. On the one hand, it equips partners to assist the United States in addressing threats targeting U.S. interests in cyberspace. On the other hand, it serves broader U.S. cyber policy, as well as economic and political goals.[56] The approach is not particular to U.S.-Latin American relations but responds to a broader strategy to commit U.S. partners to discuss state behavior in cyberspace jointly. In September 2019, Deputy Secretary of State John J. Sullivan, led an effort together with other UN members to build consensus on international law, nonbinding peacetime norms, and confidence-building measures for cyberspace.[57] In a clear signal to counteract Russia's and China's diplomatic efforts, the signatories of the joint statement pledged to work together voluntarily and to hold countries accountable when they act contrary to this internationally accepted framework. It is in this context that the United States signed bilateral agreements with Argentina, Chile, and Brazil.[58] With Argentina, in 2017, a joint commitment to an "open, interoperable, reliable, and secure cyberspace" set up the bilateral Cyber Policy Working Group. As in the U.S.-Brazil and U.S.-Chile initiatives, Washington acted through the Department of State, which led a U.S. government interagency delegation that has usually included representatives from the Department of Homeland Security, the Department of Justice, the National Security Council, the Department of Defense, and the National Institute of Standards and Technology. The counterpart delegation was led by a representative from the Argentine Ministry of Modernization, together with representatives from the Ministry of Foreign Affairs, Ministry of Security, and Ministry of Defense. Although the U.S.-Brazil Internet and Information and Communication Technology (ICT) Working Group, has convened since 2016, the agenda does not change too much from the usual topics: implementing national cyber policy frameworks, protecting networks, developing a cyber workforce, and managing cyberincidents. By way of "identifying cyber issues of mutual concern and developing new joint initiatives," Washington had high hopes of strengthening and helping protect the economic and security interests of the parties involved. With Brazil, for instance, these meetings usually include both senior representatives and other nongovernmental stakeholders. Although the discussions are meant to facilitate ongoing engagements on cyber and ICT issues in relevant regional and international fora, the dialogue further cements bilateral cyber defense and cybercrime partnerships. In this sense, the Brazilian delegation to the 2018 version of the mutual group also included officials from the Institutional Security Cabinet, Cyber Defense Command, as well as Brazilian embassy attachés from the Federal Policy, Ministry of Defense, and Brazilian Intelligence Agency.

Brazil's cyber defense policy, for example, has undoubtedly sought out some form of independence from the United States. In 2013, the plan to construct domestic Internet Exchange Points to avoid internal users going through points in the United States, Europe, and Japan was made public. The move would not only decrease the cost of broadband Internet, but would reduce the risk of cyber-espionage by a foreign government. In the wake of Edward Snowden's NSA leaks, Brazil also put its armed forces to accelerate the development of the Geostationary Satellite for Defense and Strategic Communication (SGDC), aimed at providing Brazil with autonomous control over sensitive communications. Defense-related agencies were also put under the command of the Military System for Cyber Defense (SMDC), to coordinate cybersecurity and cyberspace policy.[59]

Maybe the most enduring of the relationships examined is that between the United States and Chile. In 2018, U.S. Secretary of Defense James N. Mattis visited Santiago to sign an agreement on cybersecurity. Since 1996, U.S. and Chilean troops have trained side by side at the Rim of the Pacific Exercises, one of the most massive naval exercises in the world. Mattis's visit coincided with Chile's becoming the first Latin American nation and the first non–Five Eyes nation to serve as its combined forces, maritime component commander. Later in the month, Chile hosted the Southern Star exercise when U.S. and Chilean special operations forces trained jointly in battle scenarios monitored by regional and global observers. Mattis's trip was designed to ensure that Chile's right-wing government would honor its traditional defense cooperation, specifically on cyber operations and cyber domain, and more in-depth collaboration in science and technology. In turn, Mattis promised his support to Chile's proposal to the Conference of

Defense Ministers of the Americas for a regional cooperation mechanism to better enable humanitarian assistance amid natural disasters. "All of this, I believe, helps to solidify Chile's global reputation as a trusted security partner, providing, for example, military education, training and technical assistance in Central America and the Caribbean, while actively participating in multilevel engagements and exercises across the Indo-Pacific," Mattis stated.[60]

Group 2: Colombia, Mexico, and Venezuela

The second group of countries is comprised of Colombia, Mexico, and Venezuela. Although the first two are long-standing partners of the U.S. security and military establishment, so far their cybersecurity enterprises have not linked up officially to U.S. efforts in the region in the way the states in Group 1 have done. For the case of Venezuela, the foreseen trend is that it will remain an ally of China and Russia. The amicable link of the past between Washington and Caracas was dissolved gradually once Hugo Chavez (1999–2013) came to power. In the eyes of Washington, "Policies to seize private property and restrict media freedom eroded democratic checks and balances, while cooperation with criminal and terrorist groups as well as facilitation of activities by extra-hemispheric actors such as China, Russia, and Iran increased tensions" between them.[61] Current U.S.-Venezuela relations are marked by Washington's strong support to Juan Guaidó as the interim constitutional president since January 23, 2019, after at least fifty-three other countries denounced the reelection of Nicolas Maduro in May 2018 as unconstitutional and fraudulent.[62] Moscow has landed military assets and advisors in Venezuela, showing strong political and diplomatic support from the Kremlin to Maduro. As John E. Herbst and Jason Marczak at the Atlantic Council put it, the Kremlin has built up a mutually beneficial relationship with the regime in Caracas. "The political utility of the tie to Venezuela was evident in 2008 when Moscow sent TU-160 strategic bombers to Venezuela for a joint naval exercise in the Caribbean Sea. This served as a counterpoint to U.S. support for an increasingly pro-Western Ukraine and Georgian President Mikheil Saakashvili during Moscow's war on Georgia; Chávez offered Russia the use of a Caribbean coastal airbase in 2010," Herbst and Marczak write.[63]

The United States is pushing out foreign competitors in the areas of technology and procurement even from their non-allies in this hemisphere and globally. In the eyes of Washington, the regional ecosystem is threatened by external vendors, mostly coming from China. "Let me reiterate this point. China's national intelligence law compels any Chinese company to cooperate with China's intelligence service. With none of the checks and balances that countries like ours have—independent courts, opposition parties, free press. This means that any data transiting through China or processed by a Chinese company is potentially available to the Chinese government, by law, with total secrecy and no legal limitations on its collection or the subsequent use," argued Kimberly Breier, Assistant Secretary Bureau of Western Hemisphere Affairs.[64]

Elsewhere, U.S. diplomatic pressure has ensured that those like-minded states incorporate national security concerns whenever they develop regulations and procurement decisions into their telecommunications networks and critical infrastructure. Australia, New Zealand, Japan, and the Czech Republic conducted 5G security reviews that have kept Chinese interests at bay. While U.S. partners in Latin America are considering similar regulations—which Washington has applauded and is ready to assist—the new 5G networking infrastructure is causing much controversy. To some observers, it is uncertain who will control 5G (states or companies), but the implications for surveillance, security, and national prosperity are not. Although the 5G matrix is dependent on the Internet, it is "distinct from it and subject to much more government and private control."[65] The ongoing effort to restrict Chinese companies in the development of ICTs in the hemisphere will require much more than the continuing unilateral effort from Trump's administration. To secure cybersecurity alliances across the region, Washington will need to reassess its offer of product innovation to compete strategically with China and achieve definite ends to its policy for the hemisphere.[66]

Colombia's cyber governance is similar to those in group 1 above. Its 2011 policy guidelines for cybersecurity and cyberdefense and its main actors are supplied by a CERT team (colCERT), the Police Cyber Centre, and the armed forces' Joint Cybersecurity and Cyberdefense Command. The defense ministry is the coordinating body for cyberdefense.[67] In Mexico, the 2017 National Cybersecurity Strategy and the announcement of novel Cyberspace Operations centers around the army and the navy, to address and better coordinate defense work on cybersecurity and in cyberspace, although without mentioning partnerships with the United States. Venezuela follows the divergent path, although in that it is not unique to the region. In Trinidad and Tobago, for example, the Defense Force focuses on roles and missions in the realms of border and maritime security and counternarcotics. In 2015, Beijing gave the army light utility vehicles and other equipment and the provision of training has arrived since. The military plays a more significant role in countering criminal activity, and the government ordered new vessels from Dutch firm Damen for the coast guard. Besides the country's maritime security capacity, early in 2012 Trinidad and Tobago published its cyber strategy aimed at the defense of its critical national infrastructure.[68]

The United States has tried to have a say in these processes. At a joint U.S.-Mexico reception at the OAS before discussions at the United Nations Group of Governmental Experts (GGE) on cyber issues, OAS, through its cybersecurity team at the CICTE, provided the space for both countries to extend a dialogue. As this episode demonstrates, when the United States has not been successful in establishing a cyber working group, it has used multilateral fora as a neutral ground upon which to build new relationships. The role of OAS has helped improve cybersecurity and strengthen cyber stability in the region. Since 2002, it has worked with member states to build national cybersecurity capabilities as a part of its efforts to implement the Comprehensive Inter-American Strategy to Combat Threats to Cybersecurity.[69] A group of Chilean, Colombian, and Mexican representatives have led regional efforts to advance cyber confidence-building measures, transparency, and reduce the risk of conflict stemming from a cyberincident. Brazil chaired the GGE in 2014 and 2015. Regarding Washington's providing hands-off assistance and oversight for such ad hoc forms of governance, the State Department has claimed that "our work to date in the Americas region supports global efforts to promote responsible state behavior in cyberspace. And we all have a strong national security interest in ensuring that this important cooperation and work continues and succeeds."[70]

From the recently launched U.S.-led Joint Statement on Advancing Responsible State Behavior in Cyberspace, Colombia was the only Latin American country to be included among the initial signatory members. The statement aimed at making U.S. cyber allies, current and prospective, to engage in open discussion of their offensive cyber programs. "States must use these capabilities responsibly—whether they are in our region or not—because, in our interconnected world, all of us can be affected in unintended ways," the statement argued.[71] U.S.-led initiatives aimed at like-minded countries, although they are not necessarily allies, operate in a complementary manner to the formal "cyber dialogues" Washington currently conducts, for instance, with Japan, Estonia, the EU, South Korea, France, and the Netherlands. These initiatives include key topics for cyber deterrence and response, including multilateral and bilateral law enforcement cooperation, and joint attribution of cyber incidents.[72] The interagency focus of these "dialogues" ensures cross-government mutual collaboration and oversight on a range of cyber topics.[73]

Finally, what might worry the United States the most is that countries such as Mexico and Colombia have shown keen interest in developing their 5G networks. Washington is vigilant in assuring that the prosperity of 5G is not given to its adversaries, most notably, China. The battle over 5G and the active presence of the tech company Huawei in Latin America might lead to unprecedented consequences, for example, the United States downgrading intelligence sharing with Latin American partners or barring domestic firms from bidding on some U.S. defense contracts, privileges that the United States has extended to many countries, including, Brazil, Colombia, Chile, Mexico, and Argentina.[74] In August 2019, Robert L. Strayer, Deputy Assistant Secretary for Cyber and International Communications and Information Policy at the Department of State, was asked whether countries that choose to use Huawei or ZTE will be affected when it comes to the United States's ability to share intelligence. Strayer put it this way: "So we have important information-sharing relationships around the globe with countries. We benefit. Our partners benefit. We want to make sure we can maintain those relationships by having a vigorous flow of information, a bilateral flow of information. But if countries implement in their telecom networks that are doing 5G untrusted vendors, we're going to have to reassess how we're sharing information with them to ensure that we're not compromising that very vital data."[75]

Conclusion

States can engage in cyber alliances with the United States whenever they see that it benefits their security stance. When joining in an alliance with a superpower, countries will try to retain their voice in the decision-making arrangements so they are not stomped by the strongest. Their desire for protection dictates how they will respond to imposing preferences, and most importantly, how leaders will react to crucial conducts when it comes to the use of force. For Washington, on the other hand, making cybersecurity alliances in a unipolar world has opened the road to institutionalized and multilateral arrangements based on norms and rules, but also a secondary route of ad hoc coalitions with states believed to be exceptionally loyal and compliant.[76] The issue of cybersecurity is consistent with these two options, as some U.S. alliances commit to multilateralism, and more selective military and diplomatic alliances sometimes interconnect in several ways (see Table A.5 in the Appendix file for cybersecurity bilateral and multilateral cooperation treaties). For this matter, Washington seeks active partners that can shoulder the costs of cybersecurity.

The United States' historical linkages in the hemisphere are a constant reminder of the significance of diplomacy, alliances, partnerships, and other instruments of dialogue between nations. Across chapters 5 and 6, I have established the importance that cybersecurity has assumed in U.S.–Latin American relations and the challenges against the "American interest" that the presence of other cyber-advanced nations represents for the region's dynamics. At the outset of the chapter I asked what the prospects for U.S.–Latin American cyber partnerships were. The most discernible analysis is that Washington has affirmed the prominence of cyber resilience and enhancing cybersecurity defenses across the region. The testbed for such a move has been those countries where the United States remains an invaluable partner. The need to work together for the protection of critical areas has fostered international cooperation in cyberspace. The opportunity has also opened a more formal channel by which Washington might exert pressure upon those governments fond of Beijing. The risk escalation in cyberspace between the United States and China is worrisome due to the lack of instruments that manage crises between the superpowers.[77] Third-party countries would not want to provide proxy scenarios for cyberincidents for which there are no forms of detailed agreements or channels to avoid military conflict. The United States should engage more frequently and substantially with Beijing to clarify each other's intentions regarding cybersecurity. This way, developing countries would not be held hostage to either side's strategic policies.

Despite previous cyberincidents within the region, both chapters have also brought forward the idea that we are entering a new decade of cyber relations. Cybersecurity was tainted after the bitter NSA international espionage disclosure in which the region was revealed to have been the victim of undesired surveillance. Today, nations seem to look ahead, mindful of broader national economic and security goals. Cybersecurity has become a military and foreign policy objective on its own. Because of its comprehensive nature, governments have paired up to approach the complexity of safeguarding cyberspace. Networked approaches between governmental agencies at the national and international levels evidence the move toward a conscious effort to integrate resources from the state apparatus.

In this sense, the U.S. vision for cyberspace has echoed in group 1 (Argentina, Brazil, and Chile). For those countries in group 2 (Colombia,

Mexico, and Venezuela), only the regime in Caracas seems abstracted from U.S. influence. Authorities in Bogota and Mexico City are too deeply embedded in the current regional and global multilateral circuit not to recognize the increasing complexity of cybersecurity and the overarching influence the U.S. government has assumed in coordinating various diplomatic efforts, for instance, in the OAS and the UN. Despite U.S. diplomacy's having suffered a "devastating blow" to the efficacy of its State Department during the Trump era,[78] the cybersecurity issue has encouraged Washington's leaders to seek out ways to grow the number of governments beginning to work in concert to condemn malicious cyber activity. As well, Washington has circumscribed the cyber issue to its overall foreign and military agendas, making long-time receivers of U.S. military advice, such as Mexico and Colombia, more susceptible to engaging in formal partnerships in the short term. Time will tell if they can sustain their current state of strategic cyber autonomy. We know for sure that the U.S. diplomats and military are urgently communicating the stakes and interests at play to pursue them.

Seven

Quantifying Cybersecurity

Citizens worldwide depend on the Internet more than ever. Its use is no longer reserved for technical, academic, and research elites, but rather its mainstream access is regarded among society as a fundamental right without which we could not function. Every year, more Latin Americans are connected to the Internet. However, cybersecurity presence across the region shows different levels of development or maturity. Citizens have less cybersecurity legal protection in Nicaragua than in the Dominican Republic. National cybersecurity strategies are more advanced in Panama than in Chile, although Internet penetration is higher in the latter. The panorama is complex, and yet, we know little about the underlying factors behind the cybersecurity maturity gap.

Whether it is a consequence of regulation, policies, technological investments, or other factors, cybersecurity capacity has achieved uneven levels of readiness to address the uptake of the Internet and ICTs. To remedy the "weakest links" in cybersecurity capacity across the globe, developed and developing countries have injected resources sufficient to establish cyber knowledge at all levels of statecraft.[1] Meanwhile, the critical value of the Internet lies in its universal and voluntary interconnection; countries tend to increase regulation over such an environment, resulting in countermeasures to the associated digital risks, however, sometimes undermining what Internet culture values most highly: decentralization of its users.[2] The more the number of Internet users grows, the more networks increase in size, enabling rewards (more connections to advance the pace of development) and perils (less protection to new kinds of crime, warfare, and espionage).[3]

A denial of service attack, for example, hits a service via many network points around the Internet. While unsecured devices and applications make things worse, operators of such services need to take responsibility themselves. Many times, given the lack of national regulation, they will institute unilateral measures to protect the services they operate.[4] But of course, many new networks do not abide by either their peers' standards or any national set of policies. The literature on cybersecurity policy has so far mainly explored regulatory responses to "burgeoning" security risks. One conclusion is that these turn out to be at best futile and at worst harmful to the Internet's basic nature: its end-to-end architectural foundation. Privatization and monopolies on applications and technologies, including IoT massification, prioritize the logic of application networks that are not always open to regulation by public boundaries. The liabilities of uneven cybersecurity policies, legal frameworks, and technological advancements stress the use of the Internet without apparent constraint coming from national law.[5]

The cybersecurity maturity gap is usually reflected upon various levels. Three interrelated dimensions are found at the individual level (improving skills, knowledge, competences, and attitudes), the organizational level (developing structure, processes, and networks between organizations), and the institutional and policy level (improving the environment through legislation, policies, and strategies).[6] Policymakers lay the basis for the rule of law on the Internet to happen. This poses many governance dilemmas, such as preserving the global nature of cyberspace while respecting national legislation, while, on the other hand, fighting the misuses of the Internet while ensuring human rights protection.[7]

In emerging countries, more accurate knowledge to help governments discern the areas of cybersecurity that remain weak and exposed seems needed. Some crucial questions are: Can we survey cybersecurity policy and strategy readiness? Is cybersecurity maturity a reflection of current legal frameworks? Are cybersecurity policies ready for the rapid penetration of the Internet? Concerning Latin America and the Caribbean, none of these paradigms have been explored systematically in the literature.

This chapter uses statistical analysis to address cybersecurity maturity across the region. For implementation, the chapter purposely uses techniques that are less intricate than those available on advanced quantitative methods scholarship. I start from the assumption that not all readers have partaken statistical methods training. For different reasons, undergraduate and graduate schools do not devote the time and resources to develop and maintain quantitative training in their syllabus.[8] The idea here is to guide the reader using statistical methods. For clarity, the threefold purpose of the research is simple: account for regularities in social explanations; identify patterns of behavior separated into variables; and explain associations between variables, ideally predicting outcomes.[9]

All worked exercises are based upon the OAS Observatory of Cybersecurity in Latin America and the Caribbean dataset, available free online for download.[10] The chapter uses a selection of indicators that reflect on levels of policy, regulatory frameworks, institutional arrangements, and ICT operations. The Observatory constitutes a unique resource for international comparative work with thirty-two countries surveyed. I coded the raw data from the survey into a STATA dataset file using a partial selection of the forty-nine indicators, which were originally grouped in five dimensions: national cybersecurity policy and strategy; cyber culture and society; cybersecurity education, training, and skills; legal and regulatory frameworks; and standards, organizations, and technologies. Each indicator had a value that represented the level of maturity on an ascending scale. For methodological purposes, I coded this information using a five-point scale, with 1 = "very low," and 5 = "very high."[11] Values for variables in my dataset showed that countries differed widely in their responses regarding levels of cybersecurity maturity across the given dimensions. In sum, the countries investigated became cases, the information obtained represents variables, and the specific answers became values.

I began the analysis with an examination of the data before any statistical testing. I identified broad levels of measurement. Most ordinal variables show categories that can be arranged into order, in this case, from lowest to highest. We can say that a "very low" level of cybersecurity policy development is more extreme than "low" levels. This order followed the original coding of the Observatory dataset. Also, scalar, quantitative, or metric measurement was given to continuous data, including variables used to account for the size of the population, Internet access, Internet penetration, and secure Internet servers.

Analyzing a dataset that was collected by others for different purposes (in this case, to present an all-inclusive portrayal of cybersecurity in the region) requires specific alteration of the data in some way, to carry out the secondary analysis. I recast the dataset into a form that makes sense with the analysis so far presented in the book using data selection and data manipulation processes. I selected the cases of interest and the variables of interest for two reasons: efficiency (restricting the analysis to less than the whole dataset), and safety (restricting the analysis from unwanted cases and variables). The sample of thirty-two countries remains the same as the one collected by the Observatory. This allowed me to include in the analysis some nations that I have not significantly touched upon in the book (e.g., countries in the Caribbean and Suriname and Guyana in South America). As well, I worked only with a subset of variables under three dimensions: policy and strategy; legal and regulatory frameworks; and standards, organizations, and technologies.

Internet Penetration, Policies, and Legislation

Using numerical data helps us address questions concerned with the testing of relationships between variables in the data. Ideas, or hypotheses, about the relationships between the data, for example, the level of cybersecurity policy development and the amount of Internet penetration, or the extent of cybersecurity crime legislation and cyber defense strategies. Confirmatory statistics, also called hypothesis testing, allows testing or evaluating the validity of these ideas.

Latin American countries have different scores of Internet penetration. A larger size of the population does not equate with higher Internet penetration. For example, Brazil, the largest country in the dataset, has less than 70 percent of its population connected to the Internet. Argentina and Chile are much smaller than Brazil and have slightly higher Internet penetration. Many countries in the Caribbean, considerably smaller than Brazil, also have higher levels of Internet penetration.

Because the Internet is not so much public and free anymore, as it used to be when it was created, the privatization and monopolization of specific dominant applications have made governments react in the traditional way: regulatory regimes. While some countries have put up national firewalls and net neutrality architectures, the difficulty lies in their scope. States can only protect what lies within their national boundaries, while most of the Internet is global in its reach. While one country might protect network access very well, the utility and robustness of their efforts might be undermined by other countries with less advanced regulatory regimes. Harmonization between countries requires a common language on regulatory regimes. But nations typically have different goals, means, and resources to regulate their jurisdictions, not only regarding Internet-driven problems but national security, trade, commerce, investment, etcetera.

Cybersecurity norms show inconsistency and incoherence across the globe. While Western authorities are preoccupied with cyberattacks to their critical national infrastructure, among the Russian political elite, for example, there is more anxiety that state and nonstate adversaries will influence Russian via cyberspace than they are about their critical infrastructure vulnerability. The irony is that Russian-speaking groups are believed to be behind most of the large-scale cyber incidents against public and private corporations. Using financial trojans, these groups can, for example, target many hundreds of thousands of individuals worldwide each year.[12] In the liberal democracies, despite the growing fear of cyberincidents infiltrating voter systems and other aspects of elections, cyber operations have become more common. The point at the end of the day is that existing concerns about cybersecurity vary and so do the different countermeasures to cyberthreats.[13] In Taiwan, for example, the feeling that cyberspace might positively transform cross-straits exchanges with mainland China has failed to materialize. Taiwanese rhetoric has prompted the perception of constant cyberthreat from Beijing. They have revamped their cyber capabilities, policies, and organizational infrastructure around their military cyber command.[14]

Mutually inconsistent regulatory regimes produce divisions among nations' authorities trying to lay standard cybersecurity norms and users trying to navigate the Web and use Internet-based applications safely.[15] The perceived need for cyber sovereignty, or information sovereignty, compels countries to decide whether to abide or repel external influence, institute regulatory frameworks, or enter into binding agreements to protect their information and digital space. Many times, we mistake the Internet for a single unitary network, when in fact it is *a network of computer networks*.[16] Thus, countries can use their part of a network (their own computer networks) under their own terms. Most recently, the United States and the eleven signatories of the Trans-Pacific Partnership spent years negotiating a cross-border information flow agreement. The TPP required transparency and political harmonization in controlling Internet-related trade flows and doing business online. It also tried to establish censorship and filtering reduction rules to promote open Internet and digital rights. This aimed to break with some of the members' tougher Internet censorship and filtering (for example, in Vietnam or Malaysia). Despite President Trump's dismissal of the TPP, some countries decided to go ahead and formulate the basis of digital trade and cybersecurity cooperation across the Pacific.[17]

The phenomenon of so-called Internet Balkanization has gained momentum as a response to diverging values and sensitivities. Worldwide tensions between local jurisdictions have empowered locally splintered subnets (Iran, China, Russia, and some Middle Eastern regimes, for example). In a way, the fear that the Internet is no longer conceived of as a global network that has transcended territorial boundaries, which has encouraged its overprotection, is deemed acceptable under national customs, culture, and laws.[18]

Four decades after the Internet's inception, Internet openness is at the core of future Internet governance. Many developing countries are reaping the success and social benefits of their economies now being substantially connected to the Internet. However, an affordable Internet is not always granted. Considered a development priority, countries struggle to provide open platforms on which network providers can treat content, applications, and services equally.[19] Developed and developing countries try to install rules and regulations for net neutrality, especially as it favors the uploading and downloading of content without censorship. Many times these are seen as reflecting a security concern. The challenge here is whether policymakers and regulators can provide a regulatory framework that permits them to expand Internet penetration without compromising the fundamental principles of a free and open Internet. Overregulation and oversecurity of the Internet, ICTs and digital applications can stall attempts to increase connectivity. At the same time, underregulation and insufficient security can result in untenable conditions for all stakeholders.[20]

Liberal governments tend to support technological innovations by providing direct and indirect support, for example, via sympathetic regulatory and legal frameworks. When it comes to cybersecurity and cyberattacks, Internet technology has received full support from interested policymakers (sometimes because they have been victims of cyberincidents). In the United States, for example, public and private sectors have merged to ensure broad cooperation, for instance, between the Pentagon and Silicon Valley.[21] The defense community in the United States has partly benefited from an industrial policy favoring research and development in the cyber domain. Governments and industries as facilitators of the cybersecurity market have led some scholars to argue that the state can intervene in cybersecurity policymaking from a provider, facilitator, and promoter perspective. The case here is that for many advanced democracies, cybersecurity policymaking has not only stressed their global economic stance, but also their global security and defense policies.[22] Holistic cybersecurity policymaking calls for a balance between national security and geopolitical and development priorities.[23] The trend is visible in the United Kingdom, China, the European Union, and the United States.

A separate example worth discussing is the case of France. Continuous cyberthreats against critical infrastructure, corporate interests, and personal data pressed French authorities to launch a 2011 cybersecurity white paper in which a cybersecurity strategy was outlined. In 2013 after €150 million investments in research and development, a civil-military center of excellence in cyber defense was set up in Brittany. Among other policy goals, France's ambitions were to "become a world leader in cyber defense, protect information related to its sovereignty, safeguard critical national infrastructure, and ensure security in cyberspace educating its people, adapting legislation and incorporating new technological practices."[24] In late 2018, President Emmanuel Macron launched the Paris Call for Trust and Security in Cyberspace, hoping to revive efforts to regulate cyberspace, limit hacking, and defend against cyberweapons after the failed round of negotiations at the UN in 2017. The document, initially pushed by tech companies and launched at the UNESCO Internet Governance Forum, was signed by sixty-six countries (absent the United States, Russia, and China), 139 civil society groups, and 347 companies. Latin American signatories included Argentina, Chile, Colombia, and Panama.[25]

To further theorize on these relationships in the Latin American and Caribbean contexts, we must test some of the above observations taken from the literature using confirmatory statistics.[26] In this case, three opening research questions might be: "Among those countries with higher Internet penetration, do they differ in their level of cybersecurity policy development?" Second, "Among those countries with higher Internet penetration, do they differ in their level of cybersecurity legislation?" Third, "Among those countries with higher Internet penetration, do they differ in their level of cyber defense maturity?"

Making a statement in the form of a hypothesis suggests a relation between parameters of variables where a prediction can be expected. This could be predicting a difference or a relationship between selected variables.[27] Using the above research questions, one hypothesis might be: "Among those countries with higher Internet penetration, the mean of their cybersecurity policy development will be higher than the mean of those with a lesser degree of Internet penetration." Thus, I am hypothesizing a difference, and giving a direction to the gap, the more internet penetration in a country, the higher average on cybersecurity maturity. The null hypotheses are the logical opposite of the premises. I will aim to disprove the logical opposite of each hypothesis. Using our current examples, the Null hypothesis will be: "Among those countries with higher Internet penetration, the mean of their degree of cybersecurity policy development will be the same or less than the mean of those with a lesser degree of Internet penetration." The probability that the null hypothesis is incorrectly rejected based on the observed data is called a Type I error.[28]

For both hypotheses, I use a t-test as a confirmatory statistic method. A t-test will tell us whether the differences between the means of two groups are statistically significant. This type of statistical test is presented in terms of probability. This is the odds of accepting a false hypothesis as real. So, the smaller the size of the significance, the less likely it is that a Type I error has been made, and the more likely it is the hypothesis holds. The rule of thumb for risk level is at .05, meaning that five times out of 100, you will find a statistically significant difference even if there was none. A stricter standard of confidence cutoff is usually considered at the .001 level.[29]

The descriptive statistics (see Table A.6 in the Appendix) reveal different mean scores (although there appears to be not much difference between groups, we can confirm this using the t-test). However, there is a chance of Type I error. When comparing the scores of two groups, it is essential to examine the difference between their mean scores relative to the spread or variability of cases. The t-test examines whether the means of two groups are significantly different from one another.[30] The type of t-test conducted was an "independent sample t-test." With this type of test, there are two distinct categories for the independent variable. I recast the original measurement and created one group being those countries with a score above 1 for the variable "cybersecurity strategy development," and another group for those that scored 1, the lowest level of maturity. There are also two scalar dependent variables measured at the interval or ratio level (here, the "cyber defense" and the "cybercrime law maturity" scores). The independent samples t-test will be testing whether the means of the dependent variable for each group defined by the independent variable are significantly different.[31] The scale variables adapted from the OAS cybersecurity observatory and the ITU were:

- Cybersecurity strategy development: The level of maturity of a documented or official national cybersecurity strategy.
- Cyber defense: The level of maturity on national security policy and national defense cyber defense strategy.

- Cybercrime law: The level of maturity on substantive cybercrime legislation.
- Internet Penetration: a country's percentage of the population connected to the Internet. Internet users are individuals who have used the Internet (from any location) in the last three months.

Next, I tested the hypothesis: "Among those countries with higher Internet penetration, the mean of their cybersecurity strategy maturity will be higher than the mean of those with a lesser degree of Internet penetration." The independent samples t-test will tell us the odds or probability that the difference we see in the figures is genuine. If this is confirmed, we accept our hypothesis and reject the Null hypothesis.

My recast independent variable, Internet penetration, is now categorical with two categories: Low (scores below 63 percent) and High (scores above or equal to 63 percent). See that I selected the 63 percent cut-point to define the groups by splitting the variable into two categories. The variable has two values: 1 = below 63 percent, 2 = Above or equal to 63 percent. Cybersecurity policy (the dependent variable) is a scalar variable and was kept at the same values as recorded initially (1 = very low or the least mature; 5 = very high or the most mature).

From the descriptive statistics we knew that for countries with below-average Internet penetration, the mean score was 1.53 (standard deviation = .83), while for seventeen countries with above-average Internet levels, the mean score was 1.70 (standard deviation = .68). We want to discover if there is a significant difference between the mean scores for the two independent groups. I assume that the variance in the populations being compared is the same (see Table A.7 in the Appendix) to examine first the homogeneity of the variance between the two groups). The t-test uses Levene's Test for Equality of Variance for this purpose. In this case, $F = 1.024$ and Sig. (p) = 0.32, which indicates that $p > .05$, thus the variance between the two groups is equal. We then test the hypothesis using the t-test row of results for Equal variances not assumed. This gives us a t value ($t = -.634$) and the degrees of freedom ($df = 27.22$). We can also find a two-tailed significance (p-value) of .53. Thus, the difference between means is not significant at $p < 0.05$. We conclude that a statistically significant difference does not exist between the two groups in their strategy development scores. The research hypothesis that among those countries with higher Internet penetration, the mean of their cybersecurity strategy maturity will be higher than the mean of those with less Internet penetration is not upheld, and the Null hypothesis is accepted.

I ran another t-test exercise to test the second hypothesis: "Among those countries with higher Internet penetration, the mean of their cyber defense maturity will be higher than the mean of those with a lesser degree of Internet penetration." Table A.7 in the Appendix reports a Levene's Test for Equality of variance with the F value being 9.79 and the Sig. (p-value) at .003. This means that the variance is assumed as not equal. We then read the row of Equal Variance not assumed. This gives us a t value (t = −1.58) and the degrees of freedom (df = 25.91). We can also find a two-tailed significance (p-value) of .124. Thus, the difference between means is not significant at $p < 0.05$. We can accept the Null hypothesis and conclude that a statistically significant difference does not exist between the two groups in their cyber defense strategy development scores. The research hypothesis that among those countries with higher Internet penetration, the mean of their cyber defense strategy maturity will be higher than the mean of those with less Internet penetration is not upheld, and the Null hypothesis is accepted.

Finally, and for the third hypothesis: "Among those countries with higher Internet penetration, the mean of their cybercrime law maturity will be higher than the mean of those with lesser degree of Internet internet penetration," the t-test shows a Levene's Test for Equality of variance with the F value being .706 and the Sig. (p-value) at .407. This is far greater than 0.05 ($p > 0.005$), thus not significant, indicating that the equal variances can be assumed. We read then the row labeled Equal variances assumed to test the hypothesis. This gives us a t value (t = −2.236) and the degrees of freedom (df = 29.98). We can also find a two-tailed significance (p-value) of .003. Thus, the difference between means is significant at $p < .01$. We can reject the Null hypothesis and conclude that a statistically significant difference exists between the two groups in their cybercrime law scores. The research hypothesis that among those countries with higher Internet penetration, the mean of their cybercrime law will be higher than the mean of those with less Internet penetration is upheld, and the Null hypothesis is rejected. The chances of Type I error are made is less than 1 in 100 ($p < .01$).

Determining Relationships

Next, I used a statistical method known as general linear modeling, which includes correlation and regression statistics, to help us shed light on the relationship between two quantitative variables (interval or ratio variables) by drawing straight lines.[32] I have chosen pairs of variables where there is good reason to assume that the relationships will be either strongly positive or strongly negative.

To determine whether there is a genuine relationship between variables or not, Pearson's correlation coefficient, or r, gives us a parametric test to determine whether a significant correlation occurs between two variables, and the direction of this correlation, (positive or negative), and the strength of the relationship.[33] The correlation coefficients can take values ranging from +1.00 (a perfect positive correlation) through 0.00 (no correlation) to −1.00 (a perfect negative correlation). I have produced a matrix for a series of bivariate (relationships between two variables) correlations simultaneously.

The correlation coefficients matrix (see Table A.8 in the Appendix) helps to interpret the Pearson's r correlation coefficient more easily. Looking at each cell at a time provides information related to the direction of the relationship between variables, whether the relationship is significant, and the strength of the relationship.

I am interested in the cell showing the correlation between strategy development and cyber defense strategy. The Pearson correlation r is .524, the Sig. (2-tailed) is .002, and N = 32. The value is positive, telling us that there is a positive relationship between the two variables. A significance level of .002 indicates that we have certainty that the correlation is statistically significant. The Pearson value indicates that this is a moderate relationship. The next cell of interest is providing information on the correlation between cybercrime legislation and Internet penetration. The Pearson correlation r is .40, the Sig. (2-tailed) is .026 ($p < .05$). Again, Pearson r suggests a moderate relationship, statistically significant. The rest of the cells show weak correlations that are not statistically significant, with two-tailed correlations greater than the cutoff point of .05. Notice also that no correlation turned out to be negative.

After examining the association between variables (I have not shown that one *causes* another), we might have theoretical reasons for hypothesizing a causal link. I thus continue with regression statistics.[34] The assumptions for a regression are the same as for correlation (interval/ratio data but can tolerate ordinal variables; homoscedasticity; and a linear relationship between variables). Regression statistics assume that the independent variable (X) is, at least in part, a cause or predictor of the dependent variable (Y).[35] A regression "singles out one of the variables and treats it as potentially determined by the others and their interactions.[36] It is appropriate when the researcher has one or more causal hypotheses derived from previous studies, observations from databases, and preliminary analysis from qualitative studies. Causal questions require some knowledge of the "data-generating process"[37] For example, I have made the theoretical assumption that higher degree of cybersecurity policy development will cause more cyber defense maturity. So, cybersecurity policy development will be the independent/causal X variable, and cyber defense maturity will be the dependent/caused Y variable. The standard regression formula is $y = a + bx$. This translates as the value of the dependent variable (y) is predicted to be a constant (a) plus a coefficient (b) times the independent variable (x).[38]

I ran a bivariate regression to test the association between cyber defense strategy and cybersecurity strategy development after the r correlation test told me that they were positively correlated and had a statistically significant relationship. Interpreting a linear regression is more straightforward, for instance, than interpreting a logistic regression.

We can consider a regression as a statistical technique convenient to explain how variation in one or more variables predicts or explains the difference in another variable. This popular analytical technique is known for the following characteristics. First, it can be used to analyze experimental or nonexperimental data with multiple categorical and continuous independent variables. Second, if only one variable is used to predict or explain the variation in another variable, the technique is known as bivariate regression. The usual model includes a single regressor x that has a relationship with a response in y. Third, when more than one variable is used to predict or explain variation in another variable, the technique is generally referred to as multiple regression. In this case, more than one regressor x has a relationship with a response in y.[39] Multiple linear regression is often called an extension of simple linear regression. The latter can be considered as a different case of multiple regression because simple regression results can be obtained using the multiple regression results when the number of predictor variables equals 1.[40]

The model summary for the linear regression results (see Table A.9 in the Appendix) indicates a Pearson r is .524; R squared (r^2), or the percentage of variance shared by the variables and calculated by squaring the r-value, is .274. Statistical software will tell us the quality of the linear regression fit, indicated by the coefficient of determination, or r^2. The value of r^2 lies between 0 and 1, showing how much of the variation in y is explained by the linear relationship (between 0 and 100 percent).[41] Typically, social scientists will use the more reliable adjusted r^2 (here with a value of .25) to represent the percent of shared variance. The cyber defense strategy maturity can explain a quarter of the variance in cybersecurity development. Next, the ANOVA box could help us identify if our findings occurred from a sampling error. Here the value of F is 11.335, and the confidence value = .002 (very significant, p < .05), telling us that the results are no product of sampling error.

Regression can also be a multivariate statistical technique. The exercise can be extended to take in more than one independent causal x variable.[42] This is shown in the formula: $y = a + b_1x_1 + b_2x_2 + b_3x_3 + ... + b_nx_n$. Each independent variable X will explain some of the variances in the dependent variable Y. In other words, multiple regression is testing the extent to which each independent variable X will play a part in predicting what the most likely value of the dependent variable Y will be.[43] I apply these ideas to cybersecurity maturity levels. So far, I have played with the idea that cyber defense strategy impacts cybersecurity development. However, there might be variables other than cyber defense strategy that also influence cybersecurity policy maturity, which we can call causal factors. I have presented the importance of Internet penetration, as it may be that higher levels move authorities to protect their cyber environment through more security measures. Also, if there might be higher levels of policy organization and policy content, it might be expected to predict a robust cybersecurity policy development, as these are two essential stages leading toward policymaking. The maturity of incident response capacity might also lead to higher levels of cybersecurity development, and so on with other causal variables such as critical national infrastructure response planning, and a command-and-control center. Multiple regression allows me to shed light on these speculations. As explained by Lemercier and Zalc, "The method of multiple regression seeks to reproduce the conditions of such an experiment as closely as possible by statistical means: by observing a large number of individuals with various combinations of characteristics, one hopes to isolate the effect of certain variables on the result of interest to the investigator."[44] By this statistical method I can identify which causal variables are more relevant by explaining which of the factors (or independent, "explanatory" variables) indicate a significant amount of variance in the dependent variable.

The variable Cybersecurity strategy development (the level of maturity of a documented or official national cybersecurity strategy) is the dependent y variable. Several independent variables under four categories, drawn from the theoretical debate at the beginning of this chapter and across the book, were entered in the model to test their predictive power.

Access

- Internet Penetration: a country's percentage of the population connected to the Internet. Internet users are individuals who have used the Internet (from any location) in the last three months.

Policy and Strategy

- Organization: The level of maturity on an overarching entity for cybersecurity organization.
- Content: The level of maturity on national strategies containing contents, goals, or directives for cybersecurity.
- Cyber Defense Strategy: The level of maturity on national security policy and national defense cyber defense strategy.

Legislation

- Cybercrime Law: The level of maturity on substantive cybercrime legislation.

Technologies

- Incident Response Capacity: The level of maturity on a national incident response capacity team or personnel.
- CNI Response Planning: The level of maturity on response planning to an attack on critical national infrastructure (CNI).
- Command and Control Center: The level of maturity on a cybersecurity Command and Control Center at a national level.

The multivariate regression summary box (see Table A.10 in the Appendix) shows that with the additional predictive independent variables the Adjusted r square (the amount of explained variance in the dependent variable) has gone up from 25 percent to 80 percent. In part this is because when there is more than one independent variable in a regression, the Adjusted r square gives a more accurate indication of the amount of explained variance.[45]

Similar to the univariate regression, the ANOVA test shows an F value of 12.70 and a p-value statistically significant (Sig. = .000, $p < .0005$). The regression results did not occur by chance.

By looking at the coefficients box, we can determine the relative importance of each independent variable in accounting for variance in the cybersecurity strategy development variable. The coefficient statistics show some independent variables were statistically significant while others were not. The variable Internet penetration explains a statistically significant amount of variance in the dependent variable (Sig. = .02, p < .05). Although this time, the direction of the effect is negative.

What I have done with the multiple regression is to test eight hypotheses, and whether once the effect of the other independent variables is considered, each independent variable is having a significant impact upon the dependent variable. We also discover that some of the other independent variables have significant effects upon cybersecurity maturity (Organization; Content; and Cybercrime Law).

Once the effect of the other independent variables is considered, half of the independent variables do not exert significant effects upon the dependent variable). In more detail, the results were the following:

- Does Internet Penetration have an independent effect on cybersecurity strategy development? (Accept, significant).

- Organization has an independent effect, elevating the maturity of cybersecurity strategy development (Accepted, significant).

- Content has an independent effect, elevating the maturity of cybersecurity strategy development (Accepted, significant).

- CybDefStrategy has an independent effect, elevating the maturity of cybersecurity strategy development (Reject, not significant).

- Does CybCrimeLaw have an independent effect, elevating the maturity of cybersecurity strategy development? (Accept, barely significant).

- IncidentResponseCapacity has an independent effect, elevating the maturity of cybersecurity strategy development (Reject, not significant).

- CNIResponsePlanning has an independent effect, elevating the maturity of cybersecurity strategy development (Reject, not significant).

- CommandControlCenter has an independent effect, elevating the maturity of cybersecurity strategy development (Reject, not significant).

Discussion and Conclusion

This chapter's research has demonstrated that the variance in cybersecurity maturity among Latin American countries is substantive and a crucial subject of academic and policy concern. With the results of the descriptive statistics, I have scratched the surface of the analysis and provided evidence that more academic work is indeed essential. While it is the first step, the application of quantitative methods for cybersecurity research in social science remains uncharted territory. Despite the exciting findings in the chapter, the correlations, regression, and factor analysis revealed mostly unexpected results, as they can only be a necessary exploration of partial answers to the book's research questions. Nevertheless, the findings in this chapter and chapters 1 to 6 provide some insight into qualitative and quantitative analyses in line with what other scholars in the methodology literature would encourage.[46]

I have demonstrated that the maturity levels of cybersecurity strategy development are relatively low across countries in Central America and the Caribbean, and slightly higher in South America. As well, cyber defense strategy maturity, another essential object of study throughout the book, remains relatively low in Central America and the Caribbean and slightly higher in South America (see the additional mean estimations in Table Appendix A.11). The region's cybercrime legislation tends to have better estimates with the Caribbean leading, above South America and Central America in this order. Although this book has emphasized mostly events happening lately in six distinct nations (Argentina, Brazil, Chile, Colombia, Mexico, and Venezuela), the descriptive statistical analysis has shown that certain cyber maturity factors are common across Latin America. First, it was revealed that cybersecurity governance happens across different scales and levels of analysis any researcher can pick upon. This chapter emphasized mostly on Internet penetration, issues of policy strategy, legislation, and technologies, but other aspects remain equally interesting to explain, such as those in the categories of education (cyber education and training, corporate knowledge, and public-private standards), and cybersecurity awareness, confidence in e-government and online services, and privacy standards. These are areas for which OAS and other international organizations have also compiled usable data, which could be part of further research projects.

Second, across the chapter, I have conducted a secondary analysis of data as part of this book's effort to present a mix of methodological approaches. I believe the chapter has provided evidence to help in planning future qualitative research. Which actors would be the most likely authorized to talk on cybersecurity governance across nations? Which stakeholders would be most useful to contact in an interview-based study? What type of questions would civilian authorities and military authorities be most open to discussing with an outsider researcher? Next, this chapter's research can lead to more small-scale studies. Why is cyber defense strategy making more advanced than cybersecurity in Brazil? Why is cybercrime legislation so advanced in the Dominican Republic? Further qualitative analysis could also help explain the relationships identified in the quantitative analysis (for example, the statistically significant correlation between the extent of maturity in cyber defense and cybersecurity strategies). Depending on the emphasis given by researchers, this secondary analysis can be used to test or generate further theories on cybersecurity governance, cybercrime, public administration, military strategy, or e-government policy. In essence, however, this chapter's descriptive statistics concern only the data that were analyzed and, as other researchers would argue, should not attempt to make a statement beyond the observations.[47]

In the end, the point of doing statistical research and of all the results in this chapter, especially those derived from the variables in the regression, was to recognize the connection between the data observed and the broader events surrounding these observations.[48] Theoretical concerns throughout the book suggest that it was crucial to test the relationship between cybersecurity and cyber defense strategy maturity. The tools used in the chapter were both flexible and straightforward for others to replicate and refine. From my bivariate regression results, I can confirm that cyber defense strategy maturity had an independent effect, elevating the sophistication of cybersecurity strategy development. However, when I analyzed the relationship between the dependent variable y and several covariates simultaneously, this effect was not significant at all. Looking further, what remained statistically significant in the multivariate regression (but not in a bivariate regression) was the relationship between maturity levels of cybercrime legislation and cybersecurity strategy development (see Table A.12 in the Appendix). Both variables also scored salient loadings in the factors analysis statistics (see Tables A.13, A.14 and A.15 in the Appendix).

There is one final and fundamental conclusion to yield. The Latin American gap in cyber readiness reveals a sensitive scenario where more forward-moving energy to improve cybersecurity maturity is needed. The OAS survey, out of which my dataset was created, shows that countries advance at different speeds in building up maturity for various cyber aspects.

As I have argued in the previous chapters, cybersecurity governance is a dynamic struggle between stakeholders who bring different policy perspectives to deciding on which affairs to dedicate technological and human resources. While some countries have more superior cybersecurity national strategies, others have more sophisticated criminal legislation, while a third group outclasses in cyber defense maturity. The next chapter presents the book's critical takeaways by establishing a connection between the empirical data and the theoretical ideas discussed in the chapters.

Conclusion

States, Governance, and Cybersecurity

At the outset of this book, I set a series of questions: What drives the digital Pax Latin Americana? How do developing states govern cybersecurity? What is the plausible militarization of cybersecurity telling us? My exploration of such affairs has been partly theory-led and partly policy oriented, nevertheless, heavily based on many significant events transpiring at the global and regional levels that helped me shed light on my research outcomes. In early 2016, for example, the EU and NATO signed an agreement to increase information sharing on cyberincidents, thereby combining civilian and military responses to cyberincidents. In May 2017, computers in at least 150 countries were hit by the large-scale ransomware WannaCry cyberattack demanding payment in the Bitcoin cryptocurrency. A month later, in late June, a series of cyberattacks using the Petya malware affected Microsoft-based systems organizations.[1] In April 2018, the intelligence services in the Netherlands disrupted an alleged Russian cyber operation against the Organization for the Prohibition of Chemical Weapons headquarters in the Hague. The same month, more than thirty global technology companies, including Facebook and Microsoft, announced a joint pledge not to aid any government in undertaking offensive cyberattacks. In October, authorities from the West accused Russian intelligence officers of launching cyberattacks on investigators pursuing cases of alleged Russian misconduct around the globe. In November, French president Macron launched the Paris Call, which at the end of the day was signed by many developing countries but not by the United States, Russia, or China.[2] At the 2020 Munich Security Conference, political, military, and intelligence leaders in the transatlantic alliance vented conflicting visions on the control of 5G technology and the role China is playing in Western countries' mobile communication networks.

In the UK, fears were mounting, as Brexit will have a negative impact on cybercrime policing and intelligence sharing with the EU.[3]

In Latin America, in August 2019, OAS and the worldwide technology leader Cisco launched a joint effort to create public and private innovation councils across the region to help promote educational resources and fill professional gaps in cybersecurity. In November, hackers demanded US$5 million in Bitcoin from Mexico's state oil firm Pemex as ransomware after a hack on the company's systems. DoppelPaymer, the ransomware, was also behind attacks on Chile's Agriculture Ministry and the Texas town Edcouch in the United States.[4] In February 2020, Brazil launched its second National Cybersecurity strategy, enumerating strategic steps for cyber governance programs, critical national infrastructure protection, and encryption measures, antipiracy policies, education of the public on cybersecurity, and enhanced international cooperation through its diplomatic and military executive branches. A month earlier, prosecutors charged American attorney and journalist Glenn Greenwald with cybercrimes for "helping, encouraging, and guiding" a "criminal organization" of hackers who obtained messages between leading figures in Brazil's anticorruption investigation Lava Jato. The leaks, later published in Greenwald's investigative site The Intercept, showed collusion between then-judge Sérgio Moro and prosecutors exacerbating political bias in the case. Moro went on to beccome part of Jair Bolsonaro's cabinet as his justice minister. Greenwald had previously had a central role in exposing the secret intelligence programs leaked by Edward Snowden.[5]

What became crucial to my analysis was the fact that we were entering the second decade of the twenty-first century with an overblowing number of diplomacy and statecraft developments marking a seeming contradiction: on the one hand, the globalization of cybersecurity; on the other, fragmentation.[6] When looking for a common explanatory pattern, I proposed three research pillars that served the purpose of interpreting the events surrounding cybersecurity governance. The book has offered these main components of research to shed light on my initial questions: (1) assess the cybersecurity scenario in the developed countries and illustrate current events in the Western Hemisphere; (2) explore the governance of cybersecurity in comparison to other perceived threats to national security; and (3) exemplify the plausible militarization of cybersecurity policy. The goal of this concluding chapter is to provide possible answers to the various interconnected themes examined to suggest how cybersecurity governance has developed and the policy implications it carries.

Cybersecurity in the Advanced and Emerging Nations

I began by addressing the cybersecurity spillover from the advanced economies to the developing states, my first takeaway being that governments in different regions of the world are mirroring the "American way" of conducting cyber affairs.[7] I justify my statement based on my review of policies and governmental action toward cybersecurity. The case of what I called the digital Pax Latin Americana, and, most notably, the evidence collected from the countries that I investigated the most, inform us of a declaratory policy toward building national capacity to articulate cybersecurity tangibly, most notably through CERTs, cyber commands, centers, and other dedicated positions, agencies, and departments. The proliferation of capabilities for protecting cyberspace seems the norm across developing states, although in some cases more than in others.

In the twenty-first century, various global political elements have grown more harmonious and symmetrical.[8] The relative military power of the United States is unchallengeable; however, it has declined under pressure to shift resources from defense to other priority areas. Cybersecurity compels advanced and developing states to assume greater responsibility in a field in which small countries with less military, political, and economic power can hurt even the most secure superpower. Currently, states no longer face a single threat, and each country perceives dangers from its own national perspective. Medium-sized and smaller states protect their sovereignty and position themselves in an international field, which is no longer a rigid pattern, like in the Cold War. Under this scenario, the international community has identified potential cyber "weak states" or "safe heavens" with low-capacity building to deal with the challenge.[9]

Across the book, I have tried to identify responses from developing regions in the world as cybersecurity reflects a confluence of security, technology, and policy concerns. In an age of globalization, I argued that Latin American states seek to be seen as effective and legitimate actors that can help preserve international peace and security and win the war against underdevelopment and injustice around the world.[10] Because cyber insecurity is a consideration uppermost in the minds of military, security, and foreign policy officials across the globe, cybersecurity imperatives have grown. Increased attention is growing toward those states lagging control over their cyberspace. The dilemmas associated with the actors, drivers, and processes of managing cybersecurity in the developing world call for a more sustainable form of building cybersecurity policy based on challenging the traditional models of cooperation to help close the cyber capacity gap.[11] The technical capabilities of future artificial intelligence and its impact on human cognition might deepen the gap even further. Former U.S. secretary of state Henry Kissinger argued that the incentives of the advanced nations to push the limits of technology might be more about discoveries than human comprehension of the unknown. "And governance, insofar as it deals with the subject, is more likely to investigate AI's applications for security and intelligence than to explore the transformation of the human condition that it has begun to produce."[12] Again, the security emphasis for dealing with whatever technological jumps come next is a serious preoccupation.

The chapters in the book have drawn parallels between cybersecurity governance in the advanced democracies and the Western Hemisphere, pointing out both the similar and different issues that states confront. The abruptness of today's cybersecurity incidents has called for reactive measures that need matching. Authorities, practical capacity, and in most cases, political willingness to control and administer security are still crucial to providing any form of cybersecurity governance. Presidents, prime ministers, generals, and other state officials have actively sought more powers to address cyber insecurity.

In Latin America, countries have deepened their ties with the advanced states guided by foreign policies and military-to-military diplomacy that underpins the cyber issue. I explored Latin American states' increasing activity on cybersecurity at the international level. Countries are driven by foreign policy and strategic commitments on cyber-diplomacy, much like it happens in the advanced economies. The move mirrors the global effort to create multilateral and bilateral ties between key countries, including macroregions such as the European Union, which have developed intense networks of cyber partnerships that aim to strengthen global Internet governance.[13] What is perplexing still is the vulnerability of cybersecurity to strategic rivalries between East and West. When the UK government decided to let Huawei into its 5G networks but with restrictions, in January 2020, Senator Tom Cotton, a Republican member of the U.S. Senate Intelligence Committee, tweeted that "London has freed itself from Brussels [concerning Brexit] only to cede sovereignty to Beijing." The UK assured the rest of the world that the Chinese telecommunications firm would only be allowed to account for 35 percent of the supplying kit in a network's periphery, but not in its core, near military bases and nuclear sites. Former speaker of the House of

Representatives Newt Gingrich described in a tweet that the UK decision meant a "strategic defeat" for his country.

> British decision to accept Huawei 5G is a major defeat for the United States. How big does Huawei have to get and how many countries have to sign with Huawei for the US government to realize we are losing the Internet to China? This is becoming an enormous strategic defeat.[14]

The United States had previously suggested to like-minded allies, including those in Latin America, that using Huawei's equipment posed a risk to critical information systems and would affect vital relationships such as intelligence sharing, military-to-military diplomacy, and network security. A 5G core is the heart of the mobile network's operations, capable of authenticating subscribers, connecting people's mobiles, managing voicemail delivering SMS, routing data from Apps, and keeping track of individual users. Most of this information will now be virtualized and encrypted, meaning that who handles the data flow can potentially access sensitive data. The United States and Australia, for example, claim that the network's boundaries between its core and periphery (mobile masts, base stations, and user devices) can begin to disappear due to technological advancements, thus moving functions from the center to the edge. Huawei is currently competing for contracts in the global scene with other telecom network providers, including Finland's Nokia, Sweden's Ericsson, South Korea's Samsung, and another leading Chinese multinational, ZTE.[15]

While I was writing this book, the Brazilian telecommunications regulator deciding the faith of the country's 5G bandwidth announced that any decision on using Chinese technology would ultimately be taken by the president's national security advisor. Leonardo de Morais, who runs regulator Anatel, told Reuters that national security chief General Augusto Heleno will now set the rules. Brazil is set to become the site of the world's largest 5G spectrum auction to date.[16] The implications of the move will impact the communications functions of a myriad of sectors, including the country's defense, critical national infrastructure, commercial mobile networks, banking and finance, and other productive industries for the economy, such as agriculture, maritime trade, and innovation. As I proposed in chapters 5 and 6, Brazil's options are not set in stone just yet. Vice President Hamilton Mourao visited China to meet with Huawei chief executive Ren Zhengfei and said there was no reason to stop Huawei's investments in Brazil. On the other hand, President Bolsonaro's son, Eduardo Bolsonaro, said that involving Huawei in the 5G bid might distress military cooperation between Brazil and the United States, which under the current Trump administration has come closer.[17]

If severe technological fragmentation occurs, it can cause, mostly, the United States and its partners to decouple from the rest of the world, most notably, China. Some indications that this might happen are already evident to us. As an example, in late 2019, Beijing ordered all government offices and public institutions to remove foreign computer equipment and software within three years. The move worries international tech groups such as HP, Dell Technologies, and Microsoft. China's effort to create a controlled and secured technological environment is part of its revamped cybersecurity legislation, which will put its local vendors Lenovo, Huawei, and Xiaomi, in advantage. This kind of global trend toward protectionism in the cyber issue will have a meaningful impact on the future of the digital Pax Latin Americana. The overall regional balance of power that has favored the United States in the Western Hemisphere is gradually rolling back. Washington's global commitments as a security guarantor and commercial leader, and the further impact of the cyber endeavor, has put a massive strain on its institutional capacity.

Equally important, China's rival order and its projection of influence in the Americas have triggered key developments in U.S. regional policy since the end of the Cold War. While in the past much of Washington's efforts were put into containing the socialist bloc, today's U.S. attempts to charm traditional and formerly nonaligned states to work with Washington on critical international issues.[18] Forging deeper ties with the United States on topics ranging from cybersecurity to counterterrorism could help offset its relative decline while feeding a narrative for peace and order in the Americas. In conclusion, while assessing the cybersecurity scenario in the developed countries, I illustrated current events in the Western Hemisphere that showed at least two underlying trends.

First, the search for global cyber power is balancing Western and Eastern nations that can budget for significant economic and technological advancements. As Marcus Willet argues, several states worldwide "have kick-started their own economic development by stealing intellectual property; threatened or disrupted the financial institutions, oil industries, nuclear plants, power grids and communications routers of nations they regard as adversaries; and, in at least one case, impeded another state's capacity to develop nuclear weapons."[19]

The spread of digital technologies includes states and nonstate actors pursuing strategic gains for the protection of micro (individual) and macro (critical infrastructure) assets. Nations in the East can offer Latin America commercial and tech ventures cheaper than the Western countries. The military factor and the proliferation of cyber military technologies will demand partners to come to the aid of developing states. In this, the United States will try to secure the region and make it difficult for its adversaries to access the neighborhood.[20]

Second, the liberal international order is at a crossroads between freely elected democracies and authoritarian states. This divide hits Latin America at a moment when advanced technological channels are used more than ever to favor quick and effective political and policy decision making. The digital Pax Latin Americana is then vulnerable to the international and domestic complexities of running a country under the pressure of immediate countless social demands. Populist authorities will use the spread of technology irresponsibly and abusively. At the same time, more democratic leaders will have the laborious task of enabling new channels to oversee and ensure accountable governance of cybersecurity. Countries such as China and Russia, observers argue, have built their own national networks protracted by legal barriers impeding their citizens from reaching the wider Internet. Richard Clarke, former national security coordinator in the U.S government, encouraged the authorities in the "Internet freedom league" to restrain authoritarian states from using the Internet as the backbone of censorship, monitoring, and surveillance.[21]

The Governance of Cybersecurity

In my second research pillar, I aimed to explore the governance of cybersecurity in comparison to other perceived threats to national security in the Americas. From my research, I was able to unveil the fact that countries are prone to the dual use of technologies to confront traditional and nontraditional sources of insecurity. Software and hardware technologies apply to perceived threats ranging from transnational organized crime to cybersecurity. States are prone to implement and enforce the combined use of their limited resources to revamp their defenses, many times mixing military and nonmilitary threats. For countries seeking to handle the control of this blurred line this is worrying.

On the one hand, the burdens of terrorism, transnational trafficking of various legal and illicit goods, and violent crime oblige states to continually create and recreate security governance as a means of recasting their approach to insecurity. What the new layer of cybersecurity adds to this scenario is the ever-problematic transfer of technology to central agencies that are not accountable for human rights violations and international humanitarian law, among other parameters. Tools of cybersecurity and cyber-surveillance used to, for instance, fight guerrillas and drug traffickers, can be corrupted by weak intelligence and law enforcement agencies. State actors many times monitor and exploit communications, covertly accessing the sources of storage, processing, and transfer of ICT's data. By allocating cybersecurity to the heart of the security agencies, together with the expansion of significant data control over what happens inside the boundaries of nations, states are introducing new forms of catchall control forms that blend international security, crime, and delinquency all in one. Recent cybersecurity and cyber defense policies signed in the Americas reveal the shared interests in support of sustainable growth, prosperity, and economic security. Strengthening the government apparatus and rehabilitating its security practices (whether in cybersecurity or not) seem to be having a holistic effect wherein many agencies and institutions collide in interest and perspectives.

I argued that states in the Americas are worried about accelerating their cyber defenses as other more advanced nations hurry in the so-called cyber arms race.[22] The emerging dynamics between the international actors (including states and private firms) have pushed the twentieth-century standoff to new frontiers driven mostly by the Washington and Beijing technological duel.[23] The Pentagon is concerned with China's deepening presence in the region, most notably since their more pronounced landing starting in the mid-2000s. Military sources claim that seventeen out of thirty-one Latin American nations have joined China's Belt and Road Initiative; meanwhile, twenty-three host Chinese infrastructure projects, including fifty-six in ports across the region.

Military and foreign affairs bureaucrats in the Western Hemisphere would agree that security issues are worsening at the regional scale. Immediate collective action to clear threats to peace and security is called upon to stop the proliferation of old and new ways to disturb the fragile continental Pax. While consensus on the primacy of action exists, there is a significant factor of smaller states insisting that the United States or regional organizations do not represent their interests. Countries in the Americas have reached different rules on combating perceived threats to national security. Cybersecurity follows a similar faith. I noted that while some states walk beside Washington in setting the rules of cybersecurity and cyber defense, a subgroup is not bound to honor such a proposal. Thus, the current digital Pax Latin Americana might produce different rules of cybersecurity engagement. This outcome is commonplace when it comes to security matters.

Some states use their armed forces to fight crime and others do not. Some states have cutting-edge criminal codes to punish human trafficking while others still lag international norms. Some have legalized marijuana, while it is highly unexpected that others will overturn former boundaries in the name of iron-fisted public security. Some have a terrorism and insurgency problem. Some have major organized crime organizations running wild. States reach different paths to protect their territory and its people. Although many of these security threats, just like cybersecurity, permeate physical boundaries, countries remain averse to combined security action based in old suspicions toward one another and the more pressing issue of a diverse military and financial links with the United States, Russia, and China. The Pentagon, together with the Department of State, nudges partners in the Western Hemisphere to prepare for future threats and optimize development.[24]

Chinese and Russian influence in the affairs of cybersecurity seems more worrying than the long-present security threats in the region. Moscow and Beijing play a significant role in cybersecurity globally, including Latin America. Washington's "war on Chinese technology," as the economist Jeffrey Sachs argued, exaggerates how unsafe Chinese technologies are because officials cannot know the opposite with any certainty: How secure are they?[25] The current scramble over the tech supply chain is isolating U.S.-made technologies, on the one hand, from those developed in China, on the other. The most profound challenge to Latin American states derives from this fast-widening gap in cyber threat perceptions coming from either Beijing or Washington.

Militarized Cybersecurity

The greyish boundary between military activities and foreign intelligence in cyberspace is presumably the most questioned legal and political assumption in current cybersecurity governance.[26] Because cyberincidents take place below the threshold of conflict, as understood by international humanitarian law, the military might seem many times secondary to other actors (e.g., spies or private contractors) and demilitarization might be a suitable depiction of current disruptive events.[27] In my third pillar of research, I aimed to illustrate the issue and the likely militarization of cybersecurity. Regional, as well as national, developments have taken me to conclude that the armed forces are increasingly taking on a broad range of roles and missions. As I reviewed, governments have introduced legislative and policy changes that helped me support this claim. Also, multinational exercises and international alliances make me consider that the military will stay a strong and relevant policy actor for the time being. I draw similar conclusions to recent literature to suggest the fact that the cyber issue has come to equate with or even surpass other military traditional roles.[28] It now sits atop critical issues, including countering transnational organized crime, overseas peace support, and, finally, providing domestic security roles, for instance, in the protection of critical national infrastructure in times of social revolts. As put recently by the IISS in London, common challenges across Latin America including cybersecurity continue to lead more "defence ministries to consider the benefits from greater regional cooperation, not only through training but also in areas such as information sharing."[29]

Reshaped domestic roles should be submitted to careful analysis as civil-military relations present new challenges worldwide.[30] In Argentina, the government tried to modify a decree that restricts the military from tackling internal threats. In Chile, during the social revolution in late 2019, the armed forces took to the streets, reviving scenes from the dictatorships when armed soldiers in tanks went onto the roads in a state of emergency declaration. Later, the senate passed a decree allowing the use of the military to protect critical national infrastructure. In Mexico, President Andrés Manuel López Obrador, an old-time critic of the military, followed suit with his two predecessors and ordered the army and navy to improve their protection of oil and gas pipelines. The National Guard was set up taking forces from the traditional military police branches to help with the security situation. The scenario post–peace agreement in Colombia had led the armed forces to return gradually to internal security challenges, despite initial thoughts they would refocus in the future on roles beyond internal duties.

Popular attitudes toward the military are predetermining a monopoly on confidence that can dangerously erode civilian oversight. Conversations on how to do democratic defense policy are being conducted by parts of the top echelons of the military, more evidently in Washington after Trump's arrival in the White House, sending worrying signals to the rest of the continent.[31] As I mentioned in chapter 2, the popularity of the armed forces remains a puzzle in the study of Latin American politics. For descriptive purposes, Figure C.1 shows the means of support across the region in the period between 2004 and 2018. In some countries, the military has been more prevalent than political institutions, and vice versa. From the countries that I have studied the most across the book, the trend is evident in Brazil, Chile, and Mexico. I have provided a more detailed analysis of these relationships in Table C.1 showing the summary details of a bivariate linear regression ($r = .277$, $p = .000$). These aggregate trends show that the militarization of social and political problems is still a variable worth looking into in more detail as countries show different orientations toward the role of their armies. I have argued, nonetheless, that at least in the countries emphasized in the book (Argentina, Brazil, Chile, Colombia, Mexico, and Venezuela), the military enjoys different levels of corporate advantages at their disposal, be they financial, technological, or human. I anticipate that cybersecurity will be a sought-for field for these resources to put to work.

What is clear from the analysis is that cybersecurity sits on the fence regarding costs versus opportunities. Chapters 3 and 4 revisited the defense economics of the region. It was clear from my examination that there are severe limits to military capabilities and serious economic restraints affecting individual countries and the region. Looking at the largest defense spenders (Brazil, followed by Colombia, Mexico, Chile, and Argentina, in descending order), we saw that aging inventories are replaced by key projects toward perceived threats absorbing significant time and efforts from military planners and central government. I argue that cybersecurity affairs fit well with these rationales as cyber-technologies can be juxtaposed with criminal, espionage, military, and piracy counterefforts already happening across the region. These five countries, plus Venezuela, also represent the largest active military forces in the region. The recent recovery in commodity prices might help countries regain strength and maintain modest growth. However, the austerity wave across the region should prevent defense budgets from skyrocketing. On this topic, I argue similarly to Madeline Carr in saying that political authorities consider the impact of economic losses both to cyberincidents and in financing military resources for both traditional and nontraditional roles such as cyber.[32] States need to balance the potential risks and costs of being cyber prepared.

In this book, cybersecurity governance has been presented as an overlapping and networked policy issue. It was said and exemplified that

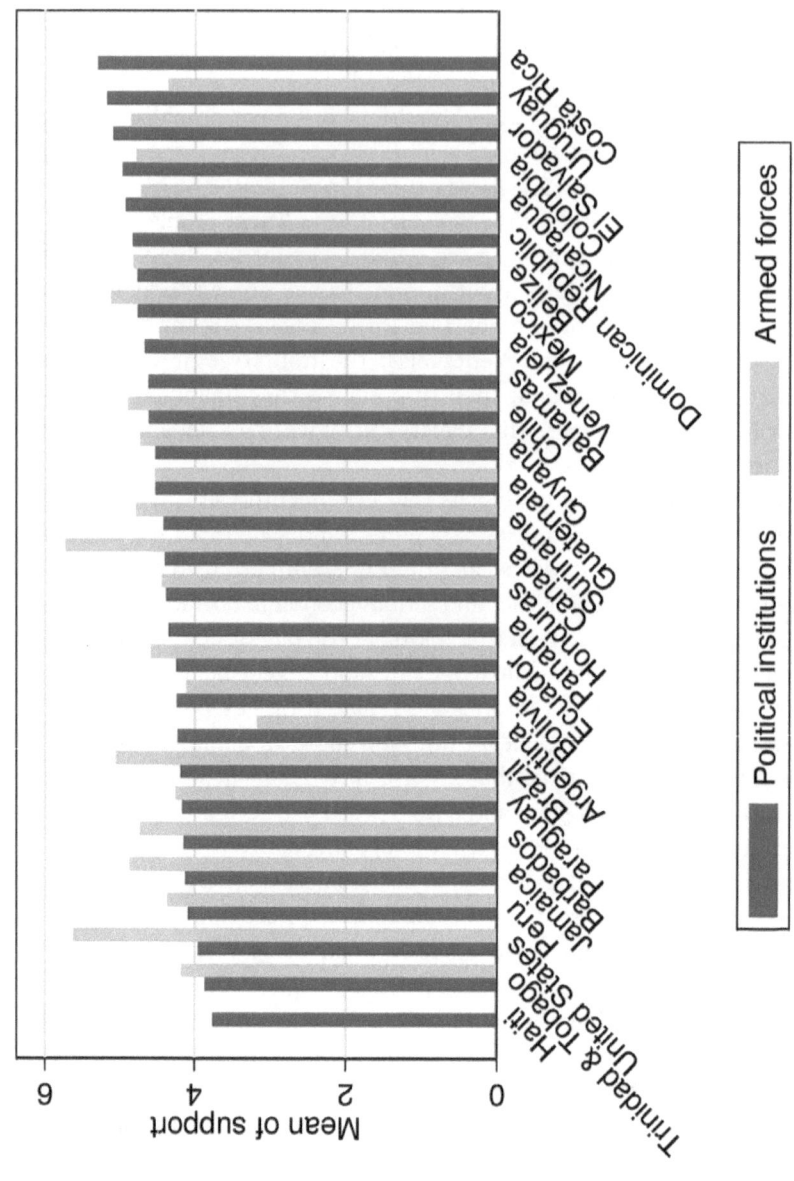

Figure C.1. Average means of institutional support in the Americas, 2004-2018. Author's construction with data from World Bank's Doing Business (2020), ITU (2019), Huawei (2020). Public domain.

those states in both the industrialized and developing regions have created, to some degree, substantial governmental structures for a cyber policy that integrates military and civilian capabilities. I presented some of the complexities in marrying these two fields, such as complicated governance, military dominance over civilians, unevenly resourced bureaucracies, different policy objectives, separate policy fields, and the degree of policy strategy maturity (as explained in chapter 7). No wonder think tanks have decided to give more credit to military cyber operations as a way to measure and assess state cyber capabilities. The most recent example is the IISS, which decided to create a systematic way of evaluating a nation's military cyber capability beginning from 2021. Its flag publication, *The Military Balance*, now carries a separate chapter on military-owned cyber capability. A few issues that I have commented on in this book arise from them, too: first, regarding how to address the cybersecurity and cyberwarfare divide.

> This complicates analysis for the Military Balance, since much cyber capability that might be used for a military purpose is civilian-owned, while some military-owned capability might be used for non-military purposes. Even in cases of cyber-attacks across borders, such as Iran's attack on Saudi Aramco in 2012 or Russia's attack on Ukraine's power grid in 2016, it is unclear in each case just how much the armed forces of these countries were involved.[33]

I have exemplified that not only in the countries with the most substantial military-owned cyber capabilities (United States, United Kingdom, China, Russia), the size and forms of their cyber assets are difficult to identify, and in other parts of the world, the intricacies of governing cybersecurity are destined for similar results. In chapter 2, for instance, I argued that in Latin America, policy developments for cybersecurity and cyber defense are meant to bring together resources and capabilities from the civilian and military sectors under a centralized command and control agency, which might be a militarized command sitting atop the pyramid of executive control. In a similar fashion to what this book has proposed, the IISS new taxonomy of analysis focuses on issues of strategy and doctrine; command and control (including intelligence, surveillance, and reconnaissance); cyber empowerment and dependence; cybersecurity and resilience; global leadership in cyberspace; and military capability for cyber coercion. It will be thought provoking to read a decade from now what these indicators tell about new ways of understanding governance, policy networks, and groups in and out of the state dealing with the cyber issue.

Policy Implications

To return to the policy implications proposed in my introductory chapter, throughout the book I mostly perceived the state as a place of conflict and diversity of opinions across policy sectors and society. Martin Smith argues that state actors and groups need each other to activate resources for a given policy process.[34] In the affairs of cybersecurity, the book has demonstrated that this idea holds. Private-sector interests need the state to step up its game on cyber legislation, policymaking, bureaucracies, and enforcement agencies. States, on the other hand, need societal groups to minimize costs, enhance knowledge, produce technologies, and lessen confrontation with particular interest groups. I have presented this debate when discussing issues of networked cybersecurity governance. After reviewing the evidence, my conclusion at this point is that the policy area of broader cyber issues is growing due to political, policy, and social pressures. Government authorities care for the sake of the state that government functions operate effectively. Policy bodies dedicated to more technical issues depend greatly on the performance of networks. Social groups see in ICTs a way to deal with the specific problems of day-to-day life.

What I have also touched upon in the book is the need for multilateral governance in the Americas to support the issue of cybersecurity with institutional and policy coherence across the. In Europe, for instance, coherent frameworks have put policy actors on the same page to deal with issues of scattered governance and independent policy areas such as law enforcement, critical information infrastructure protection, and defense.[35] In some sense, OAS is doing its part. Still, the regional body is seen with suspicion in the continent, mostly because the timing has not been auspicious for member states to devise a higher degree of cooperation in fundamental political and democratic issues. We see clear advancements in cybercrime, digital privacy, and cyber governance networks being built across the region. I have brought up questions regarding how international fora might overcome current obstacles of engagement, especially as countries seem divided in their intentions to interact with foreign actors (as I discussed in more depth in chapter 6). Here, the split between the United States and China seems more relevant. Although states can remain close to multilateral bodies that act as mediators or brokers of internet governance (again the EU is a good example of this),[36] at the end of the day governments have policy decisions to make which are adopted in a case-by-case scenario in conjunction with what seems more logical to their overall security and political agendas.

I have argued that pressures for creating cybersecurity policy networks have been both internal and external to states. While reviewing policy documents and other documents, I uncovered tensions within the same policy circles, encouraging actors to create and centralize cybersecurity efforts. I also gave proof of foreign demands to enhance cyber capabilities, join international agreements, and abide by global norms and expectations of peaceful behavior online. Although one might think that cyber policy is a closed community, in my research I have discovered that in Brazil, Argentina, Chile, Colombia, Mexico, and Venezuela many actors have a say and are ready to accept or deny engaging in multilateral cybersecurity governance.

I would be careful, nevertheless, to say that there is enough horizontal governance for each stakeholder to bring equal influence to bear on the decision-making process. In the search for whoever holds a stronger position in cybersecurity, I have arrived at the idea that state-centered pockets of power lie throughout mushrooming bureaucracies, specially designed to deal with the cybersecurity. In the countries where I observed more advanced cybersecurity maturity, I found communities with fewer empowered actors and decision makers higher up in the chain of command (e.g., Colombia). In the countries where I observed more advanced cyber defense maturity, I found a similar pattern (e.g., Brazil). Elsewhere, the question is whether we might encounter the same faith in the short term. The locus of centralized authority for cybersecurity is embedded in highly specialized actors. However, the daily practice of state authority is constantly reconceived by policymakers outside the center (i.e., bureaucracies) and reinterpreted by different societal actors (i.e., industry, universities, civic groups, etcetera).[37] As Carolina Sancho argues, "The existence of risks and threats obliges to consider cybersecurity as a condition that must be provided by the state."[38] Even if this too hard to accomplish in nonindustrialized emerging states, the book has uncovered a pathway to policy decisions that supports the idea that cybersecurity governance congregates many actors to form a policy community, but few to form a policy network at the top with enough resources to plan and implement national policy. This has always been reserved for the most powerful ministries, agencies, and branches. Here is where I found that most cyber defense commands are currently positioned. I share with other authors that to explore what sort of cyber capacity for strategic ends state can have, we need to go farther and approach the multifaceted dimension of what it means to be strong in the cyber domain.[39] For such an approach, it seems necessary to focus more systematically on the interaction of factors and actors (state and nonstate) and different governance levels (supranational, state and, substate).[40] I would add to this that scale is also a relevant governance dimension. The scale of cybersecurity governance is not the same in the United States as in Chile. When we have collaborative governance happening, we might encounter scalar problems across and inside the dimensions of the partnerships.[41] Finally, I pick up on the idea that governance as a concept implies a connection and feedback with the environment and continual adjustments of instruments and goals for policy.[42] In this sense, cybersecurity governance, as understood today, will be nothing but a series of successes and failures in need of remaking as future governments learn from past mistakes. Actions and hopes to improve governance in ensuing rounds of action will lead to "evidence-based policymaking" for cyber policy formulation.

Conclusion

In March 2019, Venezuela was hit by a power cut that affected eighteen of its twenty-three states, including the capital Caracas. The metro system was out of service for almost a week, hospitals could not keep vital machinery working, and refrigerators had no power to prevent food from rotting. The government of Nicolás Maduro blamed the Pentagon and the U.S. Southern Command for masterminding a "cyberattack against the electrical, telecommunications, and Internet systems."[43] Accounts from those familiar with the Venezuelan electric grid discredited the official version, saying that a combination of lack of maintenance, underfunding, and overexploitation of what once was Latin America's top electrical network was the most likely cause of the blackout. Considering that supply problems and drought have caused continuous electricity problems in the country, this seems a better explanation. On the other hand, one cannot dismiss the notion that the U.S military is ready to "do whatever needs to be done" in Venezuela. Admiral Faller of the Southern Command pointed to a Chinese disinformation campaign designed to blame the United States for the blackouts. When questioned about the future of Venezuela, Faller emphasized that U.S. troops "are on the balls of our feet" ready to take any action deemed necessary.[44]

Cybersecurity is both a political and a military tool that proved useful to send messages across the world. It is no secret that Beijing has offered plenty of help to the regime in Caracas with information technologies and other resources, including weapons and military supplies (as I gave evidence in the book). Action and rhetoric combine in this context. Some have tried to see the lighter side of it. "My apologies to the people of Venezuela. I must have pressed the wrong thing on the 'electronic attack' app I downloaded from Apple. My bad," commented Republican senator Marco Rubio, known for keeping track of Latin American issues and for being the anti-Madurista voice inside the Washington establishment, in a Tweet.

It is apparent from the research in this book that nations around the globe are ready to implement cyber policy decisions to align with friends and send a message to foes. This is perhaps the most crucial point to emerge from this research. States approach cybersecurity governance—the action and policies taken by civilians, the military, industry, and the private sector to safeguard their digital space—and respond to domestic and foreign pressures in different ways. What is common, though, is that countries enter the cyber era amid extraordinary political uncertainty, rising criminality, and enduring violence. For this, they are building alliances to enhance their positions, and decision makers are willingly placing cybersecurity under the domain of the military, expanding its purview beyond the four conventional battle theatres (air, sea, land, and space). Mastering the fifth realm, the cyber, comes at a moment when Eastern and Western powers test the boundaries of cyber engagement below the threshold of warfare. The implications of securing the frontier between virtual and physical networks are thus far-reaching and cannot be explored thoroughly without minding today's globally interconnected events.

Appendix

Chapter 1

METHODOLOGY

A dataset was created in the general-purpose statistical package STATA, and the measurements for all six WGI indicators and Internet penetration were entered as continuous variables. The six governance indicators recorded data across twenty years from 1996 to 2018 (data for 1997, 1999, and 2001 were not initially recorded by the WGI). The relationships between the continuous variables were analyzed, one at the time, to look at whether a high score on one of the WGI variables is associated with a high score on Internet penetration. From the literature, we have a prediction that these relationships could be in part positive, but there are serious doubts due to a lack of empirical research.

Further testing should shed light on the direction of the results. Pearson's r correlation coefficients are then reported for each variable, assuming the direction and strength and p values for the statistical significance of the relationships. Correlation coefficients vary between −1 and +1. A value of −1 indicates a perfect negative relationship; +1 a perfect positive relationship; and 0 equals to no relationship. Then, the relationship between the "effect" variable (y, the dependent or outcome variable), and the predictors (x, the independent variables) was examined.[1]

I reported the Adjusted R Squared to account for explained variance in the dependent variable.[2] When deciding among different quantitative methodologies, it is usually better to choose based on how easy it will be for someone to read and understand the statistical exercises. Often, basic results are all we want.[3]

Table A.1. Descriptive statistics, bivariate correlations coefficients, and adjusted R-squared for Internet penetration (predictor) and different governance variables (outcome).

	N	Min	Max	Mean	Std. Dev	Pearson's r	Sig.	N	Adj. R-squared
Outcome 1: Government effectiveness									
Argentina	20	−.318	.376	−.056	.175	−.213	.380	19	.011
Brazil	20	−.447	.198	−.114	.145	−.553	.014	19	.265
Chile	20	−.844	1.13	1.17	.116	−.526	.021	19	.234
Colombia	20	−.459	.071	−.131	.146	.769	.000	19	.568
Mexico	20	−.153	.364	.181	.125	−.434	.056	19	.143
Venezuela	20	−.158	.541	−1.082	.230	−.837	.000	19	.683
Outcome 2: Control of corruption									
Argentina	20	−.547	−.083	−.352	.139	−.400	.090	19	.11
Brazil	20	−.532	.166	−.103	.200	−.695	.001	19	.453
Chile	20	1.01	1.59	1.39	.168	−.480	.038	19	.185
Colombia	20	−.512	−.123	−.304	.114	−.193	.428	19	−.019
Mexico	20	−.928	−.167	−.447	.236	−.804	.000	19	.627
Venezuela	20	−.14	−.673	−1.15	.203	−.911	.000	19	.820
Outcome 3: Political stability									
Argentina	20	−.779	.205	−.063	.257	.473	.041	19	.179
Brazil	20	−.420	.330	−.162	.218	−.314	.191	19	.045
Chile	20	0.35	1.092	.530	.223	−.253	2.97	19	.009
Colombia	20	−.237	−.773	−.156	.470	.839	.000	19	.687
Mexico	20	−.923	−.056	−.578	.252	−.437	.054	19	.146
Venezuela	20	1.39	−.535	−1.10	.251	−.099	.685	19	−.048
Outcome 4: Rule of Law									
Argentina	20	−.886	.076	−.528	.276	−.33	.163	19	.059
Brazil	20	−.429	.044	−.208	.136	.417	.076	19	.125
Chile	20	1.01	1.43	1.27	.109	−.016	.948	19	−.059
Colombia	20	−.890	−.256	−.483	.191	.882	.000	19	.765
Mexico	20	−.727	−.310	−.499	.116	−.307	.188	19	.044
Venezuela	20	2.33	−.679	−1.55	.481	−.937	.000	19	.871

Outcome 5: Regulatory quality

	N	Min	Max	Mean	Std. Dev	Pearson's r	Sig.	N	Adj. R-squared
Argentina	20	−.1074	.571	−.539	.477	−.525	.021	19	.233
Brazil	20	−.313	.365	.068	.194	−.815	.000	19	.645
Chile	20	1.335	1.543	1.439	.066	−.306	.202	19	.041
Colombia	20	−.105	.496	.205	.193	.924	.000	19	.846
Mexico	20	.132	.477	.312	.109	.015	.951	19	−.055
Venezuela	20	−2.334	−.269	−1.342	.572	−.905	.000	19	.809

Outcome 6: Voice and accountability

Argentina	20	.206	.565	.373	.085	.318	.184	19	.049
Brazil	20	.242	.571	.448	.085	.538	.017	19	.248
Chile	20	.619	.129	1.028	.148	.366	.124	19	.083
Colombia	20	−.511	.191	−.169	.200	.948	.000	19	.893
Mexico	20	−.078	.347	.119	.136	−.630	.003	19	.363
Venezuela	20	−1.41	.115	−.737	.399	−.902	.000	19	.803

Source: Author's construction with data from World Bank (2021).

Chapter 5

Contingency Tables

By using a three-way contingency table, I wanted to assess the relationship between views on foreign governments and Internet usage among those that favor the Chinese or the U.S. model for future development. As with a regular contingency table, the predictor is found in the columns, the outcome, in the rows. I have given this table a title in the form of the name of the "dependent variable" (trust in a foreign government), by the "independent variable" (frequent Internet Use) controlling for the "control variable" (preferred model for development).[4] I recoded all variables to dichotomic values, with the category of interest being coded 1. I used the AmericasBarometer database for Mexico, including only those respondents who favored the Chinese or the United States as models for economic development. I found that use of Internet does affect the views of respondents about whether they trust foreign governments.

Table Appendix A.2 shows the relationship between trust in a foreign government and Internet use. I can see that frequent users of Internet are

Table A.2. Views on foreign governments, by Internet Use, Mexico

Total	China Gov	No trust	Count	115	52	167
			%within none 0, yes 1	47.3	34.2	42.3
			% of Total	29.1	13.2	42.3
		Trust	Count	128	100	228
			%within none 0, yes 1	52.7	65.8	57.7
			% of Total	32.4	25.3	57.7
	total		Count	243	152	395
			%within none 0, yes 1	100	100	100
			% of Total	61.5	38.5	100
Total	U.S. Gov	No trust	Count	340	170	510
			%within none 0, yes 1	84.8	71.4	79.8
			% of Total	53.2	26.6	79.8
		Trust	Count	61	68	129
			%within none 0, yes 1	15.2	28.6	20.2
			% of Total	9.5	10.6	20.2
	total		Count	401	238	639
			%within none 0, yes 1	100	100	100
			% of Total	62.8	37.2	100

Source: Authors' construction with data from LAPOP (2018).

more likely to support China than non-frequent users by thirteen percentage point. I am quite confident that this effect is not just by chance ($p = .009$). The gamma coefficient of .267 indicates that these two variables have a moderate relationship. The bottom half of Table 6.7 shows that non-frequent users of the Internet are thirteen percentage points more likely not to trust the U.S. government. Less than one-third of frequent Internet users trust the U.S. government. I am certainly confident that this effect is not just by chance ($p = .000$). Like before, the gamma coefficient of .381 indicates that these two variables have a moderate relationship.

Table A.3 shows the same relationship between views on foreign governments and Internet use, except this time I controlled for preferred model of development of the respondent. Among those preferring the Chinese model for development, Internet users are more likely to trust the Chinese government than nonusers of the Internet by four percentage points (79.6 percent versus 75.3 percent, respectively). Among those not preferring the Chinese model of development, Internet users are more likely to support the Chinese government than non-Internet users by seventeen percentage

Table A.3. Views on foreign governments by Internet Use in Mexico, controlling for preferred model for development

				Model Country: China			
					None	None 0, Yes 1 Frequent	Total
Others	China Gov	No trust		Count	96	42	138
				%within none 0, yes 1	57.8	40	51.3
				% of total	35.7	15.6	51.3
		Trust		Count	70	61	131
				%within none 0, yes 1	42.2	59.2	48.7
				% of total	26	22.7	48.7
	Total			Count	166	103	269
				%within none 0, yes 1	100	100	100
				% of total	61.7	38.3	100
China	China Gov	No trust		Count	19	10	29
				%within none 0, yes 1	24.7	20.4	23
				% of total	15.1	7.9	23
		Trust		Count	59	39	97
				%within none 0, yes 1	75.3	79.6	77
				% of total	46	31	77
	Total			Count	77	49	126
				%within none 0, yes 1	100	100	100
				% of total	61.1	38.9	100

continued on next page

points (59.5 percent versus 42.2 percent, respectively). Only for the group of respondents not preferring the Chinese model of development we can be certain that there is a statistically significant relationship ($p = .006$). The strength of the relationships appears to be from moderate to low (see Table A.4. showing symmetric measures).

For those preferring the United States model for development, users of the Internet are more likely to trust the U.S. government than nonusers by a difference of twenty-three percentage points (51.9 percent versus 28.4 percent, respectively). Among those preferring a different model for development, Internet users are more likely to trust the United States government than nonusers by almost twelve percentage points (22 percent versus 10.7 percent, respectively). For both groups, I am highly certain than these relationships did not occur by chance, and the relationship appears to be moderate.

Chapter 6

Cybersecurity Bilateral and Multilateral Cooperation

The UN Institute for Disarmament Research (UNIDIR) keeps an up-to-date Cyber Policy Portal that maps the cybersecurity and cybersecurity-related policy landscape. This tool provides an overview of the cyber capacity of the UN member states and a select group of intergovernmental organizations. For state profiles, it gathers information on cybersecurity policy (strategy documents and implementation frameworks), structures (national center or responsible agency; key positions, dedicated agencies and departments, National Computer Emergency Response Team [CERT] or Computer Security Incident Response Team [CSIRT]), and legal frameworks (legislation, cooperation [multilateral agreements, UN processes, and bilateral and multilateral cooperation, select activities, and membership in intergovernmental organizations]). In Table A.5, I have pointed out the most relevant cybersecurity bilateral and multilateral cooperation in the selected countries of analyses in this chapter. By bilateral and multilateral cooperation is understood the *legally binding and non–legally binding agreements, dialogues, exercises, etc., related to cyber policy matters that the state has entered into with another state and/or states.*

Table A.4. Continued

Symmetric Measures (Model Country China)

				Value	Asymptotic Standard Error[a]	Approx. T[b]	Approx. Significance
Others	Ordinal by ordinal	Gamma	Zero-Order	.332	.113	2.753	.006
	N of valid cases			269			
China	Ordinal by ordinal	Gamma	Zero-Order	.122	.218	.564	.573
	N of valid cases			126			
Total	Ordinal by ordinal	Gamma	Zero-Order	.267	.099	2.614	.009
			First-Order Partial	.304			
	N of Valid cases			395			

Symmetric Measures (Model Country United States)

				Value	Asymptotic Standard Error[a]	Approx. T[b]	Approx. Significance
Others	Ordinal by ordinal	Gamma	Zero-Order	.405	.108	3.208	.001
	N of valid cases			485			
United States	Ordinal by ordinal	Gamma	Zero-Order	.462	.139	2.815	.005
	N of valid cases			154			
Total	Ordinal by ordinal	Gamma	Zero-Order	.381	.085	3.878	.000
			First-Order Partial	.413			
	N of Valid cases			639			

Note: [a] Not assuming the null hypothesis; [b] Using the asymptotic standard error assuming the null hypothesis.
Source: Author's construction with data from LAPOP (2018).

Table A.5. Cybersecurity bilateral and multilateral cooperation in selected countries

Argentina	Brazil	Chile	Colombia	Mexico	Venezuela
U.S.-Argentina Cyber Policy Working Groups	Cooperation, EUROPOL-Brazil	Resolution on Working Group on Cooperative and Confidence Building Measures for Cyberspace, Inter-American Committee against Terrorism	Cooperation, NATO-Colombia	Workshop on Good Practices, with European Union experts	Russia-Venezuela Military-Technical Cooperation
Memorandum of understanding, Spain-Argentina	U.S.-Brazil Cyber Policy Working Group		Technical Assistance, Mission OAS-Colombia	Resolution on Working Group on Cooperative and Confidence Building Measures for Cyberspace, Inter-American Committee against Terrorism	
Cyber cooperation, Argentina Brazil	Cyber cooperation, Argentina Brazil				
OAS Cyber Security Initiative	Memorandum of Understanding, Indo-Brazil-South Africa (IBSA) Forum	OAS Cyber Security Initiative (co-sponsor)		Hemispheric Forum on International Cooperation Against Cybercrime (participant)	
Global Forum of Cyber Expertise	German-Brazil Cyber Consultations	Agreement, Ecuador-Chile		Technical Assistance Mission from the Organization of American States	
	Framework Agreement for Cooperation, Brazil-EU	Potential cooperation, Chile-Ukraine		OAS Cyber Security Initiative (co-sponsor)	
		Global Forum on Cyber Expertise, Member		Global Forum on Cyber Expertise, Member	
	Cooperation, Brazil-Turkey			Memorandum of Understanding, Honduras-Mexico	

Source: Author's construction with data from UNIDIR (2019).

Chapter 7

Factor Analysis

I used factor analysis to simplify the correlational relationships between some of the continuous variables collected by the Cybersecurity Observatory. Once a correlation is performed, it might be challenging to identify reliable patterns of association between variables. Factor analytic techniques is a data reduction technique for the descriptive analysis of statistical data.[5] Put simply, factor analysis is a useful tool for exploring rich corpora to identify salient features, a way of eliciting motifs from a mass of information.[6] It provides means to simplify relationships and identify what factors or common components of association of two variables' relationships with another, third, variable.[7] Two features of factor analytic techniques are worth revisiting. By the method of extraction, we can determine the factors underlying a collection of variables; by rotation, we can simplify the extraction technique that has identified more than one factor underlying the relationships between several variables.

From the Observatory dataset, I included the eight variables used previously in the regression in which respondents were asked to rate aspects related to the level of cybersecurity maturity (1 = the least; 5 = the highest).

The next table shows the Pearson's *r* correlation matrix for correlations between all eight variables. Most correlations are statistically significant. Some pairs of variables are statistically significant in values higher than others. High correlations between several variables might suggest that they may reflect a single underlying factor.[8] I used factor analysis to

Table A.6. Descriptive statistics for cybersecurity maturity by Internet penetration

Internet penetration	Cyber Strategy Development		Cyber Defense		Cybercrime Law	
	Mean	Std. Dev.	Mean	Std. Dev.	Mean	Std. Dev.
< 63% (N = 15)	1.53	.83	1.13	.35	2.27	.70
=> 63% (N = 17)	1.70	.68	1.41	.62	2.9	.78

Source: Author's construction with data from the Observatory of Cybersecurity in Latin America and the Caribbean (2019) and World Bank (2019).

Table A.7. Independent t-test output: Strategy development by Internet penetration

		Independent Samples Tests							
		Levene's Test for Equality of Variance		t-test for Equality of Means					
								95% Confidence Interval of Difference	
		F	Sig	t	df	Sig. (2-tailed)	Means Difference	Std. Error Difference	Lower	Upper
Cyber Strategy Development	Equal variances assumed	1.024	0.32	−0.64	30	.026	−.172	.268	−.721	.376
	Equal variances not assumed			−0.63	27.22	.53	−.172	.272	−.730	.385
Cyber Defense	Equal variances assumed	9.79	.003	−1.54	30	.135	−.278	.181	−.648	.091
	Equal variances not assumed			−1.58	25.91	.124	−.278	.176	−.638	.082
Cybercrime Law	Equal variances assumed	.706	.407	−2.33	30	.003	−.615	.264	−1.155	−.075
	Equal variances not assumed			−2.23	29.98	.003	−.615	.262	−1.151	−.079

Source: Author's construction with data from the Observatory of Cybersecurity in Latin America and the Caribbean (2019) and World Bank (2019).

Table A.8. Correlation coefficients matrix

	Cyber Strategy Development	Cyber Defense Strategy	Cybercrime Law	Internet Penetration
Cyber Strategy Development	1			
Cyber Defense Strategy	.524***	1		
Cybercrime Law	.168	.128	1	
Internet Penetration	.116	.27	.40*	1

Note: * $p < .05$, ** $p < .01$, *** $p < .001$. N= 32.

Source: Author's construction with data from the Observatory of Cybersecurity in Latin America and the Caribbean (2019) and World Bank (2019).

Table A.9. Linear regression results

	Unstandardized B	Std. Error	Sig.
Cyberdefense Strategy	.753	.244	.002
(Constant)	.661	.309	.041
N	32		
Adj. R-squared	.25		

Note: Dependent variable: Cybersecurity strategy development.

Source: Author's construction with data from the Observatory of Cybersecurity in Latin America and the Caribbean (2019) and World Bank (2019).

Table A.10. Multivariate regression results

	Unstandardized B	Std. Error	Sig.
Internet Penetration	−.005	.208	.028
Organization	.65	.211	.004
Content	.795	.244	.001
Cyberdefense Strategy	−.123	.108	.617
Cybercrime Law	.210	.155	.066
Incident Response Capacity	.140	.184	.375
CNI Response Planning	−.235	.177	.214
Command Control Center	−.284	.300	.122
(Constant)	.063	.300	.836
N	32		
Adj. R-squared	.75		

Note: Dependent variable: Cybersecurity strategy development.

Source: Author's construction with data from the Observatory of Cybersecurity in Latin America and the Caribbean (2019) and World Bank (2019).

Table A.11. Means estimation (N=32). Values from low to high (0–5).

	Mean	Std. Error	95% conf. interval	
Cyber strategy development				
Central America	1.5	.27	.95	2.045
Caribbean	1.8	.23	1.03	1.969
South America	1.83	.20	1.41	2.25
Cyber defense strategy				
Central America	1.13	.125	.87	1.37
Caribbean	1.08	.08	.91	1.25
South America	1.6	.19	1.19	1.97
Cybercrime law				
Central America	2.25	.25	1.17	2.75
Caribbean	2.8	.27	2.18	3.31
South America	2.67	.18	2.28	3.05

Note: Central America includes Belize, Costa Rica, El Salvador, Guayana, Honduras, Mexico, Nicaragua, and Panama. The Caribbean region includes Antigua and Barbuda, Bahamas, Barbados, Dominica, Dominican Republic, Grenada, Haiti, St Kitties, Saint Lucia, St. Vincent, Trinidad and Tobago, and Jamaica. South America includes Argentina, Bolivia, Brazil, Chile, Colombia, Ecuador, Guyana, Paraguay, Peru, Suriname, Uruguay, and Venezuela.

Source: Author's construction with data from the Observatory of Cybersecurity in Latin America and the Caribbean (2019).

Table A.12. Bivariate regression results

	Unstandardized B	Std. Error	Sig.
Cybercrime law	.159	.169	.357
(Constant)	1.21	.459	.013
N	32		
Adj. R^2	.029		

Note: Dependent variable: Cybersecurity strategy.

Source: Author's construction with data from the Observatory of Cybersecurity in Latin America and the Caribbean (2019).

study these variables, condensing the information in many variables into a smaller number of factors, in other words, reducing the complexity of the raw data. I also considered rotation of factors to give an idea of how the factors initially extracted differ from each other and which items are associated with each factor.

Table A.13. Correlation matrix between all variables included in the factor analysis

	Cyber Strategy Development	Cyber Defense strategy	Organization	Content	Cybercrime Law	Incident Response Capacity	CNI Response Planning	Command Control Center
Cyber Strategy Development	1							
Cyber Defense Strategy	.524***	1						
Organization	.766**	.681**	1					
Content	.748**	.713**	.671**	1				
Cybercrime Law	.168	.128	-.044	.176	1			
Incident Response Capacity	.431*	.678**	.742***	.427*	-.70	1		
CNI Response Planning	.334	.753***	.626***	.588***	-.007	.732***	1	
Command Control Center	.396*	.730***	.559***	.713***	.208	.684***	.774***	1

Note: * $p < .05$, ** $p < .01$, *** $p < .001$. $N=32$.

Source: Author's construction with data from the Observatory of Cybersecurity in Latin America and the Caribbean (2019) and World Bank (2019).

From the factor analysis initial output shown next, I measure Keiser-Meyer-Olkin (KMO) sampling adequacy and Bartlett's Test of Sphericity. The KMO index tells us how effectively the variables can be grouped into a smaller number of underlying factors (with a maximum of 1; the larger, the better the prospects are for successful factor analysis).[9] In my tests, the value of .757 is large enough.

Bartlett's Test of Sphericity tests whether the correlation matrix of the variables going into the factor analysis is significantly different from an identity matrix (a correlation matrix in which the correlation between variables other than themselves is close to zero).

If none of the variables in the factor analysis are correlated, there are no common underlying factors, and factor analysis won't work.[10] Here, the significance level is .000, meaning that the odds are very high that the matrix is not an identity matrix of close to zero correlations.

The Total Variance Explained box indicates that only two factors have Eigenvalues over 1. Thus, the exercise is suggesting we extract two factors that account for 74.688 of the variances of the relationship between variables.

Finally, I investigated the four component matrices. These contain versions of "factor loadings" of each of the variables on the two factors. Loading is the strength of each variable in defining the factor. The factors, or variables, have features in common, which I can use as clues to the conceptual function of the factor. Loadings on factors can be positive or negative (from zero to a maximum value of negative or positive one). The larger the value, the stronger the link between the variable and the factor. A negative loading indicates that the variable has an inverse relationship with the factor.[11]

Next, I concentrate in the Pattern Matrix box, specifically in the variables yielding with salient loadings (above .44). On Factor 1, the variables: StrategyDevelopment (.663); CybDefStrategy (.877); Organization (.875); Content (.791); IncidentResponseCapacity (.857); CNI Response Planning (.867); and CommanControlCenter (.835), load above .44. On Factor 2: CybCrimeLaw (.896), load above .44.

Looking at the variables, those loading on Factor 1 represent features of policy, strategy, and technologies. Factor 2 contains only one variable that seems to do with legal frameworks. My interpretation of this factor analysis would be that among countries in Latin America and the Caribbean, two main types of features are essential for the development of a documented or official national cybersecurity strategy. First, the extent to which an outline of a national cybersecurity strategy has gone beyond the initial

Table A.14. Factor Analysis Output, initial sections

KMO and Bartlett's Tests

Kaiser-Meyer-Olkin Measure of Sampling Adequacy		.757
Bartlett's Test of Sphericity	Approx. Chi-Square	192.528
	df	28
	Sig.	.000

Communalities

	Initial	Extraction
Cyber Strategy Development	1.000	.610
Cyber defense strategy	1.000	.779
Organization	1.000	.758
Content	1.000	.778
Cybercrime Law	1.000	.800
Incident Response Capacity	1.000	.774
CNI Response Planning	1.000	.758
Command Control Center	1.000	.717

continued on next page

Table A.14. Continued

Total Variance Explained

Component	Initial Eigenvalues			Extraction			Rotation Sums of Squared Loadings
	Total	% of Variance	Cumulative %	Total	% of Variance	Cumulative %	Total
1	4.819	60.241	60.241	4.819	60.241	60.241	4.809
2	1.156	14.447	74.688	1.156	14.447	74.688	1.243
3	.917	11.462	86.150				
4	.479	5.984	92.133				
5	.259	3.242	95.375				
6	.192	2.403	97.779				
7	.109	1.368	99.147				
8	.068	.853	100.000				

Note: Extraction method: Principal Component Analysis. When components are correlated, sums of squared loadings cannot be added to obtain a total variance.

Source: Author's construction with data from Observatory of Cybersecurity in Latin America and the Caribbean (2019) and World Bank (2019).

Table A.15. Factor Analysis Output, component matrices

Component Matrix[a]

	Component	
	1	2
Cyber Strategy Development	.716	.313
Cyber defense strategy	.882	-.006
Organization	.864	-.106
Content	.839	
Cybercrime Law	.120	
Incident Response Capacity	.808	
CNI Response Planning	.833	
Command Control Center	.846	

Structure Matrix

	Component	
	1	2
Cyber Strategy Development	.698	.419
Cyber defense strategy	.882	.130
Organization	.869	.028
Content	.823	.399
Cybercrime Law	.073	.894
Incident Response Capacity	.825	-.219
CNI Response Planning	.845	-.122
Command Control Center	.843	.165

Pattern Matrix[b]

	Component	
	1	2
Cyber Strategy Development	.663	.352
Cyber defense strategy	.877	.041
Organization	.875	-.061
Content	.791	.319
Cybercrime Law	-.019	.896
Incident Response Capacity	.857	-.306
CNI Response Planning	.867	-.210
Command Control Center	.835	.080

Component Correlation Matrix

Component	1	2
1	1.000	.102
2	.102	1.000

Note: Extraction Method: Principal Component Analysis. a: 2 components extracted. Rotation method: Oblimin with Kaiser Normalization. b: Rotation converged in 5 iterations.

Source: Author's construction with data from the Observatory of Cybersecurity in Latin America and the Caribbean (2019).

step of government consultation across key stakeholder groups, possibly involving privates, organized society, foreign agencies, and international cooperation. In consequence, it seems crucial to consider the extent to which the cybersecurity strategy considers any cybersecurity risks and the extent to which specific capacity building and overarching entities exist. This especially for coordination roles for incident response and the protection of critical assets mandated possibly through processes and procedures with some form of formal authority. Second, it also seems relevant to the extent to which criminal legislation exists and the presence of regulatory frameworks to some aspects of cybercrime.

Factor analysis is a descriptive procedure since it requires the researcher to describe things and attach conceptual ideas to their statistical results. For more scientific purposes, the researcher should recognize importance, first, in the variables that load highly on one factor, and second, in those that form a group that goes together conceptually as well as statistically. We can rest assure that the extraction technique presented so far fits the purpose of the exercise for two reasons: variables are not loading in more than one factor (I extracted enough factors), and variables do load in at least one factor (I have not extracted too many variables). I leave it to researchers to carry out any number of extractions using this exploratory technique. My criteria for claiming that this was a good analysis is simple: all variables loading on only one factor and low on all other factors.[12]

Conclusion

Table A.16. Bivariate regression results

	Unstandardized B	Std. Error	Sig.
Armed forces support	.277	.02	.000
(Constant)	3.24	.109	.000
N	188,407		
Adj. R-squared	.008		

Note: Dependent variable: Support for political institutions.
Source: Authors' construction with data form LAPOP (2020).

Notes

Introduction

1. These questions have attracted the attention of many scholars. A starting point for further research on the issue of U.S.-China cyber rivalry is Jon R. Lindsay, "The Impact of China on Cybersecurity: Fiction and Friction," *International Security* 39, no. 3 (Winter 2014/15): 7–47. On the Islamic State cyber jihad, see, for example, Roland Heickerö "Cyber Terrorism: Electronic Jihad," *Strategic Analysis* 38, no. 4 (2014): 554–65. To explore further the Internet and terrorism linkages, I would begin with David Benson, who argues that the Internet does not increase terrorism. States gain as many advantages as terrorist groups, leaving the status quo very much intact as it was before the Internet. See David C. Benson, "Why the Internet Is Not Increasing Terrorism," *Security Studies* 23, no. 2 (2014): 293–328.

2. Stockholm International Peace Research Institute, *Yearbook 2019: Armaments, Disarmament, and International Security*, accessed December 30, 2019, https://www.sipri.org/sites/default/files/2019-08/yb19_summary_eng_1.pdf.

3. United Nations Secretary-General, "Address at the Opening Ceremony of the Munich Security Conference," February 16, 2018, https://www.un.org/sg/en/content/sg/speeches/2018-02-16/address-opening-ceremony-munich-security-conference.

4. See a discussion on national security, privatization, and human security in cyberspace in Myriam Dunn Cavelty, "Breaking the Cyber-Security Dilemma: Aligning Security Needs and Removing Vulnerabilities," *Science and Engineering Ethics* 20, no. 3 (2014): 701–15.

5. A recent thorough publication of the topic is Alison Attrill-Smith, Chris Fullwood, Melanie Keep, and Daria J. Kuss, *The Oxford Handbook of Cyberpsychology* (Oxford: Oxford University Press, 2019).

6. Brandon Valeriano and Ryan Maness, *Cyber War versus Cyber Realities: Cyber Conflict in the International System* (Oxford: Oxford University Press, 2015), 24. This definition also entails "the physical elements because these microprocessors, mainframes, and computers are systems with a physical location."

7. Ronald J. Deibert, "Toward a Human-Centric Approach to Cybersecurity," *Ethics and International Affairs* 32, no. 4 (2018): 411–24, 411.

8. Graham Allison, *Destined for War: Can America and China Escape Thucydides' Trap?* (London: Scribe Publications, 2017), 163.

9. Jacquelyn Schneider, "The Capability/Vulnerability Paradox and Military Revolutions: Implications for Computing, Cyber, and the Onset of War," *Journal of Strategic Studies* 42, no. 6 (2019): 841–63, 843.

10. Rosa Brooks, *How Everything Became War and the Military Became Everything: Tales from the Pentagon* (New York: Simon and Schuster, 2016), 131.

11. Joseph L. Votel, David J. Julazadeh, and Weilun Lin, "Operationalizing the Information Environment: Lessons Learned from Cyber integration in the USCENTCOM AOR," *The Cyber Defense Review* 3, no. 3 (2018): 15–20.

12. McGuffin, Chris, and Paul Mitchell, "On Domains: Cyber and the Practice of Warfare," *International Journal* 69, no. 3 (2014): 394–412, 410. See also, Martin C. Libicki, "Cyberspace Is Not a Warfighting Domain," *I/S: A Journal of Law and Policy for the Information Society* 8, no. 2 (2012): 325–40, 333.

13. Thomas Rid, "Cyber War Will Not Take Place," *Journal of Strategic Studies* 35, no. 1 (2012): 5–32, 6.

14. Duncan B. Hollis and Jens David Ohlin, "What If Cyberspace Were for Fighting?" *Ethics and International Affairs* 32, no. 4 (2018): 441–56.

15. Valeriano and Maness, *Cyber War Versus Cyber Realities*, 1.

16. Robert Thompson, "The Cyber Domains: Understanding Expertise for Network Security," in *The Oxford Handbook of Expertise*, ed. Paul Ward, Jan Maarten Schraagen, Julie Gore, and Emilie M. Roth (New York: Oxford University Press, 2019), 719–39.

17. Ben Buchanan, *The Hacker and the State: Cyber Attacks and the New Normal of Geopolitics* (Cambridge: Harvard University Press, 2020).

18. Manny Fernandez, David E. Sanger, and Marina Trahan Martinez, "Ransomware Attacks Are Testing Resolve of Cities across America," *The New York Times*, August 22, 2019, https://www.nytimes.com/2019/08/22/us/ransomware-attacks-hacking.html.

19. Gary King, Robert O. Keohane, and Sidney Verba, *Designing Social Inquiry: Scientific Inference in Qualitative Research* (Princeton: Princeton University Press, 1994).

20. Susan Khazaeli and Daniel Stockemer, "The Internet: A New Route to Good Governance," *International Political Science Review* 34, no. 5 (2013): 463–82.

21. Christian Malis, "Unconventional Forms of War," in *The Oxford Handbook of War*, ed. Yves Boyer and Julian Lindley-French (New York: Oxford University Press, 2012), 186–98.

22. See Stephen Walt, *The Hell of Good Intentions: America's Foreign Policy Elite and the Decline of U.S. Primacy* (New York: Farrar, Straus and Giroux, 2019), 158–60.

23. Martin Libicki, "Is There a Cybersecurity Dilemma?" *The Cyber Defense Review* 1, no. 1 (2016): 129–40.

24. I am not the only nor the first in dealing at the same time with these concepts. See, for instance, Milton L. Mueller, *Networks and States: The Global Politics of Internet Governance* (Cambridge: MIT Press, 2010), and, Roxana Radu, *Negotiating Internet Governance* (Oxford: Oxford University Press, 2019).

25. Glenn Alexander Crowther, "The Cyber Domain," *The Cyber Defense Review* 2, no. 3 (2017): 63–78, 63.

26. John Gerring, "Case Selection for Case-Study Analysis: Qualitative and Quantitative Techniques," in *The Oxford Handbook of Political Methodology*, ed., Janet M. Box-Steffensmeier, Henry E. Brady, and David Collier (New York: Oxford University Press, 2008), 646–84.

27. King, Keohane, and Sidney, *Designing Social Inquiry*, 12.

28. I borrow this idea from Emily Goldman, *Power in Uncertain Times: Strategy in the Fog of Peace* (Stanford: Stanford University Press, 2010), 2. See also a thorough account on uncertainty, peace, and conflict in Francisco Rojas Aravena, *The Difficult Task of Peace: Crisis, Fragility, and Conflict in an Uncertain World* (New York: Palgrave, 2020).

29. See a review of Brazil's cybersecurity governance in Louise Marie Hurel and Luisa Cruz Lobato, "A Strategy for Cybersecurity Governance in Brazil," Igarapé Institute, Strategic Note 20, September 2018, https://igarape.org.br/wp-content/uploads/2019/01/A-Strategy-for-Cybersecurity-Governance-in-Brazil.pdf.

30. Lucas Kello, "Cyber Security," in *Beyond Gridlock*, ed., Thomas Hale and David Hale (Cambridge: Polity, 2017), 206–28.

31. Marina Kaljurand and Michael Miklaucic, "An Interview with Marina Kaljurand, former Minister of Foreign Affairs of Estonia," *PRISM* 7, no. 2 (2017): 116–20. Russia reacted to Estonia's decision to move a World War II memorial to Soviet soldiers from Tallinn's city center to its outskirts. Moscow orchestrated a denial-of-service attack disrupting several ICTs and Internet services for several weeks.

32. Martin C. Libicki, *Cyberdeterrence and Cyberwar* (Santa Monica: RAND Corporation, 2009).

33. I take this point from Lene Hansen and Helen Nissenbaum, "Digital Disaster, Cyber Security, and the Copenhagen School," *International Studies Quarterly* 53, no. 4 (2009): 1155–75.

34. Lucas Kello, "The Meaning of the Cyber Revolution: Perils to Theory and Statecraft," *International Security* 38, no. 2 (2013): 7–40. I argue in more detail on the term *cyberweapons* in chapter 2.

35. Thomas Rid and Peter McBurney, "Cyber-Weapons," *The RUSI Journal* 157, no. 1 (2012): 6–13.

36. Among the developing countries targeted are Algeria, Afghanistan, Belgium, Brazil, Fiji, Germany, Iran, India, Indonesia, Kiribati, Malaysia, Mexico, Pakistan, Russia, Saudi Arabia, and Syria. See "Regin: Stuxnet's Best Spying

Malware Cousin," *VR World*, November 24, 2014, http://vrworld.com/2014/11/24/regin-stuxnets-best-spying-malware-cousin/.

37. Michèle Flournoy and Michael Sulmeyer, "Battlefield Internet: A Plan for Securing Cyberspace," *Foreign Affairs* 97, no. 5 (2018): 40–46.

38. Adam Segal, "Cyber Week in Review: August 23, 2019," Council on Foreign Relations, August 23, 2019, https://www.cfr.org/blog/cyber-week-review-august-23-2019.

39. See Heather Harrison Dinniss, *Cyber Warfare and the Laws of War* (Cambridge: Cambridge University Press, 2012), 117–38.

40. Moisés Naím, "Mafia States: Organized Crime Takes Office," *Foreign Affairs* 91, no. 3 (2012): 100–11.

41. Vincent Boulanin, "Cybersecurity: A Precondition to Sustainable Information and Communication Technology–Enabled Human Development," *SIPRI Yearbook 2015: Armaments, Disarmament, and International Security*, accessed November 19, 2019, https://www.sipriyearbook.org/view/9780198787280/sipri-9780198787280-chapter-010-div1-072.xml.

42. See Libicki, "Is There a Cybersecurity Dilemma?"

43. See other international cybersecurity capacity-building instances in Vincent Boulanin, "Mapping Key Actors and Efforts in Cybersecurity for Human Development," *SIPRI Yearbook 2015: Armaments, Disarmament, and International Security*, accessed November 19, 2019, https://www.sipri.org/yearbook/2015, 397.

44. Dale C. Copeland, "Realism and Neorealism in the Study of Regional Conflict," in *International Relations Theory and Regional Transformation*, ed. T. V. Paul (Cambridge: Cambridge University Press, 2012), 49–72.

45. Joseph S. Nye Jr., "Deterrence and Dissuasion in Cyberspace," *International Security* 41, no. 3 (2016): 44–71, 44.

46. See Libicki, *Cyberdeterrence and Cyberwar*, 11–37.

47. Nye Jr., "Deterrence and Dissuasion," 46.

48. See Scott Warren Harold, Martin C. Libicki, and Astrid Stuth Cevallos, *Getting to Yes with China in Cyberspace* (Santa Monica: RAND Corporation, 2016).

49. Ronald Deibert, "Trajectories for Future Cybersecurity Research," in *The Oxford Handbook of International Security*, ed. Alexandra Gheciu and William C. Wholforth (New York: Oxford University Press, 2018), 532–46.

50. Robert M. Gates, "Helping Others Defend Themselves," *Foreign Affairs* 89, no. 3 (2010): 2–6, 2.

51. Russell Crandall, *The United States and Latin America after the Cold War* (New York: Cambridge University Press, 2008); David Mares and Francisco Rojas Aravena, *The United States and Chile: Coming in From the Cold* (New York: Routledge, 2001).

52. U.S. Department of State, "U.S.-Chile Executive Cyber Consultation," September 18, 2018, https://www.state.gov/u-s-chile-executive-cyber-consultation/.

53. Lennon Y. C. Chang and Peter Grabosky, "The Governance of Cyberspace," in *Regulatory Theory: Foundations and Applications*, ed. Peter Drahos (Canberra: ANU Press, 2017), 533–51, 534.

54. See various approaches to the governance debate in Merilee S. Grindle, *Good Governance: The Inflation of an Idea* (John F. Kennedy School of Government, Harvard University, 2010); Jan Kooiman, *Governing as Governance* (Thousand Oaks: Sage, 2003); Mark Bevir, *Democratic Governance* (Princeton: Princeton University Press, 2010); and Christopher Ansell and Jacob Torfing, *Handbook on Theories of Governance* (Cheltenham: Edward Elgar, 2016).

55. Robert Muggah and Nathan B. Thompson, "Brazil's Critical Infrastructure Faces a Growing Risk of Cyberattacks," April 10, 2018, https://www.cfr.org/blog/brazils-critical-infrastructure-faces-growing-risk-cyberattacks.

56. I borrow this approximation to the term "national security" from Derek S. Reveron, Nikolas K. Gvosdev, and John A. Cloud, "Introduction: Shape and Scope of U.S. National Security," in *The Oxford Handbook of U.S. National Security*, ed. Derek S. Reveron, Nikolas K. Gvosdev, and John A. Cloud (Oxford: Oxford University Press), 2–12.

57. Ministry of Defense, "Livro Blanco de Defesa Nacional," http://www.defesa.gov.br/arquivos/estado_e_defesa/livro_branco/lbdn_2013_ing_net.pdf

58. Presidency of the Republic, "Estratégia Nacional de Defesa, Decreto N. 6.703," http://www.planalto.gov.br/ccivil_03/_ato2007-2010/2008/Decreto/D6703.htm.

59. William Banks, "State Responsibility and Attribution of Cyber Intrusions after Tallinn 2.0," *Texas Law Review* 95, no. 7 (2017): 1488–1513, 1494.

60. See on this point, Rollin F. Tusalem, "Bringing the Military Back In: The Politicisation of the Military and Its Effect on Democratic Consolidation," *International Political Science Review* 3, no, 4 (2014): 482–501.

61. Michael Chertoff and John Michael McConnell, "A Path Forward for Cyber Defense and Security," in *Securing Cyberspace: A New Domain for National Security*, ed. Nicholas Burns and Jonathan Price (Washington, DC: Aspen Institute, 2012), 191–202.

62. For these points on cyber strategic assessment, see Brandon Valeriano and Ryan C. Maness, "The Dynamics of Cyber Conflict Between Rival Antagonists, 2001–11," *Journal of Peace Research* 51, no. 3 (2014): 347–60; and Ryan C. Maness and Brandon Valeriano, "The Impact of Cyber Conflict on International Interactions," *Armed Forces & Society* 42, no. 2 (2015): 301–23.

63. John J. Mearsheimer and Stephen M. Walt, "Leaving Theory Behind: Why Simplistic Hypothesis Testing Is Bad for International Relations," *European Journal of International Relations* 19, no. 3 (2013): 427–57.

64. Condoleezza Rice and Amy Zegart, "Managing 21st-Century Political Risk," *Harvard Business Review*, May-June 2018, https://hbr.org/2018/05/managing-21st-century-political-risk.

65. See a debate on how governance and capacity-building efforts are interlinked in the study of developing countries in Merilee S. Grindle, "Good Enough Governance Revisited," *Developmental Policy Review* 25, no. 5 (2007): 533–74, and "Governance Reform: The New Analytics of Next Steps," *Governance* 24,

no. 3 (2011): 415–18. See how governance networks affect liberal democracy in Eva Sørensen, "Democratic Network Governance," in *Handbook on Theories of Governance*, ed. Christopher Ansell and Jacob Torfing (Cheltenham: Edward Elgar, 2016), 419–27.

66. Moritz Weiss and Vytautas Jankauskas. "Securing Cyberspace: How States Design Governance Arrangements," *Governance* 32, no. 2 (2019): 259–75.

Chapter One

1. National Cyber Security Centre, "Annual Review 2018: Making the UK the Safest Place to Live and Work Online," accessed October 19, 2018, http://www.ncsc.gov.uk/annual-review-2018.

2. Some definitions of these terms by the Council on Foreign Relations "Cyber Operations Tracker" include the following:

> *Distributed denial of service*: The intentional paralyzing of a computer network by flooding it with data sent simultaneously from many individual computers.
>
> *Espionage*: The act of obtaining confidential information without the information holder's consent.
>
> *Defacement*: The unauthorized act of changing the appearance of a website or social media account.
>
> *Data Destruction*: The use of malicious software to destroy data on a computer or to render a computer inoperable.
>
> *Sabotage*: The use of malware that disrupts a physical process, such as the provision of electricity or normal function of nuclear centrifuges.
>
> *Doxing*: The act of searching and publishing private or identifying information about an individual or group on the internet, typically with malicious intent.

See fn.3 for the source of these concepts.

See Council on Foreign Relations, "Cyber Operations Tracker," accessed October 24, 2018, https://www.cfr.org/interactive/cyber-operations#Takeaways.

3. Ibid.

4. The White House, "National Security Strategy of the United States of America, December 2017," accessed October 24, 2018, https://www.whitehouse.gov/wp-content/uploads/2017/12/NSS-Final-12-18-2017-0905.pdf.

5. Ibid.

6. Joseph Hincks, "Here's What a Top General Thinks Is the Next Big Threat to the U.S.," *Time*, September 27, 2017, http://time.com/4958679/general-dunford-military-threat/.

7. Phil Stewart and Valerie Volcovichi and Reuters, "Trump Backtracks on His Idea for a Joint Cyber Security Unit with Russia after Harsh Criticism," *Time*, July 10, 2017, http://time.com/4850902/trump-russia-cyber-security-putin-criticism/.

8. Benoit Dupont, "The Cyber-Resilience of Financial Institutions: Significance and Applicability," *Journal of Cybersecurity* 5, no. 1 (2019): 1–17, 2.

9. I borrow this idea from Anne-Marie Slaughter, "America's Edge: Power in the Networked Century," *Foreign Affairs* 88, no. 1 (2009): 94–113.

10. Zoe Baird, "Governing the Internet: Engaging Government, Business, and Nonprofits," *Foreign Affairs* 81, no. 6 (2002): 15–20, 15.

11. While for others, *cyber* is a buzzword. See Brandon Valeriano and Maness, *Cyber War versus Cyber Realities*, 22.

12. Ibid., 35–36, Valeriano and Maness indicate four methods or weapons used to initiate cyber conflict: website defacement or vandalism; distributed denial of service methods, or simply put DDoS; intrusion; and infiltration.

13. Jonathan Clough, *Principles of Cybercrime* (Cambridge: Cambridge University Press, 2015), 10–11.

14. Ibid., 16–17.

15. World Bank, "World Development Indicators: States and Markets," accessed May 1, 2019, http://datatopics.worldbank.org/world-development-indicators/themes/states-and-markets.html#science-and-innovation_1.

16. Suzanne Nossel, "Google Is Handing the Future of the Internet to China," *Foreign Policy*, September 10, 2018, https://foreignpolicy.com/2018/09/10/google-is-handing-the-future-of-the-internet-to-china/.

17. "Egypt: New Media Law Aims to Silence Independent Online Media, RSF says," November 3, 2018, https://rsf.org/en/news/egypt-new-media-law-aims-silence-independent-online-media-rsf-says.

18. Anne Applebaum, "Saudi Arabia's Information War to Bury News of Jamal Khashoggi," *The Washington Post*, October 17, 2018, https://www.washingtonpost.com/opinions/global-opinions/saudi-arabias-information-war-to-bury-news-of-jamal-khashoggi/2018/10/17/e4825a5a-d227-11e8-b2d2-f397227b43f0_story.html?utm_term=.affe26290cf4.

19. Margaret Weiss, "Assad's Secretive Cyber Force," accessed October 18, 2018, https://www.washingtoninstitute.org/policy-analysis/view/assads-secretive-cyber-force.

20. Feliz Salomon, "The Thai Junta Looks Set to Tighten Control of the Internet Even Further," *Time*, December 16, 2016, http://time.com/4604521/thailand-junta-cyber-crime-internet-rights-expression/.

21. Abby Seiff and Phnom Penh, "Chinese State-Linked Hackers in Large Scale Operation to Monitor Cambodia's Upcoming Elections, Report Says," *Time*, July 18, 2018, http://time.com/5334262/chinese-hackers-cambodia-elections-report/.

22. Susan Crawford, "Net Neutrality Is Just a Gateway to the Real Issue: Internet Freedom," *Wired*, May 18, 2018, https://www.wired.com/story/net-neutrality-is-just-a-gateway-to-the-real-issue-internet-freedom/.

23. Edwin Grohe, "The Cyber Dimensions of the Syrian Civil War: Implications for Future Conflict," *Comparative Strategy* 34, no. 2 (2015): 133–48.

24. UNESCO, "World Trends in Freedom of Expression and Media Development, Global Report 2017/2018," http://unesdoc.unesco.org/images/0026/002610/261065e.pdf, 109–110.

25. Institute for Economics & Peace, "Global Peace Index 2018: Measuring Peace in a Complex World," accessed December 5, 2018, http://visionofhumanity.org/reports, 33–34.

26. Ibid., 35.

27. The Institute for Economics & Peace (2018) highlights some definitions for these four types of conflict. They are described as the following:

> *Extra systemic or extra-state armed conflict*: these occur between a state and a non-state group outside its territory, for example, colonial wars or wars of independence.
>
> *Interstate armed conflict*: these occur between two or more states.
>
> *Internal or intra-state armed conflict*: these occur between the government of a state and one or more internal opposition groups without intervention from other states.
>
> *Internationalised internal armed conflict*: these occur between the government of a state and one or more internal opposition groups with intervention from other states on one or both sides.

28. Amanda Erickson, "Our First Concern Was to Avoid a Civil War: Nicaragua's Government on Six Months of Protests," *The Washington Post*, September 26, 2018, https://www.washingtonpost.com/world/2018/09/26/our-first-concern-was-avoid-civil-war-nicaraguas-government-six-months-protests/?utm_term=.7d62b198243f.

29. Orlando J. Pérez, "Can Nicaragua's Military Prevent a Civil War?" *Foreign Policy*, July 3, 2018, https://foreignpolicy.com/2018/07/03/can-nicaraguas-military-prevent-a-civil-war/.

30. Robert Muggah, Ilona Szabo de Carvalho, and Katherine Aguirre, "Latin America Is the World's Most Dangerous Region. But There Are Signs It Is Turning a Corner," *World Economic Forum on Latin America*, March 14, 2018, https://www.weforum.org/agenda/2018/03/latin-america-is-the-worlds-most-dangerous-region-but-there-are-signs-its-turning-a-corner/.

31. Ibid.

32. Louis DiMarco, "Urban Warfare," *Oxford Bibliographies*, DOI: 10.1093/OBO/9780199791279-0171.

33. Stephen Graham, *Cities under Siege: The New Military Urbanism* (New York: Verso, 2010).

34. Andrew Cockburn, *Kill Chain: Drones, and the Rise of High-Tech Assassins* (London: Verso, 2015).

35. Diane E. Davis and Anthony W. Pereira, eds. *Irregular Armed Forces and Their Role in Politics and State Formation* (New York: Cambridge University Press, 2003); Max G. Manwaring, *Gangs, Pseudo-Militaries, And Other Modern Mercenaries: New Dynamics in Uncomfortable Wars* (Norman: University of Oklahoma Press, 2010); Thomas C. Bruneau, *Patriots for Profit: Contractors and the Military in US National Security* (Stanford: Stanford University Press, 2011).

36. Alexander Klimburg, *The Darkening Web: The War for Cyberspace* (London: Penguin, 2017).

37. William J. Astore, "Science and Technology in War," *Oxford Bibliographies*, DOI: 10.1093/OBO/9780199791279-0054.

38. Emily O. Goldman and Leslie C. Eliason, *The Diffusion of Military Technology and Ideas* (Stanford: Stanford University Press, 2003).

39. P. W. Singer, *Wired for War: The Robotics Revolution and Conflict in the 21st Century* (New York: Penguin, 2009).

40. Valeriano and Mannes, *Cyber War versus Cyber Realities*.

41. Ibid., 29.

42. Richard J. Harknett and Max Smeets, "Cyber Campaigns and Strategic Outcomes," *Journal of Strategic Studies* (2020), https://doi.org/10.1080/01402390.2020.1732354.

43. I borrow this anecdote from Astore, "Science and Technology in War."

44. The Cylance Threat Research Team, "El Machete's Malware Attacks Cut Through LATAM," *ThreatVector*, March 22, 2017, https://threatvector.cylance.com/en_us/home/el-machete-malware-attacks-cut-through-latam.html.

45. KasperskyLab, "El Machete," August 20, 2014, https://securelist.com/el-machete/66108/.

46. Kelvin Chan, "Apple CEO Backs Privacy Laws, Warns Data Being 'Weaponized,'" *The Washington Post*, October 24, 2018, https://www.washingtonpost.com/business/technology/apple-ceo-backs-privacy-laws-warns-data-being-weaponized/2018/10/24/f7485a08-d773-11e8-8384-bcc5492fef49_story.html?utm_term=.cdabba2a4c36.

47. Lily Hay Newman, "How Facebook Hackers Compromised 30 Million Accounts," *Wired*, October 12, 2018, https://www.wired.com/story/how-facebook-hackers-compromised-30-million-accounts/.

48. Carole Cadwallader and Emma Graham-Harrison, "Revealed: 50 Million Facebook Profiles Harvested from Cambridge Analytica in Major Data Breach," *Guardian*, March 17, 2018, https://www.theguardian.com/news/2018/mar/17/cambridge-analytica-facebook-influence-us-election.

49. Matthew Rosenberg and Maggie Haberman, "When Trump Phones Friends, the Chinese and the Russians Listen and Learn," *The New York Times*, October 24, 2018, https://www.nytimes.com/2018/10/24/us/politics/trump-phone-security.html.

50. United States Government Accountability Office (GAO), "Weapon Systems Cybersecurity: DOD Just Beginning to Grapple with Scale of Vulnerabilities," accessed October 26, 2018, https://www.gao.gov/assets/700/694913.pdf.

51. Elias Groll, "Many U.S. Weapons Systems Are Vulnerable to Cyberattack," *Foreign Policy*, October 9, 2018, https://foreignpolicy.com/2018/10/09/many-u-s-weapons-systems-are-vulnerable-to-cyberattack/.

52. Andrew F, Cooper and Jeremie Cornut, "The Changing Nature of Diplomacy," *Oxford Bibliographies*, DOI: 10.1093/OBO/9780199743292-0180.

53. Andrew Cottey and Anthony Forster, *Reshaping Defence Diplomacy: New Roles for Military Cooperation and Assistance* (New York: Oxford University Press, 2004).

54. Tarak Barkawi, "'Defence Diplomacy' in North-South Relations," *International Journal* 66, no. 3 (2011): 597–612.

55. Philip M. Seib, *Real-Time Diplomacy: Politics and Power in the Social Media Era* (New York: Palgrave Macmillan, 2012).

56. Internet World Stats, "World Internet Usage and Population Statistics," accessed October 24, 2018, https://www.internetworldstats.com/stats.htm.

57. Internet Traffic Report, "Global Traffic Index," accessed October 24, 2018, http://www.internettrafficreport.com/faq.htm.

58. Office for National Statistics, "Internet Users, UK: 2018," accessed October 24, 2018, https://www.ons.gov.uk/businessindustryandtrade/itandinternetindustry/bulletins/internetusers/2018.

59. Instituto Federal de Telecomunicaciones, "En México 71.3 Millones de Usuarios de Internet y 17.4 Millones de Hogares con Conexión a Este Servicio: ENDUTIH 2017 (Comunicado 015/2018)," accessed October 24, 2018, http://www.ift.org.mx/comunicacion-y-medios/comunicados-ift/es/en-mexico-713-millones-de-usuarios-de-internet-y-174-millones-de-hogares-con-conexion-este-servicio

60. Anthony Craig, "Understanding the Proliferation of Cyber Capabilities," *Council on Foreign Relations*, October 18, 2018, https://www.cfr.org/blog/understanding-proliferation-cyber-capabilities.

61. Craig (2018) explained: "A CNO unit was created in 32 of the 1041 cases (3.1 percent) in which a country had not been attacked. However, a CNO unit was also created in 19 of the 194 cases (9.8 percent) in which a country had been attacked. In other words, countries that have suffered a cyber-attack are more likely to create a CNO unit compared with countries that have not suffered a cyber-attack (The chi squared test gives a *p*-value of 0.000 suggesting the relationship is statistically significant)."

62. Craig, "Understanding the Proliferation of Cyber Capabilities."

63. Patrick Howell O'Neill, "Previously Unknown Cyber-Espionage Group Has Successfully Hacked in South America Since 2015," November 7, 2017, https://www.cyberscoop.com/previously-unknown-cyber-espionage-group-successfully-hacked-south-america-since-2015/.

64. Council on Foreign Relations, "Cyber Operations Tracker."

65. Margi Murphy and Matthew Field, "How Russia Would Wage Cyber War on Britain," *Telegraph*, April 16, 2018, https://www.telegraph.co.uk/technology/2018/04/16/russia-would-wage-cyber-war-britain/.

66. Sam Jones, "Russian Government Behind Cyber Attacks, says Security Group," *Financial Times*, October 28, 2014.

67. George Christou, *Cybersecurity in the European Union: Resilience and Adaptability in Governance Policy* (Basingstoke: Palgrave Macmillan, 2015).

68. Ian Anthony, "European Security," in *SIPRI Yearbook 2017: Armaments, Disarmament and International Security*, accessed November19, 2019, https://www.sipriyearbook.org/view/9780198811800/sipri-9780198811800-chapter-4-div1-20.xml.

69. Ian Anthony, Sam Perlo-Freeman, and Siemon T. Wezeman, "The Ukraine Conflict and its Implications," in *SIPRI Yearbook 2015: Armaments, Disarmament, and International Security*, accessed November19, 2019, https://www.sipriyearbook.org/view/9780198737810/sipri-9780198737810-chapter-3-div1-3.xml.

70. Dustin Voltz and Timothy Gardner, "In a First, U.S. Blames Russia for Cyber Attacks on Energy Grid," *Reuters*, March 15, 2018, https://www.reuters.com/article/us-usa-russia-sanctions-energygrid/in-a-first-u-s-blames-russia-for-cyber-attacks-on-energy-grid-idUSKCN1GR2G3.

71. Dam Smith, "Introduction: International Stability and Human Security in 2018," in *SIPRI Yearbook 2019: Armaments, Disarmament and International Security*, accessed November 19, 2019, https://www.sipriyearbook.org/view/9780198839996/sipri-9780198839996-chapter-1-div1-006.xml.

72. The NPR is cited in Kile, Shannon N., and Hans M. Kristensen, "World Nuclear Forces," in *SIPRI Yearbook 2019: Armaments, Disarmament, and International Security*, accessed November 19, 2019, https://www.sipriyearbook.org/view/9780198839996/sipri-9780198839996-chapter-6-div1-034.xml.

73. BBC News, "Cyber Attack on Pentagon Email," June 22, 2007, http://news.bbc.co.uk/1/hi/world/americas/6229188.stm.

74. BBC News, "China Denies Pentagon Cyber-Raid," September 4, 2007, http://news.bbc.co.uk/1/hi/world/americas/6977533.stm.

75. Isidoro Cheresky, *El Nuevo Rostro de la Democracia en América Latina: Ciudadanía, Liderazgos E Instituciones* (Buenos Aires, Argentina: Fondo de Cultura Económica, 2015).

76. C. W. Anderson, "Web 2.0," *Oxford Bibliographies*, DOI: 10.1093/OBO/9780199756841-0072; Stephen Coleman and Jay Blumler, *The Internet and Democratic Citizenship: Theory, Practice, and Policy* (Cambridge: Cambridge Univer-

sity Press, 2009); Matthew Hindman, *The Myth of Digital Democracy* (Princeton: Princeton University Press, 2008).

77. Christopher Hood, "The Tools of Government in the Information Age," in *The Oxford Handbook of Public Policy*, ed. Robert E. Goodin, Michael Moran, and Martin Rein (Oxford: Oxford University Press, 2008), 480–81.

78. I make this point from Hood (2008, 472), although I take the argument with a different connotation to reflect also on how governments can use ICTs on citizens. A thorough article revising this idea is Moisés Naím, "Why Democracies Are at a Disadvantage in Cyber Wars," *Journal of International Affairs*, The next world order: Special 70th Anniversary Issue (2017): 85–91.

79. Gretchen Helmke and Steven Levitsky, *Informal Institutions and Democracy: Lessons from Latin America* (Baltimore: Johns Hopkins University Press, 2006).

80. Guillermo O'Donnell, "On the State, Democratization, and Some Conceptual Problems: A Latin American View with Glances at Some Postcommunist Countries," *World Development* 21, no. 8 (1993): 1355–69; Guillermo O'Donnell, *Democracy, Agency, and the State: Theory with Comparative Intent* (Oxford: Oxford University Press, 2010).

81. B. Guy Peters, "What Is So Wicked About Wicked Problems? A Conceptual Analysis and a Research Program," *Policy and Society* 36, no. 3 (2017): 385–96.

82. Eduardo Dargent, *Technocracy, and Democracy in Latin America: The Experts Running Government* (New York: Cambridge University Press, 2015).

83. Carlos Solar, *Government and Governance of Security: The Politics of Organised Crime in Chile* (New York: Routledge, 2018).

84. Elke Krahmann, "Conceptualizing Security Governance," *Cooperation and Conflict* 38, no. 1 (2003): 5–26.

85. Mark Webber, Stuart Croft, Jolyon Howorth, Terry Terriff, and Elke Krahmann, "The Governance of European Security," *Review of International Studies* 30, no. 1 (2004): 3–26; James Sperling and Mark Webber, "Security Governance in Europe: A Return to System," *European Security* 23, no. 2 (2014): 126–44.

86. Peter Gill and Mark Phythian, "What Is Intelligence Studies?" *The International Journal of Intelligence, Security, and Public Affairs* 18, no. 1 (2016): 5–19.

87. Ilina Georgieva, "The Unexpected Norm-Setters: Intelligence Agencies in Cyberspace," *Contemporary Security Policy* 41, no. 1 (2020): 33–54.

88. Erik C. Nisbet, Elizabeth Stoycheff, Katy E. Pearce, "Internet Use and Democratic Demands: A Multinational, Multilevel Model of Internet Use and Citizen Attitudes about Democracy," *Journal of Communication* 62, no. 2 (2012): 249–65; Susan Khazaeli and Daniel Stockemer, "The Internet: A New Route to Good Governance," *International Political Science Review* 34, no. 5 (2013): 463–82.

89. Carl Bildt and Gordon Smith, "The One and Future Internet," *Journal of Cyber Policy* 1, no. 2 (2016): 142–56.

90. See Milton Mueller, *Networks and States: The Global Politics of Internet Governance* (London: MIT Press, 2010); Elizabeth Stoycheff and Erik C. Nisbet,

"What's the Bandwidth for Democracy? Deconstructing Internet Penetration and Citizen Attitudes about Governance," *Political Communication* 31, no. 4 (2014): 628–46; Elaine Ciulla Kamarck and Joseph S. Nye Jr. *Visions of Governance in the 21st Century* (Washington, DC: Brookings Institution Press, 2002).

91. Daniel Kaufmann, Art Kraay, and Massimo Mastruzzi, "The Worldwide Governance Indicators: Methodology and Analytical Issues," World Bank Policy Research Working Paper No. 5430 (2010), https://ssrn.com/abstract=1682130.

92. Robert I. Rotberg, "Good Governance Measures," *Governance* 27, no. 3 (2014): 511–18.

93. Francis Fukuyama, "What Is Governance?" *Governance* 26, no. 3 (2013): 347–68.

94. Ibid., 350.

95. Ibid., 347–48.

96. Bo Rothstein, *The Quality of Government: Corruption, Social Trust, and Inequality in International Perspective* (Chicago: University of Chicago Press, 2011).

97. World Bank, *World Governance Indicators, 2019*, http://www.govindicators.org.

98. Alan C. Acock, *A Gentle Introduction to Stata* (College Station: Stata Press, 2018), 119.

99. See an empirical approximation for covariates of governance in Cristoph Jindra and Ana Vaz, "Good Governance and Multidimensional Poverty: A Comparative Analysis of 71 Countries," *Governance* 32, no. 4 (2019): 1–19.

100. World Bank, "World Development Indicators: States and Markets, 2019," http://datatopics.worldbank.org/world-development-indicators.

101. See these conflictive views in Elaine Ciulla Kamarck and Joseph S. Nye Jr. *Governance.com: Democracy in the Information Age* (Washington, DC: Brookings Institution Press, 2002); Vyacheslav Polonski, "The Biggest Threat to Democracy? Your Social Media Feed," World Economic Forum, https://www.weforum.org/agenda/2016/08/the-biggest-threat-to-democracy-your-social-media-feed/; Philip N. Howard, *Pax Technica: How the Internet of Things May Set Us Free or Lock Us Up* (New Haven: Yale University Press, 2015); Karen Kornbluh, "The Internet's Lost Promise," *Foreign Affairs*, https://www.foreignaffairs.com/articles/world/2018-08-13/internets-lost-promise.

102. William H Dutton, "The Internet's Gift to Democratic Governance: The Fifth Estate," in *Can the Media Serve Democracy?* ed. Stephen Coleman, Giles Moss, and Katy Parry (Palgrave Macmillan: London, 2015). See also, Christopher Weare, "The Internet and Democracy: The Causal link Between Technology and Politics," *International Journal of Public Administration* 25, no. 5 (2002): 659–91.

103. John Gerring, "Case Selection for Case-Study Analysis: Qualitative and Quantitative Techniques," in *The Oxford Handbook of Political Methodology*, ed. Janet M. Box-Steffensmeier, Henry E. Brady, and David Collier (New York: Oxford University Press, 2009) 646–84.

104. Kaufmann, Kraay, and Mastruzzi, "The Worldwide Governance Indicators."
105. Ibid.
106. Ibid.
107. Ibid.
108. Ibid.
109. Ibid.
110. Susan Brenner, *Cyber Threats: The Emerging Fault Lines of the Nation State* (Oxford: Oxford University Press, 2009).
111. Tim Maurer, *Cyber Mercenaries: The State, Hackers, and Power* (Cambridge: Cambridge University Press, 2017).
112. Jon R. Lindsay, Tai Ming Cheung, and Derek S. Reveron, *China and Cybersecurity: Espionage, Strategy, and Politics in the Digital Domain* (Oxford: Oxford University Press, 2015).
113. P. W. Singer and Allan Friedman, *Cybersecurity and Cyberwar* (Oxford: Oxford University Press, 2014).
114. Valeriano and Maness, *Cyber War versus Cyber Realities*.
115. Libicki, *Cyberdeterrence and Cyberwar*, 28.
116. Valeriano and Maness, *Cyber War versus Cyber Realities*.
117. Lech J. Janczewski and William Caelli, *Cyber Conflicts and Small States* (New York: Routledge, 2016).
118. David J. Betz and Tim Stevens, *Cyberspace and the State: Towards a Strategy for Cyber-Power* (London: Routledge, 2012).
119. Myriam Dunn Cavelty, *Cyber-Security and Threat Politics: US Efforts to Secure the Information Age* (Abingdon: Routledge, 2008); Athina Karatzogianni, *Cyber-Conflict and Global Politics* (London and New York: Routledge, 2009).
120. Brian M. Mazanec and Bradley A. Thayer, *Deterring Cyber Warfare: Bolstering Strategic Stability in Cyberspace* (Basingstoke: Palgrave Macmillan, 2014).
121. On rethinking the internet governance issue, see, for instance, Ali Bhuiyan, *Internet Governance and the Global South: Demand for a New Framework* (Basingstoke: Palgrave Macmillan, 2016).

Chapter Two

1. See, for instance, Nazli Choucri, Stuart Madnick, and Jeremy Ferwerda, "Institutions for Cyber Security: International Responses and Global Imperatives," *Information Technology for Development* 20, no. 2 (2014): 96–121; and Ronald J. Deibert and Masashi Crete-Nishihata, "Global Governance and the Spread of Cyberspace Controls," *Global Governance* 18 (2012): 339–61.
2. Nazli Choucri, *Cyberpolitics in International Relations* (Cambridge: MIT Press, 2012); Desmond Ball, "Australian Cyber-warfare Centre," in *Australia and Cyber-Warfare*, ed. Gary Waters, Desmond Ball, and Ian Dudgeon (Canberra: ANU

Press, 2008), 119–48; Markus Fraundorfer, "Brazil's Organization of the NETmundial Meeting: Moving Forward in Global Internet Governance," *Global Governance* 23 (2017): 503–21; Andrew Clement and Jonathan A. Obar, "Canadian Internet 'Boomerang' Traffic and Mass NSA Surveillance: Responding to Privacy and Network Sovereignty Challenges," in *Law, Privacy and Surveillance in Canada in the Post-Snowden Era*, ed. Michael Geist (Ottawa: University of Ottawa Press, 2015), 13–44; Clement Guitton, "Cyber Insecurity as a National Threat: Overreaction from Germany, France, and the UK?" *European Security* 22, no. 1 (2013): 21–35; Matthew Crandall and Collin Allan, "Small States and Big Ideas: Estonia's Battle for Cybersecurity Norms," *Contemporary Security Policy* 36, no. 2 (2015): 346–68; Joe Burton, "Small States and Cyber Security: The Case of New Zealand," *Political Science* 65, no. 2 (2013): 216–38; James Andrew Lewis, "Advanced Experiences in Cybersecurity Policies and Practices: An Overview of Estonia, Israel, South Korea, and the United States," Discussion Paper No IDB-DP-457, Inter-American Development Bank, July 2016, https://publications.iadb.org/en/advanced-experiences-cybersecurity-policies-and-practices-overview-estonia-israel-south-korea-and; Jason Healey, "Eight Questions and Answers on U.S. Cyber Statecraft," in *Securing Cyberspace: A New Domain for National Security*, ed. Nicholas Burns and Jonathan Price (Washington, DC: Aspen Institute, 2012), 97–110.

    3. Joseph S. Nye Jr., "Deterrence and Dissuasion in Cyberspace."

    4. Richard A. Clarke and Robert K. Knake, *Cyber War: The Next Threat to National Security and What to Do About It* (New York: Harper Collins, 2010).

    5. Fred Kaplan, *Dark Territory: The Secret History of Cyber War* (New York: Simon and Schuster, 2016).

    6. See Johan Eriksson and Giampiero Giacomello, "Who Controls the Internet? Beyond the Obstinacy or Obsolescence of the State," *International Studies Review* 11, no. 1 (2009): 205–30.

    7. Thomas Rid, *Cyber War Will Not Take Place* (Oxford: Oxford University Press, 2013).

    8. Daniel Abebe, "Cyberwar, International Politics, and Institutional Design," *The University of Chicago Law Review* 83, no. 1 (2016): 1–22, 3.

    9. Jean-Marie Guéhenno, "10 Conflicts to Watch in 2017," *Foreign Policy*, January 5, 2017, http://foreignpolicy.com/2017/01/05/10-conflicts-to-watch-in-2017/.

    10. See, for instance, David Pion-Berlin, *Military Missions in Democratic Latin America* (New York: Palgrave Macmillan, 2016).

    11. Carlos Solar "Civil-military Relations and Human Security in a Post-Dictatorship," *Journal of Strategic Studies* 42, no. 3–4 (2019): 507–31; Vincent Boulanin, "Cybersecurity: A Precondition to Sustainable Information and Communication Technology–Enabled Human Development," *SIPRI Yearbook 2015: Armaments, Disarmament, and International Security*, SIPRI, accessed November 19, 2019, https://www.sipriyearbook.org/view/9780198787280/sipri-9780198787280-chapter-010-div1-072.xml.

12. Choucri, *Cyberpolitics in International Relations*, 32.

13. For a discussion on the idea of middle powers' security stance see, among a broad literature, Carsten Holbraad, *Middle Powers in International Politics* (London: Macmillan, 1984); Andrew F. Cooper, *Niche Diplomacy: Middle Powers after the Cold War* (New York: St. Martin's Press, 1997); Ronald M. Behringer, "Middle Power Leadership on the Human Security Agenda," *Cooperation and Conflict* 40, no. 3 (2005): 305–42; Andrew F. Cooper, "Testing Middle Power's Collective Action in a World of Diffuse Power," *International Journal* 71, no. 4 (2016): 529–44.

14. Anna Hayes, "Human (In)security," in *The Oxford Handbook of US National Security*, ed. Derek S. Reveron, Nikolas K. Gvosdev, and John A. Cloud (New York: Oxford University Press, 2018), 457–73.

15. See Hughes, "A Treaty for Cyberspace."

16. Choucri, *Cyberpolitics in International Relations*.

17. Chile and Mexico are the only Latin American countries in the OECD.

18. Dmitry Adamsky, "The Israeli Odyssey toward its National Cyber Security Strategy," *The Washington Quarterly* 40, no. 2 (2017): 113–27.

19. Ministerio del Interior y Seguridad de Chile, "Política Nacional de Ciberseguridad," April 27, 2017, https://biblioteca.digital.gob.cl/handle/123456789/738.

20. Chile's legal status was also updated to comply with the Convention on Cybercrime of the Council of Europe (as it has happened across the Americas, for example, in Panama). However, changing legal statuses is not as straightforward as desired, especially when multiple and independent domestic codes need to be tailored to comply between multiple and now interdependent regulatory institutions.

21. Rachel Glickhouse, "Explainer: Fighting Cybercrime in Latin America," Americas Society/Council of the Americas, November 14, 2013, http://www.as-coa.org/articles/explainer-fighting-cybercrime-latin-america.

22. Observatory of Cybersecurity in Latin America and the Caribbean, "Home," last accessed February 14, 2022, https://observatoriociberseguridad.org/#/home.

23. Observatory of Cybersecurity in Latin America and the Caribbean, "Home."

24. Martín Natalevich, "Uruguay sufrió 15 ciberataques de alta severidad durante 2016," January 23, 2017, http://www.elobservador.com.uy/uruguay-sufrio-15-ciberataques-alta-severidad-2016-n1022628.

25. See Miguel Centeno, Atul Kohli, and Deborah Yashar, *States in the Developing World* (Cambridge: Cambridge University Press, 2017).

26. See Carlos Solar, "Governance of Defense and Policymaking in Chile," *Latin American Policy* 6, no. 2 (2015): 205–25. On the middle power's role in promoting a human security agenda see Ronald M. Behringer, *The Human Security Agenda* (London: Continuum, 2012). On the overall role of "power" in the international system, see Moisés Naím, Katherine Pollock, and Sabin Ray, "The Decay of Power in Domestic and International Politics," *The Brown Journal of World Affairs* 21, no. 1 (2014): 245–51.

27. If we zoom in on the case of Mexico, for example, we can see Carlos Solar, "State, Violence and Security in Mexico: Developments and Consequences

for Democracy," *Mexican Studies*, 30, no. 1 (2014): 241–55; Mónica Serrano, "Del 'Momento Mexicano' A la Realidad de la Violencia Político-Criminal," *Foro Internacional* 60, no. 2 (2020): 791–852; and "States of Violence: State-Crime Relations in Mexico," In *Violence, Coercion, and State Making in Twentieth Century Mexico*, ed. Wil G. Pansters (Stanford, Stanford: University Press, 2012), 135–58.

28. See Elisa Lopez-Lucia, "Regional Powers and Regional Security Governance: An Interpretive Perspective on the Policies of Nigeria and Brazil," *International Relations* 29, no. 3 (2015): 348–62.

29. See Carlos Solar, "Police Bribery: Is Corruption Fostering Dissatisfaction with the Political System?" *Democracy and Security* 11, no. 4 (2015): 373–94.

30. Milton Mueller, Andreas Schmidt, and Brenden Kuerbis, "Internet Security and Networked Governance in International Relations," *International Studies Review* 15, no. 1 (2013): 86–104.

31. See Jacob Torfing, B. Guy Peters, Jon Pierre, and Eva Sorensen, *Interactive Governance: Advancing the Paradigm* (Oxford: Oxford University Press, 2012).

32. For a recent review of the situation in civil-military relations in Latin America, see David Pion-Berlin and Rafael Martínez, *Soldiers, Politicians, and Citizens: Reforming Civil-Military Relations in Latin America* (New York: Cambridge University Press, 2017).

33. For an updated revision of the securitization theory and the case of cybersecurity, see Thierry Balzacq, Sarah Léonard, and Jan Ruzicka, "'Securitization' Revisited: Theory and Cases," *International Relations* 30, no. 4 (2016): 494–531.

34. See Berlin, *Military Missions*; and, Solar, "Civil-Military Relations and Human Security."

35. Countries included in the AmericasBarometer 2018/19 round were Mexico, Guatemala, El Salvador, Honduras, Nicaragua, Panama, Colombia, Ecuador, Bolivia, Peru, Paraguay, Chile Uruguay, Brazil, Argentina, Dominican Republic, Jamaica, United States, and Canada. See Latin American Public Opinion Project (LAPOP), "AmericasBarometer Data Sets 2018/19," The AmericasBarometer, accessed November 29, 2020, https://www.vanderbilt.edu/lapop/raw-data.php.

36. Observatory of Cybersecurity in Latin America and the Caribbean, "Home."

37. Frankie E. Catota, M. Granger Morgan, and Douglas C. Sicker, "Cybersecurity Education in a Developing Nation: The Ecuadorian Environment," *Journal of Cybersecurity* 5, no. 1 (2019): 1–19.

38. See Rebecca Slayton, "What Is the Cyber Offense-Defense Balance? Conceptions, Causes, and Assessment," *International Security* 41, no. 3 (2016): 72–109.

39. Herbert Lin, "A Virtual Necessity: Some Modest Steps toward Greater Cybersecurity," *Bulletin of the Atomic Scientists* 68, no. 5 (2012): 75–87, 76.

40. See Paul Cornish and Kingsley Donaldson, *2020 World of War* (London: Hodder and Stoughton, 2017), 11–36, 24.

41. Paulo Vicente Alves, "Why the Chinese Are Getting Aggressive in South America," July 6, 2015, http://thehill.com/blogs/congress-blog/foreign-policy/245208-why-the-chinese-are-getting-aggressive-in-south-america.

42. "Maduro Contra la Pared," *Semana*, September 17, 2017, 11, https://www.semana.com/china-y-rusia-quieren-demostrar-que-amenazas-de-trump-son-vacias/540741/.

43. Nikolas K. Gvosdev, "The Bear Goes Digital: Russia and Its Cyber Capabilities," in *Cyberspace and National Security: Threats, Opportunities, and Power in a Virtual World*, ed. Derek S. Reveron (Washington, DC: Georgetown University Press, 2012), 173–89.

44. For a specific account on Latin America and U.S. influence see Joseph S. Tulchin, *Latin America in International Politics: Challenging US Hegemony* (Boulder: Lynne Rienner, 2016). For a broader conversation on the influence of the superpowers in the modern international system, see, for example, Michael Cox, "Power Shifts, Economic Change and the Decline of the West?" *International Relations* 26, no. 4 (2012): 369–88.

45. Organization of American States and Symantec, *Tendencias de Seguridad Cibernética en America Latina y el Caribe*, accessed May 1, 2019, http://www.symantec.com/content/es/mx/enterprise/other_resources/b-cyber-security-trends-report-lamc.pdf.

46. David R. Mares, *Latin America and the Illusion of Peace* (Abingdon: Routledge, 2012).

47. Derek S. Reveron, "An Introduction to National Security and Cyberspace," in *Cyberspace and National Security: Threats, Opportunities, and Power in a Virtual World*, ed. Derek S. Reveron (Washington, DC: Georgetown University Press, 2012), 3–19, 3.

48. See, for instance, Andrew Clement and Jonathan A. Obar, "Canadian Internet 'Boomerang' Traffic and Mass NSA Surveillance: Responding to Privacy and Network Sovereignty Challenges," in *Law, Privacy, and Surveillance in Canada in the Post-Snowden Era*, ed. Michael Geist (Ottawa: University of Ottawa Press, 2015), 13–44; and Eneken Tikk, "Ten Rules for Cyber Security," *Survival* 53, no. 3 (2011): 119–32.

49. Glickhouse, "Explainer: Fighting Cybercrime in Latin America."

50. Robert Muggah and Misha Glenny, "Why Brazil Put Its Military in Charge of Cyber Security," January 13, 2015, http://www.defenseone.com/technology/2015/01/why-brazil-put-its-military-charge-cyber-security/102756/.

51. Jay H. Anson, "United to Counter Cyber Crime," May 19, 2017, https://dialogo-americas.com/en/articles/united-counter-cyber-crime.

52. Organization of American States and Symantec, *Tendencias de Seguridad Cibernética*.

53. BBC Mundo, "¿Por qué Aumentan los Delitos Cibernéticos en América Latina?" November 24, 2015, http://www.bbc.com/mundo/noticias/2015/11/151118_tecnologia_cibercrimen_ciberdelito_aumento_america_latina_lb.

54. Organization of American States and Symantec, *Tendencias de Seguridad Cibernética*.

55. Jeanne Shaheen, "The Russian Company That Is a Danger to Our Security," September 4, 2017, https://www.nytimes.com/2017/09/04/opinion/kaspersky-russia-cybersecurity.html?_r=0.

56. U.S. Immigration and Customs Enforcement, "Russian Cyber-Criminal Pleads Guilty to Role in Multimillion Dollar Online Identity Theft Ring," August 9, 2017, https://www.ice.gov/news/releases/russian-cyber-criminal-pleads-guilty-role-multimillion-dollar-online-identity-theft.

57. Israel Rodríguez, "Reporta Condusef Récord de Presuntos Fraudes," July 18, 2017, http://www.jornada.com.mx/ultimas/2017/07/18/reporta-condusef-record-de-presuntos-fraudes.

58. "Colombia También es Víctima del Ataque Global Informático," *El Tiempo*, May 13, 2017, http://www.eltiempo.com/tecnosfera/novedades-tecnologia/ataque-cibernetico-afecta-redes-de-74-paises-87390.

59. Benjamin Mattern, "Cyber Security and Hacktivism in Latin America: Past and Future," July 24, 2014, http://www.coha.org/cyber-security-and-hacktivism-in-latin-america-past-and-future/.

60. Interpol, "Hackers Reportedly Linked to 'Anonymous' Group Targeted in Global Operation Supported," February 28, 2012, https://www.interpol.int/News-and-media/News/2012/PR014.

61. Governments that reported no increase in budgets for cybersecurity were Argentina, Barbados, Brazil, Colombia, Dominica, Ecuador, El Salvador, Grenada, Guatemala, Mexico, Nicaragua, Panama, Paraguay, Peru, and the United States. See, Organization of American States, *Report on Cybersecurity and Critical Infrastructure in the Americas*, https://www.sites.oas.org/cyber/Documents/2015%20-%20OAS%20Trend%20Micro%20Report%20on%20Cybersecurity%20and%20CIP%20in%20the%20Americas.pdf.

62. See "Wannacry Cyberattack Numbers in Latin America," *Latin American Post*, May 16, 2017, http://latinamericanpost.com/index.php/living-people/living-future/15225-wannacry-cyberattack-numbers-in-latin-america; and Sam Jones, "Timeline: How the WannaCry Cyber-attack Spread," *Financial Times*, May 14, 2017, https://www.ft.com/content/82b01aca-38b7-11e7-821a-6027b8a20f23.

63. "Latinoamérica Está Expuesta a Ataques Cibernéticos," *Notimex*, June 3, 2017, http://www.excelsior.com.mx/hacker/2017/06/03/1167503.

64. William Hague, "Arming the Information Highway Patrol," *The World Today* 68, no. 7 (2012): 35–36, 35. In 2016, the UK launched its primary National Cyber Security Strategy, available at https://assets.publishing.service.gov.uk/government/uploads/system/uploads/attachment_data/file/567242/national_cyber_security_strategy_2016.pdf.

65. It was reported that the ransomware WannaCry had once been a spying tool known as EternalBlue, belonging to the U.S. National Security Agency (NSA). See Jones, "Timeline: How the WannaCry Cyber Attack Spread."

66. Abby Ohlheiser, "Brazil and Mexico Ask the U.S. to Explain Reports of NSA Spying on Their Leaders," *The Atlantic*, September 2, 2013, https://www.theatlantic.com/international/archive/2013/09/brazil-and-mexico-want-us-explain-reports-nsa-spying-their-presidents/311546/.

67. Lily Hay Newman, "The US Gives Cyber Command the Status it Deserves," *Wired*, August 19, 2017, https://www.wired.com/story/cyber-command-elevated/.

68. James Bamford, "Commentary: The World's Best Cyber Army Doesn't Belong to Russia," August 4, 2016, http://www.reuters.com/article/us-election-intelligence-commentary/commentary-the-worlds-best-cyber-army-doesnt-belong-to-russia-idUSKCN10F1H5.

69. Scott Shane, "To Sway Vote, Russia Used Army of Fake Americans," *The New York Times*, September 8, 2017, A1.

70. Sepúlveda was charged for the use of malicious software, conspiracy, violation of personal data, and espionage. See Jordan Robertson, Michael Riley, and Andrew Willis, "How to Hack an Election," *Bloomberg*, March 31, 2016, https://www.bloomberg.com/features/2016-how-to-hack-an-election/.

71. See this discussion in Joshua Rovner and Tyler Moore, "Does the Internet Need an Hegemon," *Journal of Global Security Analysis* 2, no. 3 (2017): 184–203; and James A. Green, *Cyber Warfare: A Multidisciplinary Analysis* (Abingdon and New York: Routledge, 2015).

72. John Lee, "Cyber Kleptomaniacs: Why China Steals Our Secrets," *World Affairs* 176, no. 3 (2013): 73–79.

73. "Russia Hacked Danish Defense for Two Years, Minister Tells Newspaper," *Reuters*, April 23, 2017, http://www.reuters.com/article/us-denmark-security-russia/russia-hacked-danish-defense-for-two-years-minister-tells-newspaper-idUSKBN-17P0NR.

74. Ellen Nakashima, "Chinese Hackers Who Breached Google Gained Access to Sensitive Data, U.S. Officials Say," *The Washington Post*, May 20, 2013, https://www.washingtonpost.com/world/national-security/chinese-hackers-who-breached-google-gained-access-to-sensitive-data-us-officials-say/2013/05/20/51330428-be34-11e2-89c9-3be8095fe767_story.html?utm_term=.a19c779acaad.

75. See Ethan Gutmann, "Hacker Nation: China's Cyber Assault," *World Affairs* 173, no. 1 (2010): 70–79.

76. Mark Bromley, Kolja Brockmann, and Giovanna Maletta. "Controls on Intangible Transfers of Technology and Additive Manufacturing," *SIPRI Yearbook 2018: Armaments, Disarmament, and International Security*, accessed May 1, 2019, https://www.sipriyearbook.org/view/9780198821557/sipri-9780198821557.xml.

77. Dan Smith "Introduction: International Security, Armaments, and Disarmament," *SIPRI Yearbook 2017: Armaments, Disarmament, and International Security*, accessed November 19, 2019, https://www.sipriyearbook.org/view/9780198811800/sipri-9780198811800-chapter-1-div1-6.xml.

78. Ibid.

79. Ibid.

80. LulzSecPeru was indicated as a home-grown version of the U.S. and UK-based LulzSec, a collective dedicated to hacktivist hits that have targeted public and private organizations worldwide.

81. Kyra Gurney, "Infiltration of Chile Air Force Emails Highlights LatAm Cyber Threats," *Insight Crime*, August 15, 2014, http://www.insightcrime.org/news-briefs/chile-air-force-peru-hackers-cyber-threat.

82. Most recently, Chile also had bilateral rivalries with Argentina, for example, over the Beagle Channel dispute in 1977–1984 when both military dictatorships almost went to war. However, interstate relations have improved considerably since the return to democracy and defense partnerships have been set up, unlike Chile's relations with neighbors Peru and Bolivia. See, Andrés Villar Gertner, "The Beagle Channel Frontier Dispute between Argentina and Chile: Converging Domestic and International Conflicts," *International Relations* 28, no. 2 (2014): 207–27; and, Solar, "Civil-Military Relations."

83. Miriam Wells, "Brazil Investigating Hack of Military Police Data," September 17, 2013, *Insight Crime*, http://www.insightcrime.org/news-briefs/hackers-publish-personal-details-of-50000-rio-police.

84. "Hackearon la Página del Ejército Argentino con Supuestas Amenazas de ISIS," *Infobae*, June 19, 2017, http://www.infobae.com/politica/2017/06/19/hackearon-la-pagina-del-ejercito-argentino-con-supuestas-amenazas-de-isis/.

85. Organization of American States, "Report on Cybersecurity and Critical Infrastructure in the Americas," accessed November 19, 2019, https://www.sites.oas.org/cyber/Documents/2015%20-%20OAS%20Trend%20Micro%20Report%20on%20Cybersecurity%20and%20CIP%20in%20the%20Americas.pdf.

86. Roland Heickero, "Focus on Cyber Security, Cyber Terrorism: Electronic Jihad," *Strategic Analysis* 38, no. 4 (2014): 554–65.

87. Gabriel Weimann, "Cyberterrorism: The Sum of All Fears?" *Studies in Conflict & Terrorism* 28, no. 2 (2005): 129–49. See also, Myriam Dunn Cavelty, "Cyber-Terror—Looming Threat or Phantom Menace? The Framing of the US Cyber-Threat Debate," *Journal of Information Technology & Politics* 4, no, 1 (2008): 19–36.

88. Wendy H. Wong and Peter A. Brown, "E-Bandits in Global Activism: WikiLeaks, Anonymous, and the Politics of No One," *Perspectives on Politics* 11, no. 4 (2013): 1015–33.

89. Jay Weaver, "Will Ex-President Be Sent Home to Panama to Face Charges? His Fate up to Miami Judge," *Miami Herald*, August 3, 2017, http://www.miamiherald.com/news/nation-world/article165290732.html.

90. "Growing Scandal in Latin America over "Pegasus" Spy-hacking Program," *Univision*, June 21, 2017, http://www.univision.com/univision-news/latin-america/growing-scandal-in-latin-america-over-pegasus-spy-hacking-program.

91. Organization of Americas States and Inter-American Development Bank, *Cybersecurity: Are We Ready in Latin America and the Caribbean?* accessed November

19, 2019, available at https://publications.iadb.org/en/cybersecurity-are-we-ready-latin-america-and-caribbean.

92. See Dave Clemente, "International Security: Cyber Security as a Wicked Problem," *The World Today* 67, no. 10 (2011): 15–17.

93. U.S. Department of State, "Joint Statement on U.S.-Argentina Partnership on Cyber Policy," April 27, 2017, https://www.state.gov/r/pa/prs/ps/2017/04/270496.htm.

94. Myriam Ortega, "Colombian Armed Forces Counter Cybercrime," November 9, 2017, https://dialogo-americas.com/articles/colombian-armed-forces-counter-cybercrime/#.YguHzC10d0s.

95. Gobierno Bolivariano de Venezuela, "Ciberdefensa: Comando Operacional Estratégico," accessed November 19, 2019, https://ceofanb.mil.ve/direcciones/direccion-conjunta-de-ciberdefensa/.

96. Matt Spetalnick, "Russian Deployment in Venezuela Includes 'Cybersecurity Personnel': U.S. Official," March 26, 2019, https://www.reuters.com/article/us-venezuela-politics-russians/russian-deployment-in-venezuela-includes-cybersecurity-personnel-u-s-official-idUSKCN1R72FX?feedType=RSS&feedName=worldNews.

97. Observatory of Cybersecurity in Latin America and the Caribbean, "Home."

98. Carlos Solar, "The Inter-Institutional Governance of Money Laundering: An In-Depth Look at Chile Following Re-Democratisation," *Global Crime* 16, no. 4 (2015): 328–50.

99. Tim Stevens, "A Cyberwar of Ideas? Deterrence and Norms in Cyberspace," *Contemporary Security Policy* 33, no. 1 (2012): 148–70.

100. Markus Fraundorfer, "Brazil's Organization of the NETmundial Meeting: Moving Forward in Global Internet Governance," *Global Governance* 23, no. 3 (2017): 503–21.

101. Ibid., 504–509; see also Paul Cornish, "Governing Cyberspace through Constructive Ambiguity," *Survival* 57, no. 3 (2015): 153–76.

102. Aaron Franklin Brantly, "The Cyber Losers," *Democracy and Security* 10, no. 2 (2014): 132–55, 132.

103. Victor Platt, "Still the Fire-Proof House? An Analysis of Canada's Cyber Security Strategy," *International Journal* 67, no. 1 (2011): 155–67.

104. See Nigel Inkster, "China in Cyberspace," in *Cyberspace and National Security: Threats, Opportunities, and Power in a Virtual World*, ed. Derek S. Reveron (Georgetown University Press, 2012) 191–205; Jinghan Zeng, Tim Stevens, Yaru Chen, "China's Solution to Global Cyber Governance: Unpacking the Domestic Discourse of 'Internet Sovereignty,'" *Politics & Policy* 45, no. 3 (2017): 432–64; and, Jon R. Lindsay, "The Impact of China on Cybersecurity Fiction and Friction," *International Security* 39, no. 3 (2014): 7–47.

105. Eric Schmidt and Jared Cohen, "The Digital Disruption: Connectivity and the Diffusion of Power," *Foreign Affairs* 89, no. 6 (2010): 75–85, 76.

106. See what going from "analogue" to "digital" cybersecurity thinking demands in, for example, David J. Betz and Tim Stevens, "Analogical Reasoning and Cyber Security," *Security Dialogue* 44, no. 2 (2013): 147–64.

107. Saltzman, "Cyber Posturing."

108. Gary McGraw, "Cyber War Is Inevitable (Unless We Build Security In)," *Journal of Strategic Studies* 36, no. 1 (2013): 109–19.

109. Richard J. Harknett and James A. Stever, "The New Policy World of Cybersecurity," *Public Administration Review* 71, no. 3 (2011): 455–60.

110. Nye Jr., "Deterrence and Dissuasion."

111. Ibid.; Slayton, "What is Cyber Offense-Defense Balance?"; Gartzke, "The Myth of Cyberwar."

112. Henry Farrell and Martha Finnemore, "The End of Hypocrisy: American Foreign Policy in the Age of Leaks," *Foreign Affairs* 92, no. 6 (2013): 22–26.

113. Samantha Bradshaw and Philip N. Howard, "The Global Disinformation Order," accessed March 1, 2020, https://comprop.oii.ox.ac.uk/wp-content/uploads/sites/93/2019/09/CyberTroop-Report19.pdf.

114. See Robert O. Keohane and Joseph S. Nye Jr., "Power and Interdependence in the Information Age," *Foreign Affairs* 77, no. 5 (1998): 81–94.

115. Richard B. Andres, "The Emerging Structure of Strategic Cyber Offense, Cyber Defense, and Cyber Deterrence," in *Cyberspace and National Security: Threats, Opportunities, and Power in a Virtual World*, ed. Derek S. Reveron (Washington, DC: Georgetown University Press, 2012), 89–104.

116. Mark Galeotti, "The Cyber Menace," *The World Today* 68, no. 7 (2012): 32–35.

Chapter Three

1. Deb Reichmann and Associated Press, "President Trump Reduces Restrictions on U.S. Responding Offensively to Cyberattacks," October 31, 2018, http://time.com/5441193/trump-offensive-cyberattacks-bolton/.

2. Samuel P. Huntington, "Conventional Deterrence and Conventional Retaliation in Europe," *International Security* 8, no. 3 (1983/1984): 32–56.

3. Thomas Rid and Peter McBurney, "Cyber-Weapons," *The RUSI Journal* 157, no. 1 (2012): 6–13.

4. See Jeffrey Carr, "The Misunderstood Acronym: Why Cyber Weapons Aren't WMD," *Bulletin of the Atomic Scientists* 69, no. 5 (2013): 32–37.

5. Kello, "The Meaning of the Cyber Revolution," 23.

6. Ibid.

7. See this fivefold taxonomy in Ioannis Agrafiotis, Jason R. C. Nurse, Michael Goldsmith, Sadie Creese, and David Upton, "A Taxonomy of Cyber-Harms:

Defining the Impacts of Cyber-Attacks and Understanding How They Propagate," *Journal of Cybersecurity* 4, no. 1 (2018): 1–15.

8. John R. Lindsay and Eric Gartzke, "Politics by Many Other Means: The Comparative Strategic Advantages of Operations Domains," *Journal of Strategic Studies* (2020), https://doi.org/10.1080/01402390.2020.1768372.

9. See a particularly useful discussion on conventional and nonconventional weapons use in Rebecca Johnson, "Arms Control and Disarmament Diplomacy," in *The Oxford Handbook of Modern Diplomacy*, ed. Andrew F. Cooper, Jorge Heine, and Ramesh Thakur (New York: Oxford University Press, 2013), 594–613.

10. Valeriano and Maness, *Cyber War versus Cyber Realities*, 21.

11. I make this claim throughout the book based on Thomas C. Bruneau and Florina Cristiana Matei, "Towards a New Conceptualization of Democratization and Civil-Military Relations," *Democratization* 15, no. 5 (2008): 909–29; and Thomas Bruneau and Aurel Croissant, "Civil-Military Relations: Why Control Is Not Enough," in *Civil Military Relations: Control and Effectiveness Across Regimes*, ed. Thomas C. Bruneau and Aurel Croissant (Boulder: Lynne Rienner, 2019), 1–18.

12. Joel F. Brenner, "Eyes Wide Shut: The Growing Threat of Cyber Attacks on Industrial Control Systems," *Bulletin of the Atomic Scientists* 69, no. 5(2013): 15–20, 16.

13. Raul Gouvea, "US–Latin America's Security: Moving through an Inflection Point?" *Defense & Security Analysis* 33, no. 3(2017): 223–41.

14. Daron Acemoglu and Pierre Yared, "Political Limits to Globalization," *American Economic Review: Papers and Proceedings* 100 (2010): 83–88.

15. See ibid.

16. I used the World Bank's grouping of countries under each of these regions.

17. Brazil accounted for 51 percent of total military expenditure in South America. See Nan Tian, Alexandra Kuimova, Diego Lopes da Silva, Pieter D. Wezeman, and Siemon T. Wezeman, "Trends in World Military Expenditure," SIPRI Fact Sheet, accessed November 29, 2020, https://www.sipri.org/sites/default/files/2020-04/fs_2020_04_milex_0_0.pdf.

18. Here the literature is extensive. One article worth pointing out is Abraham F. Lowenthal and Sergio Bitar, "Getting to Democracy: Lessons from Successful Transitions," *Foreign Affairs* 95, no. 1 (2016): 134–44.

19. See Valeriano and Maness, *Cyber War versus Cyber Realities*, 26.

20. Ibid., 25.

21. Todd S. Sechser, Neil Narang, and Caitlin Talmadge, "Emerging Technologies and Strategic Stability in Peacetime, Crisis, And War," *Journal of Strategic Studies* 42, no. 6 (2019): 727–35.

22. Ibid.

23. Stephen Biddle and Robert Zirkle, "Technology, Civil-Military Relations, and Warfare in The Developing World," *Journal of Strategic Studies* 19, no. 2 (1996): 171–212.

24. Marc R. DeVore, "Commentary on the Value of Domestic Arms Industries: Security of Supply or Military Adaptation?" *Defence Studies* 17, no. 3 (2017): 242–59. See pages 243–44.

25. Çağlar Kurç and Stephanie G. Neuman, "Defence Industries in the 21st Century: A Comparative Analysis," *Defence Studies* 17, no. 3 (2017): 219–27. See pages 219–20.

26. Robert Muggah and John P. Sullivan, "The Coming Crime Wars," September 21, 2018, https://foreignpolicy.com/2018/09/21/the-coming-crime-wars/.

27. Ibid.

28. Kello, "The Meaning of the Cyber Revolution."

29. Valeriano and Maness, *Cyber War versus Cyber Realities*, 21–22.

30. *The Economist*, "Defence Companies Target the Cyber-Security Market," July 26, 2018, https://www.economist.com/business/2018/07/26/defence-companies-target-the-cyber-security-market.

31. Ibid.

32. Heinz Schulte, "Industry and War," in *The Oxford Handbook of War*, ed. Ives Boyer and Julian Lindley-French (New York: Oxford University Press, 2012), 518–29.

33. Ibid.

34. Ibid., 529.

35. Joseph S. Nye Jr., "Nuclear Lessons for Cyber Security?" *Strategic Studies Quarterly* 5, no. 4 (2011): 18–38, 19.

36. Valeriano and Maness, *Cyber War versus Cyber Realities*, 23.

37. Ibid., 27–28.

38. Richard K. Betts, "The Lost Logic of Deterrence: What the Strategy That Won the Cold War Can—and Can't—Do Now," *Foreign Affairs* 92, no. 2 (2013): 87–99, 88.

39. Lawrence Freedman, "Deterrence: A Reply," *Journal of Strategic Studies* 28, no. 5 (2005): 789–801.

40. Carlos Solar, "Reassessing Chilean International Security," *International Politics* 56, no. 5 (2019): 569–84; and Monica Herz, "Concepts of Security in South America," *International Peacekeeping* 17, no. 5 (2010): 598–612, 601, 604.

41. John Chipman and James Lockhart Smith, "South America: Framing Regional Security," *Survival* 51, no. 6 (2009): 77–104.

42. J. Paul Dunne, Sam Perlo-Freeman, and Ron P. Smith, "The Demand for Military Expenditure in Developing Countries: Hostility Versus Capability," *Defence and Peace Economics* 19, no. 4 (2008): 293–302.

43. Valeriano and Maness, *Cyber War versus Cyber Realities*, 20–21.

44. Ibid., 20–21.

45. J. P. Clark, "Military Operations and the Defense Department," in *The Oxford Handbook of U.S. National Security*, ed. Derek Reveron, Nikolas K. Gvosdev, and John A. Cloud (New York: Oxford University Press, 2018), 292–302.

46. I rescue this idea from Andres Eduardo Fernandez-Osorio, Fabio Nelson Cufiño-Gutierrez, Cesar Andres Gomez-Diaz, and Gustavo Andres Tovar-Cabrera, "Dynamics of State Modernization in Colombia: The Virtuous Cycle of Military Transformation," *Democracy and Security* 15, no. 1 (2019): 75–104.

47. J. P. Clark, "Military Operations and the Defense Department," 291.

48. Fernandez-Osorio, Cufiño-Gutierrez, Gomez-Diaz, and Tovar-Cabrera, "Dynamics of State Modernization in Colombia."

49. See Bruneau and Matei, "Towards a New Conceptualization," and Thomas C. Bruneau, "Civilians and the Military in Latin America: The Absence of Incentives," *Latin American Politics and Society* 55, no. 4 (2013): 143–60.

50. Valeriano and Maness, *Cyber War versus Cyber Realities*, 36.

51. Joseph S. Nye Jr., "Hard, Soft, and Smart Power," in *The Oxford Handbook of Modern Diplomacy*, ed. Andrew F. Cooper, Jorge Heine, and Ramesh Thakur (New York: Oxford University Press, 2013), 560–72.

52. Dan Smith, "Dual-Use and Arms Trade Controls," *SIPRI Yearbook 2018: Armaments, Disarmament, and International Security*, accessed November 12, 2019, https://www.sipriyearbook.org/view/9780198821557/sipri-9780198821557-chapter-10-div1-019.xml.

53. Smith, "Dual-Use and Arms Trade Controls."

54. Monica Herrera and Ron Matthews, "Latin America in Step with Global Defence Offset Phenomenon," *The RUSI Journal* 159, no. 6 (2014): 50–57, 53.

55. Jose Carlos Albano Amarante and Patrice Franko, "Defense Transformation in Latin America: Will It Transform the Technological Base?" *Democracy and Security* 13, no. 3 (2017): 173–95, 173.

56. Albano Amarante and Franko, "Defense Transformation," 174–75.

57. Raul Gouvea, "Brazil's New Defense Paradigm," *Defense & Security Analysis* 31, no. 2 (2015): 137–51, see 143–47.

58. Karl R. DeRouen Jr., "Defense Spending and Economic Growth in Latin America: The Externalities Effects," *International Interactions* 19, no. 3 (1994): 193–212, see 197–98.

59. Jose Miguel Benavente, Gustavo Crespi, Jorge Katz, and Giovanni Stumpo, "New Problems and Opportunities for Industrial Development in Latin America," *Oxford Development Studies* 25, no. 3 (1997): 261–77, 263.

60. J. Paul Dunne, Sam Perlo-Freeman, and Aylin Soydan, "Military Expenditure and Debt in South America," *Defence & Peace Economics* 15, no. 2 (2004): 173–87.

61. See a thoughtful discussion on the topic in Thomas C. Bruneau, *Patriots for Profit: Contractors and the Military in U.S. National Security* (Stanford: Stanford University Press, 2011).

62. Joseph Cerami and John David Young, "Military Science," *Oxford Bibliographies*, DOI: 10.1093/OBO/9780199743292-0052.

63. Bruce Berkowitz, *The New Face of War: How War Will Be Fought in the 21st Century* (New York: Free Press, 2003).

64. See Ofer Fridman, *Russian "Hybrid Warfare": Resurgence and Politicisation* (New York: Oxford University Press, 2018).
65. Clark, "Military Operations," 293–94.
66. Ibid., 294.
67. Smith, "Introduction: International Security, Armaments and Disarmament."
68. See this point in Kurç and Neuman, "Defence Industries," 223.
69. Kenneth E. Sharpe, "The Military, The Drug War, and Democracy in Latin America: What Would Clausewitz Tell Us?" *Small Wars & Insurgencies* 4, no. 3 (1993): 56–86, 263.
70. Mark Bromley and Alfredo Malaret, "ATT-Related Activities in Latin America and the Caribbean: Identifying Gaps and Improving Cooperation," February, 2017, https://www.sipri.org/sites/default/files/ATT-related-activities-Latin-America-and-Caribbean.pdf.
71. See this discussion through the Colombian example in Fabio Andres Diaz and Syed Mansoob Murshed, "'Give War a Chance': All-Out War as a Means of Ending Conflict in the Cases of Sri Lanka and Colombia," *Civil Wars* 15, no. 3 (2013): 281–305.
72. Valeriano and Maness, *Cyber War versus Cyber Realities*, 37–38.
73. For some regions the most recent values in all three indicators correspond to 2017, but not all regions have up-to-date figures. No data were recorded for researchers in R&D in Africa. See World Bank (2019).
74. Valeriano and Maness, *Cyber War versus Cyber Realities*, 23.
75. Kurç and Neuman, "Defence Industries," 221–22.
76. See Tyler Moore and Ross Anderson, "Internet Security," in *The Oxford Handbook of the Digital Economy*, ed. Martin Peitz and Joel Waldfogel (New York: Oxford University Press), 572–600.
77. Richard Shirreff, "Conducting Joint Operations," in *The Oxford Handbook of War*, ed. Yves Boyer and Julian Lindley-French (New York: Oxford University Press), 374–84.
78. Nye Jr. "Deterrence and Dissuasion," 63.
79. Elisabeth Braw and Gary Brown. "Personalised Deterrence of Cyber Aggression." *RUSI Journal* 154, no. 2 (2020): 48–54, 5.
80. I take these ideas from Huntington, "Conventional Deterrence," 36.
81. Richard K. Betts, "The Lost Logic of Deterrence: What the Strategy That Won the Cold War Can—and Can't—Do Now," *Foreign Affairs* 92, no. 2 (2013): 87–99, 88.
82. Slayton, "What Is the Cyber Offense-Defense Balance?," 82.
83. Sadie Creese, Ruth Shillair, Maria Bada, and Willian H. Dutton, "Building the Cybersecurity Capacity of Nations," in *Society and the Internet: How Networks of Information and Communication are Changing Our Lives*, ed. Mark Graham and William H. Dutton (Oxford: Oxford University Press, 2019), 166–79.
84. Business Monitor International, "Defence & Security Report, Americas," July 7, 2015, EBSCOhost.

85. Business Monitor International, "Latin America Security Risk Ratings," July 7, 2014, EBSCOhost.
86. Slayton, 'What Is the Cyber Offense-Defense Balance?" 83.

Chapter Four

1. Hal Brands, "Crime, Irregular Warfare, and Institutional Failure in Latin America: Guatemala as a Case Study," *Studies in Conflict & Terrorism* 34, no. 3 (2011): 228–47.
2. James Bargent, "Top Paraguay Official Reveals Arms Trafficking Modus Operandi," *Insight Crime*, March 10, 2016, https://www.insightcrime.org/news/brief/top-paraguay-official-reveals-modus-operandi-arms-trafficking/.
3. Mark Pyman, "Addressing Corruption in Military Institutions," *Public Integrity* 19, no. 5 (2017): 513–28.
4. C. H. Gardiner, "Rising Brazil Border Seizures Troubling Sign for the Region," *Insight Crime*, March 5, 2018, https://www.insightcrime.org/news/brief/brazil-frontier-region-busts/.
5. Nir Kshetri, *Cybercrime and Cybersecurity in the Global South* (London: Palgrave Macmillan, 2013), 135.
6. Phil Williams and Sandra Quincoses, *The Evolution of Threat Networks in Latin America*, accessed February 20, 2022, https://gordoninstitute.fiu.edu/research/publications/evolution-of-threat-networks-in-latam.pdf, 3.
7. Anastasia Austin, "Latin America Under Threat of Cybercrime amid Coronavirus," *Insight Crime*, April 8, 2020, https://insightcrime.org/news/analysis/threat-cyber-crime-coronavirus/.
8. I discuss the MINUSTAH further in Carlos Solar, "Chile's Peacekeeping and the Post-UN Intervention Scenario in Haiti," *International Studies* 56, no. 4 (2019): 272–91. See also, Arturo Sotomayor, *The Myth of the Democratic Peacekeeper: Civil-Military Relations and the United Nations* (Baltimore: Johns Hopkins University Press, 2014)
9. Business Monitor International, "Latin America Security Overview," July 7, 2015, EBSCOhost.
10. Valeriano and Maness, *Cyber War versus Cyber Realities*, 32.
11. I find that the 2017 U.S. National Security Strategy encompassed many of the aspects that Washington would like to see reflected in Latin American cybersecurity governance.
12. W. Alejandro Sanchez Nieto, "Whatever Happened to South America's Splendid Little Wars," *Small Wars & Insurgencies* 22, No. 2 (2011): 329–35.
13. Arend Lijphart, "The Comparable Cases Strategy in Comparative Research," *Comparative Political Studies* 8, no. 2 (1975): 158–77.
14. James Mahoney, "Qualitative Methodology and Comparative Politics," *Comparative Political Studies* 40, no. 2 (2007): 122–44.

15. Ibid., 129.

16. "Argentina Buys Five Used Fighter Jets from French Navy," *Reuters*, May 11, 2018, https://www.reuters.com/article/us-argentina-military-france-idUKKB-N1IC1OP.

17. Juan Battaleme, and Francisco de Santibañes, "Argentina's Defence Deficit," *Survival* 61, no. 4 (2019): 63–78, 63–64. According to the authors, Argentina's strategic technologies, including nuclear power and satellites, "make[s] the decline of its military so incongruous and significant."

18. Business Wire, "Future of the Argentina Defense Industry to 2022—Market Attractiveness, Competitive Landscape, and Forecasts—Research and Markets," June 13, 2017, https://www.businesswire.com/news/home/20170713005850/en/Future-Argentina-Defense-Industry-2022—Market.

19. Ben Wolfgang, "U.S., to Provide Military Security for the G-20 Summit in Buenos Aires," *The Washington Post*, August 16, 2018, https://www.washingtontimes.com/news/2018/aug/16/us-provide-military-security-g-20-summit-argentina/.

20. U.S. Department of State, "Joint Statement on U.S.-Argentina Partnership."

21. See a thorough account of U.S.-Argentina Relations during the second half of the twentieth century in Deborah L. Norden, *Argentina and the United States: Changing Relations in a Changing World* (New York: Routledge Press, 2002).

22. RT, "Argentina, Brazil Agree on a Cyber-Defense Alliance Against US Espionage," September 15, 2013, https://www.rt.com/news/brazil-argentina-cyber-defense-879/.

23. Privacy International, "State of Privacy: Argentina," January 23, 2019, https://privacyinternational.org/state-privacy/57/state-privacy-argentina.

24. Business Monitor International, "BMI Defence & Security Report, Brazil," February 3, 2015, EBSCOhost.

25. Ibid.

26. U.S. Department of State, "U.S-Brazil Bilateral Cooperation on Cyber and Internet Policy," May 28, 2018, https://www.state.gov/r/pa/prs/ps/2018/05/282074.htm.

27. Frank Mora and Brian Fonseca, "Latin America's High-Tech Warriors," *Americas Quarterly*, May 7, 2015, https://www.americasquarterly.org/content/latin-americas-high-tech-warriors.

28. Ibid.

29. I rescue this idea from Mathieu Poujol, "Key Cyber Security Issues in the Aerospace and Defense Industry," *Teknowlogy*, September 11, 2018, https://www.pac-online.com/key-cyber-security-issues-aerospace-defense-industry.

30. Aravind Krishnan, "LAAD 2017 Defence and Security: The Biggest Arms Fair in Latin America," April 9, 2017, https://medium.com/@aravind17/laad-2017-defence-security-the-biggest-arms-fair-in-latin-america-81b9e268156d.

31. Taciana Moury, "Brazilian Army Invests in Cyber Defense," *Dialogo*, May 12, 2017, https://dialogo-americas.com/en/articles/brazilian-army-invests-cyber-defense.

32. International Institute for Strategic Studies, "Chapter Eight: Latin America and the Caribbean," *The Military Balance* 119, no. 1 (2019): 380–437.

33. Business Monitor International, "BMI Defence and Security Report, Chile," January 1, 2018, EBSCOhost.

34. Ministerio de Defensa Nacional de Chile, "Balance De Gestión Integral Año 2017, Subsecretaría de Defensa," March 2018, http://www.dipres.gob.cl/597/articles-172590_doc_pdf.pdf.

35. U.S. Department of State, "U.S.-Chile Executive Cyber Consultation," September 18, 2018, https://www.state.gov/r/pa/prs/ps/2018/09/285999.htm.

36. Ibid.

37. Paulina Silva, "Chile: Presidential Instructive on Cybersecurity," November 5, 2018, https://www.hipervinculos.cl/en/presidential-instructive-on-cybersecurity/.

38. "Chile y Estados Unidos Colaboraran en Seguridad Cibernética," *Agencia EFE*, August 16, 2018, https://www.efe.com/efe/america/politica/chile-y-estados-unidos-colaboraran-en-seguridad-cibernetica/20000035-3721877.

39. More resources available at http://www.ssdefensa.gov.cl/n4942_18-05-2016.html; https://www.alianzaciberseguridad.cl; https://www.ciberseguridad.gob.cl.

40. Suldery Vargas Vásquez, "Ejercito Pone en Marcha el Primer Batallón de Prevención y Atención de Desastres," January, 2009, https://www.ejercito.mil.co/?idcategoria=246906.

41. Business Monitor International, "Defence and Security Report, Security, Colombia, 2017: Market Overview," January 1, 2017.

42. Business Monitor International, "BMI Defence and Security Report, Colombia," January 1, 2017, EBSCOhost.

43. Ibid.

44. Ibid.

45. Indra, "Indra and CODALTEC to Develop the First Air Defense System in Colombia, Fully Manufactured in Latin America," June 26, 2018, https://www.indracompany.com/en/noticia/indra-codaltec-develop-first-air-defense-system-colombia-fully-manufactured-latin-america.

46. Ibid.

47. Zachary Volkert, "Microsoft to Bring ICT Innovation to Colombia," *BN Americas*, December 6, 2013, http://www.bnamericas.com/news/technology/microsoft-to-bring-ict-innovation-to-colombia?lang=en.

48. Miriam Ortega, "Colombian Armed Forces Counter Cybercrime," *Dialogo*, November 9, 2017, https://dialogo-americas.com/en/articles/colombian-armed-forces-counter-cybercrime.

49. Ibid.

50. Martin Arostegui, "Colombia Probes Voter Registration Cyberattacks Traced to Russia's Allies," March 15, 2018, *Voice of America News*, https://www.voanews.com/a/colombia-voter-registration-cyberattacks-russia-allies/4300571.html.

51. Business Monitor International, "BMI Defence and Security Report, Mexico," January 1, 2018, EBSCOhost.

52. Ibid.

53. John Vidal, "Mexico's Zapatista Rebels, 24 Years on and Defiant in Mountain Strongholds," *Guardian*, February 17, 2018, https://www.theguardian.com/global-development/2018/feb/17/mexico-zapatistas-rebels-24-years-mountain-strongholds.

54. Tara Copp, "Deployed Border Troops Are Preparing for Militias Stealing Their Gear, Protester Violence, Documents Show," October 31, 2018, https://www.militarytimes.com/news/your-military/2018/10/31/deployed-border-troops-are-preparing-for-militias-stealing-their-gear-protester-violence-documents-show/.

55. According to a recent report by Business Monitor International (BMI): "Mexico has included its vast Economic Exclusive Zone in its national economic plan and intends to develop the local maritime and naval industries significantly, therefore further procurement for inshore and offshore patrol vessels is expected in coming years."

56. According to BMI: "Mexico is the largest military and commercial aircraft market in Latin America, and US manufacturers control 80 percent of the domestic market. Sweden, France, Italy, and the UK and Russia have managed to periodically penetrate the market into the Mexican military aircraft sector."

57. Business Monitor International. "BMI Defence and Security Report, Mexico."

58. Gobierno de México, "Estrategia Nacional de Ciberseguridad, 2017," https://www.gob.mx/cms/uploads/attachment/file/271884/Estrategia_Nacional_Ciberseguridad.pdf.

59. Kavitha Surana, "Spyware Sold to Mexican Government Was Used to Target Experts Investigating Missing Students," *Foreign Policy*, July 10, 2017, https://foreignpolicy.com/2017/07/10/spyware-sold-to-mexican-government-was-used-to-target-experts-investigating-missing-students-cyberwar-ayotzinapa-hacking/.

60. Elia Groll, "Spy vs. Spy, El Chapo Edition," *Foreign Policy*, January 11, 2019, https://foreignpolicy.com/2019/01/11/spy-vs-spy-el-chapo-edition-guzman-narco-flexispy-hacking/.

61. John P. Sullivan and Robert J. Bunker, "Mexican Cartel Strategic Note No. 18: Narcodrones on the Border and Beyond," *Small Wars Journal*, 2016, https://smallwarsjournal.com/jrnl/art/mexican-cartel-strategic-note-no-18-narcodrones-on-the-border-and-beyond.

62. Ibid.

63. Emilio Godoy, "Fuerzas Armadas, Vulnerable a Ciberataque," *Proceso*, June 10, 2017, https://www.proceso.com.mx/490521/fuerzas-armadas-vulnerables-a-ciberataques. Two recent articles by Deborah L. Norden give proof of Venezuela's state of (in)security and the role of the military. See Deborah L. Norden, "Sowing Conflict in Venezuela: Political Violence and Economic Policy," in *Economic Development Strategies and the Evolution of Violence in Latin America*, ed. William Ascher and Natalia S. Mirovitskaya (New York: Palgrave Macmillan, 2012), https://doi.org/10.1057/9781137272690_6; and Deborah L. Norden, "The Making of Socialist Soldiers: Radical Populism and Civil-Military Relations in Venezuela,

Ecuador and Bolivia," in *Debating Civil-Military Relations in Latin America*, ed. David R. Mares and Rafael Martinez (Sussex: Sussex Academic Press, 2014), 155–80.

64. Business Monitor International, "BMI Defence and Security Report, Venezuela," January 1, 2018, EBSCOhost.

65. UNIDIR, "Poverty and Death: Disaster Mortality, 1996–2015," accessed November 1, 2019, https://www.unisdr.org/we/inform/publications/50589.

66. Ariel Cohen and Ray Walser, "The Russia-Venezuela Axis: Using Energy for Geopolitical Advantage," July 21, 2008, https://www.heritage.org/europe/report/the-russia-venezuela-axis-using-energy-geopolitical-advantage.

67. Business Monitor International, "BMI Defence and Security Report, Venezuela."

68. Spetalnick, "Russian Deployment in Venezuela."

69. Ibid.

70. Ibid.

71. Gobierno Bolivariano de Venezuela, "Ciberdefensa," Comando Estratégico Operacional, accessed November 21, 2019, https://ceofanb.mil.ve/direcciones/direccion-conjunta-de-ciberdefensa/.

72. "Venezuela Crea Un Comando Cibernético Para Defenderse Contra La 'Conjura Mediática,'" *RT*, June 7, 2014, https://actualidad.rt.com/actualidad/view/130410-maduro-comando-cibernetico-venezuela.

73. "Venezuela Cyberattack Targets Government Websites," *DW*, August 8, 2017, https://www.dw.com/en/venezuela-cyberattack-targets-government-websites/a-40002475.

74. Lorenzo Franceschi-Bicchierai, "Venezuela's Government Appears to Be Trying to Hack Activists with Phishing Pages," *Vice*, February 15, 2019, https://motherboard.vice.com/en_us/article/d3mdxm/venezuela-government-hack-activists-phishing.

75. Oliver Marguelas, "Venezuela Cybersecurity Profile: Seeking Stability on an Unstable Medium," News, The Henry M. Jackson School of International Studies, University of Washington, August 12, 2016, https://jsis.washington.edu/news/venezuela-cybersecurity-profile-seeking-stability-unstable-medium/

76. "El Gobierno crea comando militar para garantizar servicios públicos tras apagón," *El Universal*, March 15, 2019, http://www.eluniversal.com/economia/35501/el-gobierno-crea-comando-militar-para-garantizar-servicios-publicos-tras-apagon.

77. "FANB Desplegada en el marco del Ejercicio de Acción Integral 'Ana Karina Rote 2019,'" Comando Estratégico Operacional Fuerza Armada Nacional Bolivariana, March 17, 2019, https://ceofanb.mil.ve/fanb-desplegada-en-el-marco-de-los-ejercicios-de-accion-integral-ana-karina-rote-2019/.

78. Valeriano and Maness, *Cyber War versus Cyber Realities*, 24–25.

79. Ibid., 26.

80. Ibid.

81. Paul Ashby, "U.S. National Security in the Western Hemisphere," in *The Oxford Handbook of U.S. National Security*, ed. Derek S. Reveron, Nikolas K. Gvosdev, and John A. Cloud (New York: Oxford University Press, 2018) 611–38.

82. The White House, "National Security Strategy," 12–13.
83. Lora Saalman, "New Domains of Crossover and Concern in Cyberspace," *SIPRI Commentary*, July 26, 2017, https://www.sipri.org/commentary/topical-backgrounder/2017/new-domains-crossover-and-concern-cyberspace.

Chapter Five

1. James Andrew Lewis, *Rethinking Cybersecurity: Strategy, Mass Effect, and States* (Lanham: Rowman and Littlefield, 2018), 2.

2. See a progression of the idea from the early writings of R. Evan Ellis in *China in Latin America: The Whats and Wherefores* (Boulder: Lynne Rienner, 2009) to Ashby, "U.S. National Security in the Western Hemisphere."

3. G. John Ikenberry, "Between the Eagle and the Dragon: America, China, and Middle State Strategies in East Asia," *Political Science Quarterly* 131, no. 1 (2015): 9–43; Amitav Acharya, *Constructing a Security Community in Southeast Asia: ASEAN and the Problem of Regional Order* (Basingstoke: Routledge, 2009)

4. See Mac Margolis, "Obama's Low-Key Legacy of Success in Latin America," *Bloomberg*, January 3, 2017, https://www.bloomberg.com/view/articles/2017-01-03/obama-s-low-key-legacy-of-success-in-latin-america; Michael Reid, "Obama and Latin America: A Promising Day in the Neighborhood," *Foreign Affairs* (September/October), 2015, https://www.foreignaffairs.com/articles/americas/obama-and-latin-america; Abraham F. Lowenthal, Ted Piccone, Laurence Whitehead, *Shifting the Balance: Obama and the Americas* (Washington, DC: Brookings Institute Press, 2010); and Hanna Samir Kassab and Jonathan D. Rosen, *The Obama Doctrine in the Americas* (Lanham, MD: Lexington Books, 2017).

5. See a list of officials and their roles at the bureau in, "On Trump's Latin America Team: Tomás Regalado appointed as New Head of Radio and TV Martí," *The Global Americans*, May 13, 2018, https://theglobalamericans.org/2018/05/trumps-latin-america-team/.

6. See R. Evan Ellis, *China on the Ground in Latin America: Challenges for the Chinese and Impacts on the Region* (New York: Palgrave Macmillan, 2014), and David B.H. Denoon, *China, The United States, and the Future of Latin America* (New York: New York University Press, 2017).

7. See "Xi, Morales Hold Talks, Agree to Establish China-Bolivia Strategic Partnership," *Xinhua*, June 19, 2018, http://www.xinhuanet.com/english/2018-06/19/c_137265305.htm.

8. Maha S. Kamel, "China's Belt and Road Initiative: Implications for the Middle East," *Cambridge Review of International Affairs* 31, no. 1 (2018): 76–95, 76.

9. Charlotte Gao, "China Says Latin America 'Eager' to Join Belt and Road," *The Diplomat*, January, 24, 2018, https://thediplomat.com/2018/01/china-says-latin-america-eager-to- join-belt-and-road/.

10. Francisco Panizza, *Contemporary Latin America: Development and Democracy beyond the Washington Consensus* (London: Zed Books, 2009); and Kevin M. Morrison, "The Washington Consensus and the New Political Economy of Economic Reform," in *The Oxford Handbook of the Politics of Development*, ed. Carol Lancaster and Nicolas van de Walle (Oxford: Oxford University Press, 2018), 73–87.

11. "Argentina," The Observatory of Economic Complexity (OEC), accessed October 29, 2019, https://oec.world/en/profile/country/arg.

12. Shannon K. O'Neil, "Argentina's IMF Package Could Trigger Ugly Blowback," *Council on Foreign Relations*, June 14, 2018, https://www.cfr.org/blog/argentinas-imf-package-could-trigger-ugly-blowback?sp_mid=56809101&sp_rid=Y-2Flc29sYXJAZ21haWwuY29tS0.

13. Cynthia J. Arson, "U.S.-Argentine Relations and the Visit of President Mauricio Macri," April 27, 2017, https://www.wilsoncenter.org/article/us-argentine-relations-and-the-visit-president-mauricio-macri; see also, "Argentine and Russian Presidents Macri and Putin Sign Strategy Accord," *Deutsche Welle*, January 23, 2018, http://www.dw.com/en/argentine-and-russian-presidents-macri-and-putin-sign-strategy-accord/a-42274790.

14. See, for example, Joshua Rovner, "Cyber War as an Intelligence Contest," September 16, 2019, https://warontherocks.com/2019/09/cyber-war-as-an-intelligence-contest/.

15. See Hal Brands, *American Grand Strategy in the Age of Trump* (Washington, DC: Brookings Institution Press, 2018); Derek S. Reveron and Nikolas K. Gvosdev, "National Interests and Grand Strategy," in *The Oxford Handbook of U.S. National Security*, ed. Derek S. Reveron, Nikolas K. Gvosdev, and John A. Cloud (Oxford: Oxford University Press, 2018), 35–56.; and, Colin Dueck, *The Obama Doctrine: American Grand Strategy Today*, Oxford: Oxford University Press, 2015.

16. Suisheng Zhao, "A New Model of Big Power Relations? China-US Strategic Rivalry and Balance of Power in the Asia–Pacific," *Journal of Contemporary China* 24, no. 93 (2015): 377–97.

17. Bruce Gilley and Andrew O'Neil, "China's Rise through the Prism of Middle Powers," in *Middle Powers and the Rise of China*, ed. Bruce Gilley and Andrew O'Neil (Washington: Georgetown University Press, 2014), 1–22, 2.

18. See Carsten Holbraad, "The Role of Middle Powers," *Cooperation and Conflict* 6, no. 1 (1991): 77–90.

19. Anthony Peter Spanakos and Joseph Marques, "Brazil's Rise as a Middle Power: The Chinese Contribution," in *Middle Powers and the Rise of China*, ed. Bruce Gilley and Andrew O'Neil (Washington: Georgetown University Press, 2014), 213–36, 213.

20. Gilley and O'Neil, "China's Rise through the Prism of Middle Powers," 13. See also, Alison Brysk, *Global Good Samaritans: Human Rights as Foreign Policy* (Oxford: Oxford University Press, 2009).

21. James Manicom and Jeffrey Reeves, "Locating Middle Powers in International Relations Theory and Power," in *Middle Powers and the Rise of China*, ed. Bruce Gilley and Andrew O'Neil (Washington: Georgetown University Press, 2014), 23–44, 34–38. For China's White Paper on the region see, "Full Text of China's Policy Paper on Latin America and the Caribbean," Xinhua, November 24, 2016, http://english.gov.cn/archive/white_paper/2016/11/24/content_281475499069158.htm.

22. See Paul Coyer, "China's Pivot to Latin America: Beijing's Growing Security Presence in America's Backyard," *Forbes*, February 20, 2016, https://www.forbes.com/sites/paulcoyer/2016/02/20/chinas-pivot-to-latin-america-beijings-growing-security-presence-in-americas-backyard/2/#70eb48951426. For China's Defense White Paper see, "China's Military Strategy," *Ministry of National Defense*, May 26, 2015, http://eng.mod.gov.cn/Press/2015-05/26/content_4586805.htm.

23. "Factbox: Venezuela's Jailed, Exiled or Barred Opposition Politicians," *Reuters*, February 19, 2018, https://www.reuters.com/article/us-venezuela-politics-factbox/factbox-venezuelas-jailed-exiled-or-barred-opposition-politicians-idUSKCN1G31WU.

24. For a review of U.S. participation in regional forums see Daniel P. Erikson, "Make the Summit of the Americas Great Again," *Foreign Policy*, April 20, 2018, http://foreignpolicy.com/2018/04/20/make-the-summit-of-the-americas-great-again/.

25. "Trump Has a Chance to Correct Obama's Mistake on Venezuela," *The Washington Post*, March 17, 2017, https://www.washingtonpost.com/opinions/global-opinions/trumps-chance-to-correct-obamas-mistake-on-venezuela/2017/03/17/8389943a-09aa-11e7-b77c-0047d15a24e0_story.html?noredirect=on&utm_term=.d88d6583ea18.

26. For a review of the Americas' hot spots of conflict see, David R. Mares, *Latin America and the Illusion of Peace* (London: Routledge, 2012). For two different takes on peace in the region written two decades apart see Francisco Rojas Aravena, "Williamsburg: ¿Un Giro Definitivo en Las Relaciones Hemisféricas de Seguridad?" *Estudios Internacionales* 29, no. 114 (1996): 139–64; and Mónica Serrano, "¿Promesas Idealistas? América Latina en Tiempos de la Responsabilidad de Proteger," *Foro Internacional* 57, no. 1 (2017): 55–108.

27. Brigitte Weiffen, Leslie Wehner, and Detlef Nolte, "Overlapping Regional Security Institutions in South America: The Case of OAS and UNASUR," *International Area Studies Review* 16, no. 4 (2013): 370–89.

28. On the USSOUTHCOM's areas of responsibility and "lines of effort," see http://www.southcom.mil/About/Area-of-Responsibility/.

29. Department of the U.S. Navy, "US 4th Fleet Participates in SIFOREX 2018," April 19, 2018, http://www.navy.mil/submit/display.asp?story_id=105190.

30. Navy Live, "Building Partnerships During 56th Annual UNITAS Exercise," October 15, 2015, http://navylive.dodlive.mil/2015/10/16/building-partnerships-during-56th-annual-unitas-exercise/.

31. La Tercera, "Chadwick Señala que 'No Está Contemplado por Ahora' Visita de Piñera a Oficinas de Huawei," April 14, 2019, https://www.latercera.com/politica/noticia/chadwick-senala-no-esta-contemplado-ahora-visita-pinera-oficinas-huawei/614168/.

32. Oxford Analytica, "Huawei's Fate Is Inseparable from Europe's Geostrategy," March 7, 2019, https://dailybrief.oxan.com/Analysis/GA242262/Huaweis-fate-is-inseparable-from-Europes-geostrategy.

33. Mark Stokes, "The Chinese People's Liberation Army Computer Network Operations Infrastructure," in *China and Cybersecurity: Espionage, Strategy, and Politics in the Digital Domain*, ed. Jon R. Lindsay, Tai Ming Cheung, and Derek S. Reveron (Oxford: Oxford University Press, 2015): 164–77.

34. Gabriela Mello, "Huawei Partners with Local Retailers to Sell Flagship Smartphones in Brazil," April 30, 2019, https://uk.reuters.com/article/uk-huawei-tech-brazil/huawei-partners-with-local-retailers-to-sell-flagship-smartphones-in-brazil-idUKKCN1S62TN.

35. Huawei, "A New Era for Brazil's Digital Economy Powered by ICT Innovation," accessed May 15, 2020, https://www.huawei.com/en/about-huawei/executives/articles/brazil-digital-economy-ict-innovation.

36. The Business Year, "Connecting Country: Colombia 2018," accessed May 15, 2020, https://www.thebusinessyear.com/colombia-2018/connecting-country/interview.

37. Daniel Fajardo Cabello, "No Solo Amazon: Huawei Estudia Instalar En Chile Su Primer Data Center Para América Latina," *La Tercera*, February 21, 2019, https://www.latercera.com/pulso/noticia/no-solo-amazon-huawei-tambien-estudia-instalar-chile-gran-data-center-america-latina/537873/.

38. *The Economist*, "The Tax Heaven at the End of the World," July 16, 2016, https://www.economist.com/the-americas/2016/07/16/the-tax-haven-at-the-end-of-the-world.

39. Oliver Stuenkel. "Huawei Heads South," *Foreign Affairs*, May 10, 2019, https://www.foreignaffairs.com/articles/brazil/2019-05-10/huawei-heads-south.

40. Access Now, "Argentina Province in Talks with Huawei over Acquiring Facial Recognition Surveillance Technology," July 12, 2018, https://www.accessnow.org/mendoza-surveillance-tech/.

41. Fernanda Uriegas, "The Chinese Brands Latin Americans Love," April 15, 2019, https://www.americasquarterly.org/content/chinese-brands-latin-americans-love.

42. Cassandra Garrison, "On South America's Largest Solar Farm, Chinese Power Radiates," April 29, 2019, https://www.weforum.org/agenda/2019/04/on-south-americas-largest-solar-farm-chinese-power-radiates/.

43. BBC, "Meng Wanzhou: Huawei Executive Suffers US Extradition Blow." May 27, 2020, https://www.bbc.co.uk/news/world-us-canada-52793343.

44. Nigel Inkster, "The Huawei Affair and China's Technology Ambitions," *Survival* 61, no. 1 (2019): 105–11, 108.

45. See Jefferson P. Marquis, Jennifer D. P. Moroney, Justin Beck, Derek Eaton, Scott Hiromoto, David R. Howell, Janet Lewis, Charlotte Lynch, Michael J. Neumann, and Cathryn Quantic Thurston, *Developing an Army Strategy for Building Partner Capacity for Stability Operations* (Santa Monica: Rand Corporation, 2010), 60–62.

46. U.S. Southern Command, "Theater Strategy," May 19, 2017, http://www.southcom.mil/Portals/7/Documents/USSOUTHCOM_Theater_Strategy_Final.pdf?ver=2017-05-19-120652-483.

47. Ibid., 2.

48. Jeff Colgan, "Venezuela and Military Expenditure Data," *Journal of Peace Research* 48, no. 4 (2011): 547–56.

49. Nan Tian and Diego Lopes da Silva, "Improving South American Military Expenditure Data," *SIPRI*, September 4, 2017, https://www.sipri.org/commentary/topical-backgrounder/2017/improving-south-american-military-expenditure-data.

50. Stockholm International Peace Research Institute. "SIPRI Military Expenditure Database," accessed November 30, 2020., https://www.sipri.org/databases/milex.

51. For analysis of the statistics of Chinese arms transfers to the region, see Tom O'Connor, "Russian Military Sells New Missile Defense System to Brazil," *Newsweek*, April 4, 2017, http://www.newsweek.com/russia-military-influence-expand-brazil-missile-system-579156; Allan Nixon, "China's Growing Arms Sales to Latin America," *The Diplomat*, August 24, 2016, https://thediplomat.com/2016/08/chinas-growing-arms-sales-to-latin-america/.

52. Richard Connolly and Cecilie Sendstad, "Russia's Role as an Arms Exporter: The Strategic and Economic Importance of Arms Exports for Russia," *Chatham House*, March, 2017, http://www.chathamhouse.org/sites/default/files/publications/research/2017-03-20-russia-arms-exporter-connolly-sendstad.pdf.

53. Carlos Torralba, "Mexico Triples its Arms Purchases in Space of Five Years," *El País*, February 21, 2017, https://elpais.com/elpais/2017/02/21/inenglish/1487673374_904061.html.

54. Peter Wazeman, Aude Fleurant, Alexandra Kuimova, Nan Tian, and Siemon T Wezeman, "Trends in International Arms Transfers, 2017," *SIPRI*, March, 2018, https://www.sipri.org/sites/default/files/2018-03/fssipri_at2017_0.pdf.

55. See Todd Sandler and Justin George, "Military Expenditure Trends for 1960–2014 and What They Reveal," *Global Policy* 7, no. 2 (2016): 174–84, https//doi: 10.1111/1758-5899.12328.

56. Yong Wang, "Offensive for Defensive: The Belt and Road Initiative and China's New Grand Strategy," *The Pacific Review* 29, no. 3 (2016): 455–63, 461.

57. For a useful timeline of events in the Americas involving U.S. interventions, see Richard W. Slatta, "Time Line of US-Latin American Relations," accessed May 15, 2020, https://faculty.chass.ncsu.edu/slatta/hi216/hi453time.htm.

58. Nicholas Khoo, "Interstate Rivalry in East Asia," in *The Oxford Handbook of U.S. National Security*, ed. Derek S. Reveron, Nikolas K. Gvosdev, and John A. Cloud (Oxford: Oxford University Press, 2018), 573–90.

59. Lisa Ferdinando, "U.S. Wants to Remain 'Partner of Choice' in Latin America," April 6, 2017, http://www.southcom.mil/MEDIA/NEWS-ARTICLES/Article/1144692/us-wants-to-remain-partner-of-choice-in-latin-america/.

60. Adam Isacson and Sara Kinosian, "A First Glance at the 2017 Foreign Aid Request for Latin America," *WOLA*, February 9, 2016, https://www.wola.org/analysis/a-first-glance-at-the-2017-foreign-aid-request-for-latin-america/.

61. Gian Carlo Delgado-Ramos and Silvina María Romano, "Political-Economic Factors in U.S. Foreign Policy: The Colombia Plan, the Mérida Initiative, and the Obama Administration," *Latin American Perspectives* 38, no. 4 (2011): 93–108, https://doi.org/10.1177/0094582X11406208.

62. Carolina Contreras, "Chilean Armed Forces Support Peace in Colombia," *Dialogo*, September 15, 2016, https://dialogo-americas.com/en/articles/chilean-armed-forces-support-peace-colombia.

63. Jairo Muñoz, "Will China Make the Most of Colombia's Peace Deal with the FARC," *The Diplomat*, September 13, 2016, https://thediplomat.com/2016/09/will-china-make-the-most-of-colombias-peace-deal-with-farc/.

64. Gregory B. Weeks, *U.S. and Latin American Relations* (Chichester: Wiley Blackwell, 2015) 207–31.

65. Michael Shifter, "The US and Latin America Through the Lens of Empire," *Current History* 103, no. 670 (2004), http://archive.thedialogue.org/page.cfm?pageID=32&pubID=1044.

66. "What Latin Americans Think of the United States," *The Economist*, March 29, 2016, https://www.economist.com/graphic-detail/2016/03/29/what-latin-americans-think-of-the-united-states.

67. See Carlos Solar, *Government and Governance of Security: The Politics of Organised Crime in Chile* (New York: Routledge, 2018).

68. Weifeng Zhou and Mario Esteban, "Beyond Balancing: China's Approach towards the Belt and Road Initiative," *Journal of Contemporary China* 27, no. 112 (2018): 487–501, 492.

69. See Astrid H. M. Nordin and Mikael Weissmann, "Will Trump Make China Great Again? The Belt and Road Initiative and International Order," *International Affairs* 94, no. 2 (2018): 231–49.

70. Alex Cuadros, "Open Talk of a Military Coup Unsettles Brazil," *The New Yorker*, October 13, 2017, https://www.newyorker.com/news/news-desk/open-talk-of-a-military-coup-unsettles-brazil.

71. See Lourdes Casanova, "The Challenges of Chinese Investment in Latin America," *World Economic Forum*, March 15, 2018, https://www.weforum.org/agenda/2018/03/latin-america-china-investment-brazil-private-public/.

72. See "China's Chinalco Starts $1.3 Billion Expansion of Peru Copper Mine," *Reuters*, June 4, 2018, https://www.reuters.com/article/us-peru-copper-china/chinas-chinalco-starts-1-3-billion-expansion-of-peru-copper-mine-idUSKCN1J00CI; "Chinas' COSCO to Build $2 Billion Shipping Port in Peru: Ambassador," *Reuters*, June 1, 2018, https://www.reuters.com/article/us-peru-china/chinas-cosco-to-build-2-billion-shipping-port-in-peru-ambassador-idUSKCN1IX5WW.

73. Riordan Roett and Guadalupe Paz, *Latin America and the Asian Giants: Evolving Ties with China and India* (Washington, DC: Brookings Institution Press, 2016); Kevin P. Gallagher and Roberto Porzecanski, "China and the Future of Latin American Economic Development," in *The Oxford Handbook of Latin American Economics*, ed. José Antonio Ocampo and Jaime Ros (Oxford: Oxford University Press, 2011), 461–87.

74. Yanran Xu, *China's Strategic Partnerships in Latin America: Case studies of China's Oil Diplomacy in Argentina, Brazil, Mexico, and Venezuela, 1991–2015* (London: Lexington Books, 2017).

75. Andrea Bianculli, "Latin America," in *The Oxford Handbook of Comparative Regionalism*, ed. Tanja A. Börzel and Thomas Risse (Oxford: Oxford University Press, 2016), 154–77.

76. "Ambassadors Expect More China, Latin America Cooperation," *Xinhua*, October 26, 2016, http://english.gov.cn/news/international_exchanges/2016/10/26/content_281475475859018.htm.

77. See Lisa Ferdinando, "U.S. Wants to Remain 'Partner of Choice' in Latin America," U.S. Southcom, April 6, 2017, http://www.southcom.mil/MEDIA/NEWS-ARTICLES/Article/1144692/us-wants-to-remain-partner-of-choice-in-latin-america/.

78. Nadège Rolland, "China's 'Belt and Road Initiative': Underwhelming or Game-Changer?" *The Washington Quarterly* 40, no. 1 (2017): 127–42.

79. Meia Nouwens, "Global Defence-Industry League: Where Is China?" August 28, 2018, https://www.iiss.org/blogs/military-balance/2018/08/china-global-defence-industry-league.

80. Ibid.

81. Nan Tian, "Military Expenditure," *SIPRI Yearbook 2017: Armaments, Disarmament, and International Security*, accessed November 12, 2019, https://www.sipriyearbook.org/view/9780198811800/sipri-9780198811800-chapter-9-div1-45.xml.

82. Susan T. Jackson, "Arms Production," *SIPRI Yearbook 2010: Armaments, Disarmament, and International Security*, accessed November 12, 2019, accessed November 12, 2019, https://www.sipriyearbook.org/view/9780199581122/sipri-9780199581122-div1-53.xml.

83. Mikael Grinbaum and Susan T. Jackson. "Arms Production and Military Services," *SIPRI Yearbook 2012: Armaments, Disarmament, and International Security*, accessed November 12, 2019, https://www.sipriyearbook.org/view/9780199650583/sipri-9780199650583-div1-30.xml.

84. Cameron Ortis and Paul Evans, "The Internet and Asia-Pacific Security: Old Conflicts and New Behaviour," *The Pacific Review* 16, no. 4 (2003): 549–50, 553–54.

85. Nouwens, "Global Defence-Industry League."

86. Ortis and Evans, "The Internet and Asia-Pacific Security," 553–54.

87. Ian Bowers, "The Use and Utility of Hybrid Warfare on the Korean Peninsula," *The Pacific Review* 31, no. 6 (2018): 762–86, 765–66.

88. Council on Foreign Relations, "Cyber Operations Tracker."

89. Bowers, "The Use and Utility of Hybrid Warfare," 771.

90. James Andrew Lewis, "Five Myths about Chinese Hackers," *The Washington Post*, March 22, 2013, accessed at http://articles.washingtonpost.com/2013-03-22/opinions/37923854_1_chinese-hackers-cyberattacks-cold-war.

91. James Andrew Lewis, *Rethinking Cybersecurity*, 7

92. Bowers, "The Use and Utility of Hybrid Warfare," 772.

93. John Keller, "2017 DOD Budget to Roll Out Next Week: Has Pentagon Spending Finally Bottomed Out?" February 2, 2016, https://www.militaryaerospace.com/computers/article/16715133/2017-dod-budget-to-roll-out-next-week-has-pentagon-spending-finally-bottomed-out

94. Mackenzie Eaglen, "Pentagon Should Take the Lead on Tech," September 4, 2019, https://www.wsj.com/articles/pentagon-should-take-the-lead-on-tech-11567636730.

95. Jae Ho Chung, "How America and China See Each Other: Charting National Views and Official Perceptions," *The Pacific Review* 32, no. 2 (2019): 188–209, 202.

96. Ibid.

97. James Andrew Lewis, *Rethinking Cybersecurity*, 8.

98. China's "peaceful" foreign policy under President Xi Jinping has been the object of study by Eastern scholars and thoroughly covered, among other papers and monographs, in Jian Zhang, "China's New Foreign Policy Under Xi Jinping: Towards 'Peaceful Rise 2.0'? *Global Change, Peace & Security* 27, no. 1 (2015): 5–19.

99. Foreign Ministry of the People's Republic of China (FMPRC), "China's Defence Expenditure," April 7, 2011, available at http://www.fmprc.gov.cn/mfa_eng/wjb_663304/zzjg_663340/jks_665232/ kjlc_665236/gjjk_665238/t410723.shtml

100. Latin American Public Opinion Project (LAPOP), "AmericasBarometer Data Sets 2016/17," The AmericasBarometer, accessed May 10, 2018, https://www.vanderbilt.edu/lapop/raw-data.php.

101. There are various statistical formulae to calculate the coefficient of correlation, r. Across the book I have used Pearson's r coefficient. See Roger Pierce, *Research Methods in Politics* (London: Sage, 2011), 206–19.

102. See Hari Das, "Statistical Analysis," in *Encyclopedia of Case Study Research*, ed. Albert J. Mills, Gabrielle Durepos, and Elden Wiebe (Thousand Oaks: Sage, 2012), 889–92, 889.

103. Jacob Cohen, Patricia Cohen, Stephen G. West, and Leonna S. Aiken, *Applied Multiple Regression/Correlation Analysis for the Behavioral Sciences* (New York: Routledge, 2013), 32.

104. Scott Menard, "Logistic Regression," in *Encyclopedia of Research Design*, ed. Neil J. Salkind (Thousand Oaks: Sage, 2012), 731–35.

105. Judea Pearl, "The Foundations of Causal Inference," *Sociological Methodology* 40, no. 1 (2010): 75–149, 79.

106. Cohen, Cohen, West, and Aiken, *Applied Multiple Regression/Correlation*, 7.

107. Ferdinando, "U.S. Wants to Remain 'Partner of Choice' in Latin America."

108. See "Posture Statement of Admiral Kurt W. Tidd Commander, United States Southern Command Before the 115th Congress Senate Armed Services Committee," U.S. Southcom, February 15, 2018, http://www.southcom.mil/Portals/7/Documents/Posture%20Statements/SOUTHCOM_2018_Posture_Statement_FINAL.PDF?ver=2018-02-15-090330-243.

109. Barbara Stallings, "Does Asia Matter? The Political Economy of Latin America's International Relations," in *The Oxford Handbook of Latin American Political Economy*, ed. Javier Santiso and Jeff Dayton-Johnson (Oxford: Oxford University Press, 2012), 210–32.

110. Hongying Wang, "The Missing Link in Sino–Latin American Relations," *Journal of Contemporary China* 24, no. 95 (2015): 922–42.

111. Jian Zhang, "China's New Foreign Policy under Xi Jinping: Towards 'Peaceful Rise 2.0'?" *Global Change, Peace & Security* 27, no. 1 (2015): 5–19; David M. Lampton, "China: Challenger or Challenged?" *The Washington Quarterly* 39, no 3 (2016): 107–19.

112. June Teufel Dreyer, "Managing Sino-American Relations." *Orbis* 61, no 1 (2017): 83–90, 89.

Chapter Six

1. U.S. Department of State, "Secretary of State Michael R. Pompeo and Argentine Foreign Minister Jorge Faurie at a Press Availability at the Western Hemisphere Counterterrorism Ministerial Plenary," July 19, 2019, https://2017-2021.state.gov/secretary-of-state-michael-r-pompeo-and-argentine-foreign-minister-jorge-faurie-at-a-press-availability-at-the-western-hemisphere-counterterrorism-ministerial-plenary/index.html.

2. Reuters, "Pompeo to Visit Four Latin American Nations in Security, Migration Push," July 16, 2019, https://www.reuters.com/article/uk-usa-pompeo-idUKKCN1UB2Q7; U.S. Department of State, "Secretary Pompeo's Meeting with Argentine President Macri," July 19, 2019, https://ar.usembassy.gov/secretary-pompeos-meeting-with-argentine-president-macri/.

3. Stephen M. Walt, "Alliances in a Unipolar World," *World Politics* 61, no. 1 (2009): 86–120, 86.

4. U.S. Department of Defense, "Cyber Strategy," September 18, 2018, https://media.defense.gov/2018/Sep/18/2002041658/-1/-1/1/CYBER_STRATEGY_SUMMARY_FINAL.PDF.

5. Mark Pomerleau, "Here's How Cyber Command Is Using 'Defend Forward,'" November 12, 2019, https://www.fifthdomain.com/smr/cybercon/2019/11/12/heres-how-cyber-command-is-using-defend-forward/?utm_source=clavis.

6. U.S. Southern Command Strategy. "United States Southern Command Strategy: 'Enduring Promise for the Americas,'" May 8, 2019, https://www.southcom.mil/Portals/7/Documents/SOUTHCOM_Strategy_2019.pdf?ver=2019-05-15-131647-.

7. U.S. Department of Defense, "Southcom Chief Outlines Keys for Success in South America," May 24, 2019, https://www.defense.gov/explore/story/Article/1857725/southcom-chief-outlines-keys-for-success-in-south-america/.

8. Department of Defense, "Press Briefing by Admiral Kurt Tidd," March 5, 2018, https://www.defense.gov/Newsroom/Transcripts/Transcript/Article/1458541/.

9. Walt, "Alliances in a Unipolar World," 88–89.

10. *Financial Times*, "Latin America Resists US Pressure to Exclude Huawei," June 9, 2019, https://www.ft.com/content/38257b66-83c5-11e9-b592-5fe435b57a3b.

11. Stephen M. Walt, "Alliance Formation and the Balance of World Power," *International Security* 9, no. 4 (1985): 4–7.

12. U.S. Southern Command, "United States Southern Command Strategy."

13. Walt argues that these factors can reflect a state's perception of a level of risk based on the impact of aggregate power, proximity, offensive capability, and offensive intentions. See Walt, "Alliance Formation," 8–9.

14. Luis Simon, Alexander Lanoszka and Hugo Meijer, "Nodal Defense: The Changing Structure of U.S. Alliance Systems in Europe and East Asia," *Journal of Strategic Studies* 44, no. 3 (2021): 350–88.

15. Ibid.

16. Larry Wortzel, "Change Partners: Who Are America's Military and Economic Allies in the 21st Century?" June 6, 2005, https://www.heritage.org/defense/report/change-partners-who-are-americas-military-and-economic-allies-the-21st-century; Michael Wesley, *Global Allies: Comparing US alliances in the 21st Century* (Canberra: Australian National University Press, 2017).

17. Rafael Hernández, "Commentary: Alliances and Dis-alliances between the United States and Latin America and the Caribbean," *Latin American Perspectives* 38, no. 4 (2011): 131–36.

18. Mark Pomerleau, "In Era of 'Defend Forward,' What Does Success Look Like?," April 24, 2019, https://www.fifthdomain.com/dod/2019/04/24/in-era-of-defend-forward-what-does-success-look-like.

19. Walt, "Alliances in a Unipolar World," 93.

20. The White House, "Advancing Our Interests: Actions in Support of the President's National Security Strategy," May 27, 2010, https://obamawhitehouse.archives.gov/the-press-office/advancing-our-interests-actions-support-presidents-national-security-strategy.

21. Alex Grigsby, "The Brazil-U.S. Cyber Relationship Is Back on Track," July 1, 2015, https://www.cfr.org/blog/brazil-us-cyber-relationship-back-track.

22. Ibid.

23. The White House, "National Security Strategy."

24. The White House, "Foreign Policy," accessed November 29, 2020, https://obamawhitehouse.archives.gov/issues/foreign-policy/.

25. UNIDIR, "Cybersecurity, and Cyberwarfare: Preliminary Assessment of National Doctrine and Organization, 2011," accessed October 29, 2020, https://www.unidir.org/files/publications/pdfs/cybersecurity-and-cyberwarfare-preliminary-assessment-of-national-doctrine-and-organization-380.pdf.

26. Ibid.

27. The White House, "Factsheet: Cybersecurity National Action Plan," February 9, 2016, https://obamawhitehouse.archives.gov/the-press-office/2016/02/09/fact-sheet-cybersecurity-national-action-plan.

28. BBC News, "Mike Pompeo Warns UK over Huawei 'Security Risks,'" May 8, 2019, https://www.bbc.co.uk/news/uk-politics-48198932.

29. *Washington Examiner*, "China Selling High-Tech Tyranny to Latin America, Stoking US Concern," April 10, 2019, https://www.washingtonexaminer.com/policy/defense-national-security/china-selling-high-tech-tyranny-to-latin-america-stoking-us-concern.

30. Wesley, *Global Allies*, 2, 4.

31. Joseph L. Lengyel, "Securing the Nation One Partnership at a Time," *Strategic Studies Quarterly* 12, no. 3 (2018): 3–9.

32. Mark Pomerleau, "Two Years In, How Has a New Strategy Changed Cyber Operations?" November 11, 2019, https://www.fifthdomain.com/dod/2019/11/11/two-years-in-how-has-a-new-strategy-changed-cyber-operations/.

33. U.S. Senate, "Statement of General Paul M. Nakasone Commander United States Cyber Command Before the Senate Committee on Armed Services," February 14, 2019, https://www.armed-services.senate.gov/imo/media/doc/Nakasone_02-14-19.pdf.

34. Ibid., 13.

35. Ibid., 2.

36. Ibid.

37. Ibid. 4.

38. Robbie Gramer, "Can State's New Cyber Bureau Hack It?" *Foreign Policy*, January 18, 2019, https://foreignpolicy.com/2019/01/18/state-department-cyber-security-cyber-threats-russia-china-diplomacy-capitol-hill-lawmakers-pompeo/.

39. Ibid.

40. Paul Marks, "The Cybersecurity 202: International Cooperation On China Hacking Could Be Signal for 2019," December 21, 2018, https://www.washingtonpost.com/news/powerpost/paloma/the-cybersecurity-202/2018/12/21/the-cybersecurity-202-international-cooperation-on-china-hacking-could-be-signal-for-2019/5c1be9491b326b6a59d7b21c/; Paul Marks, "U.S. to Try New Approach to Punish Hacking Nations: Working with Allies," March 7, 2019, https://www.washingtonpost.com/news/powerpost/paloma/the-cybersecurity-202/2019/03/07/the-cybersecurity-202-u-s-to-try-new-approach-to-punish-hacking-nations-working-with-allies/5c80132b1b326b2d177d5ff1.

41. Justin Lynch, "North Korea 'Hacking the Hell Out of Latin America,'" June 29, 2018, https://www.fifthdomain.com/critical-infrastructure/2018/06/29/north-korea-hacking-the-hell-out-of-latin-america/.

42. Inter-American Development Bank and OAS, "Cybersecurity: Are We Ready in Latin America and the Caribbean?"

43. U.S. Department of State, "Co-Chairs' Statement on the Inaugural ASEAN-U.S. Cyber Policy Dialogue," October 3, 2019, https://2017-2021.state.gov/co-chairs-statement-on-the-inaugural-asean-u-s-cyber-policy-dialogue/index.html.

44. Gehrke, Joel, "Pompeo to Latin America: 'We Love You, China Doesn't,'" *Washington Examiner*, April 12, 2019, https://www.washingtonexaminer.com/policy/defense-national-security/pompeo-to-latin-america-we-love-you-china-doesn't.

45. U.S. Department of State, "Historic Accomplishments During Diplomatic Engagements in the Western Hemisphere July 18–21," July 22, 2019, https://uy.usembassy.gov/historic-accomplishments-during-diplomatic-engagements-in-the-western-hemisphere/.

46. Elias Groll, "The U.S.-Iran Standoff Is Militarizing Cyberspace," September 27, 2019, https://foreignpolicy.com/2019/09/27/the-u-s-iran-standoff-is-militarizing-cyberspace/.

47. The White House, "Strategy to Combat Transnational Organized Crime, July 2011," accessed January 15, 2019, https://obamawhitehouse.archives.gov/sites/default/files/microsites/2011-strategy-combat-transnational-organized-crime.pdf.

48. International Institute for Strategic Studies. "Chapter Three: North America." *The Military Balance* 118, no. 1 (2018): 27–64, 29.

49. The Canadian Forces Network Operation Centre is the "national operational cyber defense unit" permanently assigned to support Canadian forces' operations. The armed forces' Information Management Group (IMG) is responsible for electronic warfare and network defense. The Canadian Force Information Operations Group, under the IMG, commands the Canadian Forces Information Operations Group Headquarters; the Canadian Forces Electronic Warfare Centre; the Canadian Forces Network Operation Centre, which is the "national operational cyber defense unit" permanently assigned to support Canadian Forces operations; and other units. See International Institute for Strategic Studies, "Chapter Three: North America," 45.

50. International Institute for Strategic Studies, "Chapter Five: Russia and Eurasia," *The Military Balance* 118, no. 1 (2018): 169–218.

51. International Institute for Strategic Studies, "Chapter Six: Asia," *The Military Balance* 118, no. 1 (2018): 219–314, 244–81.

52. International Institute for Strategic Studies, "Chapter Eight: Latin America."

53. U.S. Department of State, "Recommendations to the President on Protecting American Cyber Interests through International Engagement, Office of the Coordinator for Cyber Issues," May 31, 2018, https://www.state.gov/wp-content/uploads/2019/04/Recommendations-to-the-President-on-Protecting-American-Cyber-Interests-Through-International-Engagement.pdf.

54. Ibid.

55. Walt, "Alliances in a Unipolar World," 94.

56. U.S. Department of State, "Recommendations to the President."

57. The joint statement was signed by Australia, Belgium, Canada, Colombia, the Czech Republic, Denmark, Estonia, Finland, France, Germany, Hungary, Iceland, Italy, Japan, Latvia, Lithuania, the Netherlands, New Zealand, Norway, Poland, the Republic of Korea, Romania, Slovakia, Spain, Sweden, the United Kingdom, and the United States. See U.S. Department of State, "The United States Co-Hosts Second Ministerial Meeting on Advancing Responsible State Behavior in Cyberspace," September 23, 2019, https://www.state.gov/the-united-states-co-hosts-second-ministerial-meeting-on-advancing-responsible-state-behavior-in-cyberspace/.

58. As I mentioned in chapter 5, a Chilean government delegation visited Washington, D.C., in August 2018, to discuss bilateral cooperation on cyber issues, including government capacity to address emerging challenges and shared threats in cyberspace. Chile has openly sought collaboration with U.S. agencies to "continue close collaboration on cybersecurity, critical infrastructure protection, incident response, data protection, information and communication technology procurement, international security in cyberspace, and military and law enforcement cooperation by establishing strong channels for open communication about cyber issues of concern." Undersecretary of Defense Cristián de la Maza led the Chilean delegation. Chile's Joint Staff coordinates cybersecurity policies for the Ministry of National Defense and the armed forces. Each service has a cybersecurity organization. In 2017, Chile announced the creation of a Joint Cyberdefense Command and defense CERT teams. See International Institute for Strategic Studies, "Chapter Eight: Latin America," 406.

59. Ibid., 363.

60. U.S. Department of Defense, "Remarks by Secretary Mattis on Cybersecurity in Santiago, Chile," August 16, 2018, https://www.defense.gov/Newsroom/Transcripts/Transcript/Article/1608504/remarks-by-secretary-mattis-on-cybersecurity-in-santiago-chile/.

61. U.S. Department of State, "U.S. Relations with Venezuela," July 6, 2021, https://www.state.gov/u-s-relations-with-venezuela/.

62. Ibid.

63. John E. Herbst and Jason Marczak, "Russia's Intervention in Venezuela: What's at Stake?" September 12, 2019, https://www.atlanticcouncil.org/in-depth-research-reports/report/russias-intervention-in-venezuela-whats-at-stake/. See also Mason Shuya, "Russian Influence in Latin America," *Journal of Strategic Security* 12, no. 2 (2019): 17–41.

64. U.S. Department of State, "Remarks Kimberly Breier, Assistant Secretary Bureau of Western Hemisphere Affairs. China's New Road in the Americas: Beyond Silk and Silver," April 29, 2019, https://cl.usembassy.gov/chinas-new-road-in-the-americas-beyond-silk-and-silver//.

65. Scott Malcomson, "The Real Fight for the Future of 5G: Who Will Patrol the Borders of a New Network?" *Foreign Affairs*, November 14, 2019, https://www.foreignaffairs.com/articles/2019-11-14/real-fight-future-5g.

66. Kurt M. Campbell, and Jake Sullivan, "Competition Without Catastrophe: How America Can Both Challenge and Coexist with China," *Foreign Affairs* 98, no. 5 (2019): 96–110.

67. International Institute for Strategic Studies, "Chapter Eight: Latin America," 400.

68. International Institute for Strategic Studies, "Chapter Eight: Latin America and the Caribbean," *The Military Balance* 117, no. 1 (2017): 417–78, 471.

69. According to a recent report, OAS cybersecurity efforts have produced crucial results such as: national CSIRTs have increased from four to twenty-one since 2004; eleven countries have adopted national cybersecurity strategies; a hemispheric network (CSIRTAmericas.org) has been established that facilitates communication, information, and knowledge sharing; and specialized technical training has been provided to thousands of government officials and other cybersecurity stakeholders in the region. See U.S. Department of State, "Remarks of Deputy Assistant Secretary Robert L. Strayer at a Joint U.S.-Mexico Reception at the OAS for Participants of the UN GGE Regional Consultation," August 15, 2019, https://2017-2021.state.gov/remarks-of-deputy-assistant-secretary-robert-l-strayer-at-a-joint-u-s-mexico-reception-at-the-oas-for-participants-of-the-un-gge-regional-consultation/index.html.

70. Ibid.

71. Ibid.

72. For a recent discussion of cyberincidents attribution among public actors, see Florian J. Egloff, "Contested Public Attributions of Cyber Incidents and the Role of Academia," *Contemporary Security Policy* 41, no. 1 (2020): 55–81.

73. See, for instance, U.S. Department of State, "Joint Statement on the Inaugural U.S.-Dutch Cyber Dialogue," May 20, 2019, https://2017-2021.state.gov/joint-statement-on-the-inaugural-u-s-dutch-cyber-dialogue/index.html.

74. Oliver Stuenkel, "Huawei Heads South: The Battle Over 5G Comes to Latin America," *Foreign Affairs*, May 10, 2019, https://www.foreignaffairs.com/articles/brazil/2019-05-10/huawei-heads-south.

75. U.S. Department of State, "U.S. Policy on 5G Technology," August 28, 2019, https://2017-2021.state.gov/US-Policy-On-5g-Technology/index.html.
76. Walt, "Alliances in a Unipolar World," 116–17.
77. Campbell and Sullivan, "Competition without Catastrophe."
78. William J. Burns, "The Demolition of U.S. Diplomacy," *Foreign Affairs*, October 14, 2019, https://www.foreignaffairs.com/articles/2019-10-14/demolition-us-diplomacy.

Chapter Seven

1. Patryk Pawlak and Panagiota-Nayia Barmpaliou, "Politics of Cybersecurity Capacity Building: Conundrum and Opportunity," *Journal of Cyber Policy* 2, no. 1 (2017): 123–44.
2. Andrew Sullivan, "Avoiding Lamentation: To Build a Future Internet," *Journal of Cyber Policy* 2, no. 3 (2017): 323–37.
3. John Naughton, "The Evolution of the Internet: From Military Experiment to General Purpose Technology," *Journal of Cyber Policy* 1, no. 1 (2016): 5–28.
4. Sullivan, "Avoiding Lamentation."
5. Ibid.
6. Pawlak and Barmpaliou, "Politics of Cybersecurity."
7. Susan Ariel Aaronson, "What Might Have Been and Could Still Be: the Trans-Pacific Partnership's Potential to Encourage an Open Internet and Digital Rights," *Journal of Cyber Policy* 2, no. 2 (2017): 232–54, 235.
8. For a recent call to integrate quantitative methods training in the classroom see Matthew Alan Morehouse, Tomáš Lovecký, Hannah Read, Marcus Woodman, "Quantify? or, Wanna Cry? Integrating Methods Training in the IR Classroom," *International Studies Perspectives* 18, no. 2 (2017): 225–45.
9. I borrow the features of statistical research from Geoff Payne and Judy Payne, *Key Concepts in Social Research* (London: Sage, 2011), 181–86.
10. *Observatory of Cybersecurity in Latin America and the Caribbean*, "Home."
11. The original coding given by the *Observatory* was 1 = Start-up; 2 = Formative; 3 = Established; 4 = Strategic; and 5 = Dynamic.
12. Jamie Saunders, "Tackling Cybercrime—the UK response," *Journal of Cyber Policy* 2, no. 1 (2017): 4–15.
13. Lincoln Pigman, "Russia's Vision of Cyberspace: A Danger to Regime Security, Public Safety, and Societal Norms and Cohesion," *Journal of Cyber Policy* 4, no. 1 (2019): 22–34.
14. Hon-min Yau, "A Critical Strategy for Taiwan's Cybersecurity: A Perspective from Critical Security Studies," *Journal of Cyber Policy* 4, no. 1 (2019): 35–55.
15. Sullivan, "Avoiding Lamentation."

16. John Naughton, "The Evolution of The Internet: From Military Experiment to General Purpose Technology," *Journal of Cyber Policy* 1, no. 1 (2016): 5–28, 8.

17. Aaronson, "What Might Have Been."

18. Naughton, "The Evolution of the Internet," 21.

19. Nanjira Sambuli, "Challenges and Opportunities for Advancing Internet Access In Developing Countries While Upholding Net Neutrality," *Journal of Cyber Policy* 1, no. 1 (2016): 61–74.

20. Ibid., 62.

21. Vinod K. Aggarwal and Andrew W. Reddie, "Comparative Industrial Policy and Cybersecurity: The US Case," *Journal of Cyber Policy* 3, no. 3 (2018): 445–66.

22. Benjamin Bartlett, "Government as Facilitator: How Japan Is Building Its Cybersecurity Market," *Journal of Cyber Policy* 3, no. 3 (2018), 327–43.

23. Tai Ming Cheung, "The Rise of China as A Cybersecurity Industrial Power: Balancing National Security, Geopolitical, and Development Priorities," *Journal of Cyber Policy* 3, no. 3 (2018): 306–26.

24. Danilo D'Elia, "Industrial Policy: The Holy Grail of French Cybersecurity Strategy?" *Journal of Cyber Policy* 3, no. 3 (2018): 385–406.

25. Ministry for Europe and Foreign Affairs, "Paris Call for Trust and Security in Cyberspace," accessed May 22, 2019, https://www.diplomatie.gouv.fr/en/french-foreign-policy/digital-diplomacy/france-and-cyber-security/article/cybersecurity-paris-call-of-12-november-2018-for-trust-and-security-in.

26. To make my hypotheses most useful, I based them on the theoretical parts of this chapter; I establish that there is a predictor (what we think of as a cause) and an outcome; and that the hypotheses are testable. See Theresa Marchant-Shapiro, *Statistics for Political Analysis: Understanding the Numbers* (London: Sage, 2017), 179–221.

27. Robert Miller and Ciaran Acton. *SPSS for Social Scientists* (London: Palgrave Macmillan, 2009), 123.

28. It is standard practice to decide to reject null hypotheses if p values fall below a certain threshold, such as .05 or .01, so that the probability of a Type I error is small. See Michelle Lacey, "Concept of Significance Level," in *Encyclopedia of Research Design*, ed. Neil J. Salkind (Thousand Oaks: Sage, 2012), 1367.

29. Miller and Acton, *SPSS*, 126.

30. Ibid., 127.

31. Ibid., 128.

32. Ibid., 199.

33. Ibid., 204.

34. See Christof Wolf and Helen Best, "Linear Regression," in *The SAGE Handbook of Regression Analysis and Causal Inference*, ed. Henning Best and Christof Wolf (London: Sage, 2013), 57–82.

35. Miller and Acton, *SPSS*, 209.

36. Claire Lemercier and Claire Zalc, *Quantitative Methods in the Humanities: An Introduction* (Charlottesville: University of Virginia Press, 2019), 73.

37. Judea Pearl, "The Foundations of Causal Inference," 76.
38. Miller and Acton, *SPSS*, 212.
39. See Matthew W. Reynolds, "Bivariate Regression," in *Encyclopedia of Research Design*, ed. Neil J. Salkind (Thousand Oaks: Sage, 2012), 87–91, and Douglas C. Montgomery, Elizabeth A. Peck, G. Geoffrey Vining, *Introduction to Linear Regression Analysis* (London: John Wiley, 2012).
40. Samprit Chateerjee and Alis S. Hadi, *Regression Analysis by Example* (London: John Wiley, 2012), 58.
41. Lemercier and Claire Zalc, *Quantitative Methods*, 76.
42. Miller and Acton, *SPSS*, 212.
43. Ibid.
44. Lemercier and Zalc, *Quantittive Methods*, 74. See also, Marchant-Shapiro, *Statistics for Political Analysis*, 343–76.
45. Miller and Acton, *SPSS*, 216.
46. Manfred Max Bergman, "Combining Different Types of Data for Quantitative Analysis," in *The SAGE Handbook of Social Research Methods*, ed. Pertti Alasuutari, Leonard Bickman, and Julia Brannen (London: Sage), 585–601.
47. Sean Gilmard, *Statistical Modeling and Inference for Social Science* (New York: Cambridge University Press, 2014).
48. Ibid.

Conclusion

1. Ian Davis, "Annex C. Chronology 2017," in *SIPRI Yearbook 2018: Armaments, Disarmament and International Security*, accessed November 12, 2019, https://www.sipriyearbook.org/view/9780198821557/sipri-9780198821557-chapter-13.xml.
2. Ian Davis, "Annex C. Chronology 2018," *SIPRI Yearbook 2019: Armaments, Disarmament and International Security*, November 8, 2019, https://www.sipriyearbook.org/view/9780198839996/sipri-9780198839996-chapter-13.xml.
3. Tim Stevens and Kevin O'Brien, "Brexit and Cyber Security." *RUSI Journal* 164, no. 3 (2019): 22–30.
4. Adriana Barrera and Raphael Satter, "Hackers Demand $5 Million from Mexico's Pemex in Cyberattack." *Reuters*, November 13, 2019, https://www.reuters.com/article/us-mexico-pemex/hackers-demand-5-million-from-mexicos-pemex-in-cyberattack-idUSKBN1XN03A.
5. Sam Cowie, "Brazilian Prosecutors Charge Journalist Glenn Greenwald with Cybercrimes." *Guardian*, January 21, 2020, https://www.theguardian.com/world/2020/jan/21/glenn-greenwald-charged-cybercrime-brazil.
6. The fragmentation of cyberspace has received a critical approach lately, which proposes its territoriality as mutable and dynamic involving the constant renegotiation of space in direction and depth. See Daniel Lambach, "The Territorialization of Cyberspace," *International Studies Review* 22, no. 3 (2020): 482–506.

7. See David C. Gompert, and Martin Libicki, "Waging Cyber War the American Way," *Survival* 57, no. 4 (2015): 7–28.

8. I borrow this idea from Henry Kissinger, *Diplomacy* (New York: Simon and Schuster, 1994).

9. Pawlak and Barmpaliou, "Politics of Cybersecurity."

10. See Stewart Patrick, *Weak Links: Fragile States, Global Threats, and International Security* (New York: Oxford University Press, 2011).

11. See Ben Buchanan, *The Cybersecurity Dilemma* (London: Hurst, 2016), and Pawlak and Barmpaliou, "Politics of Cybersecurity."

12. Henry Kissinger, "How the Enlightment Ends," *The Atlantic*, June, 2018, https://www.theatlantic.com/magazine/archive/2018/06/henry-kissinger-ai-could-mean-the-end-of-human-history/559124/.

13. See this discussion, for instance, in Thomas Renard, "EU Cyber Partnerships: Assessing the EU Strategic Partnerships with Third Countries in The Cyber Domain," *European Politics and Society* 19, no. 3 (2018): 321–37.

14. Leo Kelion, "Huawei Set for Limited Role in UK 5G Networks," *BBC News*, January 28, 2020, https://www.bbc.co.uk/news/technology-51283059.

15. Ibid.

16. Anthony Boadle, "Huawei Role in Brazil 5G Up to National Security Chief: Regulator," *Reuters*, February 18, 2020, https://uk.reuters.com/article/us-brazil-telecoms/huawei-role-in-brazil-5g-up-to-national-security-chief-regulator-idUKKBN20C1U1.

17. Ibid.

18. Hal Brands, "American Grand Strategy in the Post–Cold War Era," in *New Directions in Strategic Thinking 2.0*, ed. Russell W. Glenn (Canberra: ANU Press, 2018), 133–48.

19. Marcus Willett, "Assessing Cyber Power," *Survival* 61, no. 1(2019): 85–90, 85.

20. See, for instance, a similar analysis in Hans Binnendijk. *Friends, Foes, and Future Directions: U.S. Partnerships in a Turbulent World: Strategic Rethink* (Santa Monica: RAND Corporation, 2016).

21. Richard A. Clarke, and Rob Knake, "The Internet Freedom League: How to Push Back Against the Authoritarian Assault on the Web," *Foreign Affairs* 98, no. 5(2019): 184.

22. See Jarno Limnéll, "The Cyber Arms Race Is Accelerating—What Are The Consequences?" *Journal of Cyber Policy* 1, no. 1 (2016): 50–60.

23. Hunter Marshton, "The U.S.-China Cold War Is a Myth," *Foreign Policy*, September 6, 2019, https://foreignpolicy.com/2019/09/06/the-u-s-china-cold-war-is-a-myth/.

24. See a relevant analysis in Desha Girod, "How to Win Friends and Influence Development: Optimising US Foreign Assistance," *Survival* 61, no. 6 (2019): 99–114. For a general overview of themes on U.S.-Latin American security links,

see David Mares and Arie Kacowicz, *Routledge Handbook of Latin American Security* (London: Routledge, 2016).

25. Jeffrey S. Sachs, "America's War on Chinese Technology," *Project Syndicate*, November 7, 2019, https://www.project-syndicate.org/commentary/cheney-doctrine-us-war-on-chinese-technology-by-jeffrey-d-sachs-2019-11.

26. As I have mentioned throughout the book, the gray areas of military and cybersecurity policy making are an outcome of the overlapping areas between civilian and military roles in national security. See on the matter Deborah L. Norden, "Latin American Militaries in the 21st Century: Civil-Military Relations in the Era of Disappearing Boundaries," in *Routledge Handbook of Latin American Security*, ed. David Mares and Arie Kacowicz (London: Routledge, 2016), 242–53.

27. See this point in particular in Sergei Boeke and Dennis Broeders, "The Demilitarisation of Cyber Conflict," *Survival* 60, no. 6 (2018): 73–90.

28. Damien Van Puyvelde and Aaron F. Brantley, *Cybersecurity: Politics, Governance, and Conflict in Cyberspace* (Cambridge: Polity, 2019).

29. International Institute for Strategic Studies, "Chapter Eight: Latin America," 382–83.

30. See Thomas Bruneau and Aurel Croissant, *Civil-Military Relations: Control and Effectiveness across Regimes* (Boulder: Lynne Rienner, 2019).

31. Mara E. Karlin and Alice H. Friend, "Military Worship Hurts U.S. Democracy," *Foreign Policy*, September 21, 2018, https://foreignpolicy.com/2018/09/21/military-worship-hurts-us-democracy-civilian-trump/.

32. Madeline Carr, *US Power and the Internet in International Relations* (Basingstoke: Palgrave Macmillan, 2016).

33. See International Institute for Strategic Studies, "Chapter Ten: Military Cyber Capabilities," *The Military Balance* 120, no. 1 (2020): 515–18.

34. Over the years, I have reflected on policy networks and institutional ideas presented by Martin Smith since the publication of his 1993 book *Pressure, Power & Policy* (Pittsburgh University Press). I am indebted to Martin and Adam White for their academic support and especially for passing on to me their teachings on governance and public policy.

35. Helena Carrapico and Andre Barrinha, "European Union Cyber Security as an Emerging Research and Policy Field," *European Politics and Society* 19, no. 3 (2018): 299–303.

36. George Christou and Seamus Simpson, "The European Union, Multilateralism, and The Global Governance of the Internet," *Journal of European Public Policy* 18, no. 2 (2011): 241–57.

37. Adam White reflected upon this idea based on the state-in-society approach coined by Joel Migdal. See Adam White, "Introduction: A State-in-Society Agenda," in *The Everyday Life of the State*, ed. Adam White (Seattle: University of Washington Press, 2013), 3–12.

38. Carolina Sancho, "Ciberseguridad: Presentación del Dossier," *Urvio* 20 (2018): 8–18, 14.

39. See, for instance, Christopher Whyte and Brian Mazanec, *Understanding Cyber Warfare: Politics, Policy, and Strategy* (New York: Routledge, 2018), and Herbet Lin and Amy Zegart, *Bytes, Bombs, and Spies: The Strategic Dimension of Offensive Cyber Operations* (Washington, DC: Brookings, 2018).

40. See Myriam Dunn Cavelty, "Europe's Cyber-Power," *European Politics and Society* 19, no. 3 (2018): 304–20.

41. See the issue of scale in governance in Chris Ansell and Jacob Torfing, *How Does Collaborative Governance Scale* (Bristol: Polity Press, 2018).

42. B. Guy Peters and Jon Pierre, *Comparative Governance: Rediscovering the Functional Dimension of Governing* (New York: Cambridge University Press, 2016).

43. Sam Jones, "Venezuela Blackout: What Caused It and What Happens Next?" *Guardian*, March 13, 2019, https://www.theguardian.com/world/2019/mar/13/venezuela-blackout-what-caused-it-and-what-happens-next.

44. Lara Seligman, "U.S. Military Wary of China's Foothold in Venezuela," *Foreign Policy*, April 8, 2019, https://foreignpolicy.com/2019/04/08/us-military-wary-of-chinas-foothold-in-venezuela-maduro-faller-guaido-trump-pentagon/.

Appendix

1. The following cutoff points to measure Pearson's *r* were used: +/–.1 = weak; +/–.3 = modest; +/–.5 = moderate; +/–.8 = strong.

2. I used the following cutoff points for the Adjusted R Square values: 0.1 = poor fit; 0.11–0.3 = modest fit; 0.31–0.5 = moderate fit; 0.5 = strong fit. See, Daniel Mujis, *Doing Quantitative Research in Education with SPSS* (London: Sage, 2011).

3. Acock, *A Gentle Introduction*.

4. See how to control for a third variable in Marchant-Shapiro, *Statistics for Political Analysis*, 319–42.

5. Miller and Acton, *SPSS*, 241.

6. Lemercier and Zalc, *Quantitative Methods in the Humanities*, 87.

7. Miller and Acton, *SPSS*, 241.

8. Ibid., 242.

9. Ibid., 246.

10. Ibid., 247.

11. Ibid., 251.

12. Ibid., 255.

Bibliography

Aaronson, Susan Ariel. "What Might Have Been and Could Still Be: The Trans-Pacific Partnership's Potential to Encourage an Open Internet and Digital Rights." *Journal of Cyber Policy* 2, no. 2 (2017): 232–54.
Abebe, Daniel. "Cyberwar, International Politics, and Institutional Design." *The University of Chicago Law Review* 83, no. 1 (2016): 1–22.
Access Now. "Argentina Province in Talks with Huawei over Acquiring Facial Recognition Surveillance Technology." July 12, 2018. https://www.accessnow.org/mendoza-surveillance-tech/.
Acemoglu, Daron, and Pierre Yared. "Political Limits to Globalization." *American Economic Review: Papers and Proceedings* 100 (2010): 83–88.
Acharya, Amitav. *Constructing a Security Community in Southeast Asia: ASEAN and the Problem of Regional Order*. Basingstoke: Routledge, 2009.
Acock, Alan C. *A Gentle Introduction to Stata*. College Station: Stata Press, 2018.
Adamsky, Dmitry. "The Israeli Odyssey toward Its National Cyber Security Strategy." *The Washington Quarterly* 40, no. 2 (2017): 113–27.
Agencia EFE. "Chile y Estados Unidos Colaboraran En Seguridad Cibernética." August 16, 2018. https://www.efe.com/efe/america/politica/chile-y-estados-unidos-colaboraran-en-seguridad-cibernetica/20000035-3721877.
Aggarwal, Vinod K., and Andrew W. Reddie. "Comparative Industrial Policy and Cybersecurity: The US Case." *Journal of Cyber Policy* 3, no. 3 (2018): 445–66.
Agrafiotis, Ioannis, Jason R. C. Nurse, Michael Goldsmith, Sadie Creese, and David Upton. "A Taxonomy of Cyber-harms: Defining the Impacts of Cyber-attacks and Understanding How They Propagate." *Journal of Cybersecurity* 4, no. 1 (2018): 1–15.
Albano Amarante, Jose Carlos, and Patrice Franko. "Defense Transformation in Latin America: Will It Transform the Technological Base?" *Democracy and Security* 13, no. 3 (2017): 173–95.
Allison, Graham. *Destined for War: Can America and China Scape the Thucydides' Trap?* London: Scribe Publications, 2017.

Alves, Paulo Vicente. "Why the Chinese Are Getting Aggressive in South America." July 6, 2015. http://thehill.com/blogs/congress-blog/foreign-policy/245208-why-the-chinese-are-getting-aggressive-in-south-america.

Amanda Erickson, "Our First Concern Was to Avoid a Civil War: Nicaragua's Government on Six Months of Protests." *The Washington Post*, September 26, 2018. https://www.washingtonpost.com/world/2018/09/26/our-first-concern-was-avoid-civil-war-nicaraguas-government-six-months-protests/?utm_term=.7d62b198243f.

Anderson, C. W. "Web 2.0," *Oxford Bibliographies*. DOI: 10.1093/OBO/9780199756841-0072.

Andres, Richard B. "The Emerging Structure of Strategic Cyber Offense, Cyber Defense, and Cyber Deterrence." In *Cyberspace, in Cyberspace and National Security: Threats, Opportunities, and Power in a Virtual World*, edited by Derek S. Reveron, 89–104. Washington, DC: Georgetown University Press, 2012.

Ansell, Chris, and Jacob Torfing. *How Does Collaborative Governance Scale?* Bristol: Polity Press, 2018.

Ansell, Christopher, and Jacob Torfing. *Handbook on Theories of Governance*. Cheltenham: Edward Elgar, 2016.

Anson, Jay H. "United to Counter Cyber Crime." *Dialogo*, May 19, 2017. https://dialogo-americas.com/en/articles/united-counter-cyber-crime.

Anthony, Ian. "European Security." In *SIPRI Yearbook 2017: Armaments, Disarmament and International Security*. https://www.sipriyearbook.org/view/9780198811800/sipri-9780198811800-chapter-4-div1-20.xml.

Anthony, Ian, Sam Perlo-Freeman, and Siemon T. Wezeman. "The Ukraine Conflict and its Implications." In *SIPRI Yearbook 2015: Armaments, Disarmament, and International Security*. https://www.sipriyearbook.org/view/9780198737810/sipri-9780198737810-chapter-3-div1-3.xml.

Applebaum, Anne. "Saudi Arabia's Information War to Bury News of Jamal Khashoggi." *Washington Post*, October 17, 2018. https://www.washingtonpost.com/opinions/global-opinions/saudi-arabias-information-war-to-bury-news-of-jamal-khashoggi/2018/10/17/e4825a5a-d227-11e8-b2d2-f397227b43f0_story.html?utm_term=.affe26290cf4.

Arostegui, Martin. "Colombia Probes Voter Registration Cyberattacks Traced to Russia's Allies." *Voice of America News*, March 15, 2018. https://www.voanews.com/a/colombia-voter-registration-cyberattacks-russia-allies/4300571.html.

Arson, Cynthia J. "U.S.-Argentine Relations and the Visit of President Mauricio Macri." *Wilson Center*, April 27, 2017. https://www.wilsoncenter.org/article/us-argentine-relations-and-the-visit-president-mauricio-macri.

Ashby, Paul. "U.S. National Security in the Western Hemisphere." In *The Oxford Handbook of U.S. National Security*, edited by Derek S. Reveron, Nikolas K. Gvosdev, and John A. Cloud, 611–83. New York: Oxford University Press, 2018.

Astore, William J. "Science and Technology in War." *Oxford Bibliographies*. DOI: 10.1093/OBO/9780199791279-0054.
Attrill-Smith, Alison, and Chris Fullwood, Melanie Keep, and Daria J. Kuss. *The Oxford Handbook of Cyberpsychology*. Oxford: Oxford University Press, 2019.
Austin, Anastasia. "Latin America under Threat of Cybercrime amid Coronavirus." *Insight Crime*, April 8, 2020. https://insightcrime.org/news/analysis/threat-cyber-crime-coronavirus/.
Baird, Zoe. "Governing the Internet: Engaging Government, Business, and Nonprofits." *Foreign Affairs* 81, no. 6 (2002): 15–20.
Ball, Desmond. "Australian Cyber-warfare Centre." In *Australia and Cyber-Warfare*, edited by Gary Waters, Desmond Ball, and Ian Dudgeon, 119–48. Canberra: ANU Press, 2008.
Balzacq, Thierry, Sarah Léonard, and Jan Ruzicka. "'Securitization' Revisited: Theory and Cases." *International Relations* 30, no. 4 (2016): 494–531.
Bamford, James. "Commentary: The World's Best Cyber Army Doesn't Belong to Russia." *Reuters*, August 4, 2016. http://www.reuters.com/article/us-election-intelligence-commentary/commentary-the-worlds-best-cyber-army-doesnt-belong-to-russia-idUSKCN10F1H5.
Banks, William. "State Responsibility and Attribution of Cyber intrusions After Tallinn 2.0." *Texas Law Review* 95, no. 7 (2017): 1488–513.
Bargent, James. "Top Paraguay Official Reveals Arms Trafficking Modus Operandi." *Insight Crime*, March 10, 2016. https://www.insightcrime.org/news/brief/top-paraguay-official-reveals-modus-operandi-arms-trafficking/.
Barkawi, Tarak. "'Defence Diplomacy' in North-South Relations." *International Journal* 66, no. 3 (2011): 597–612.
Barrera, Adriana, and Raphael Satter. "Hackers Demand $5 Million from Mexico's Pemex in Cyberattack." *Reuters*, November 13, 2019. https://www.reuters.com/article/us-mexico-pemex/hackers-demand-5-million-from-mexicos-pemex-in-cyber attack-idUSKBN1XN03A.
Bartlett, Benjamin. "Government as Facilitator: How Japan Is Building Its Cybersecurity Market." *Journal of Cyber Policy* 3, no. 3 (2018): 327–43.
Battaleme, Juan, and Francisco de Santibañes. "Argentina's Defence Deficit." *Survival* 61, No. 4 (2019): 63–78.
BBC Mundo. "¿Por qué Aumentan los Delitos Cibernéticos en América Latina?" November 24, 2015. http://www.bbc.com/mundo/noticias/2015/11/151118_tecnologia_cibercrimen_ciberdelito_aumento_america_latina_lb.
BBC News. "China Denies Pentagon Cyber-Raid." September 4, 2007, http://news.bbc.co.uk/1/hi/world/americas/6977533.stm.
BBC News. "Cyber Attack on Pentagon Email." June 22, 2007, http://news.bbc.co.uk/1/hi/world/americas/6229188.stm.
BBC News. "Meng Wanzhou: Huawei Executive Suffers US Extradition Blow." May 27, 2020. https://www.bbc.co.uk/news/world-us-canada-52793343.

BBC News. "Mike Pompeo Warns UK Over Huawei 'Security Risks.'" May 8, 2019. https://www.bbc.com/news/uk-politics-48198932.

Behringer, Ronald M. "Middle Power Leadership on the Human Security Agenda." *Cooperation and Conflict* 40, no. 3 (2005): 305–42.

Behringer, Ronald M. *The Human Security Agenda.* London: Continuum, 2012.

Benavente, José Miguel, Gustavo Crespi, Jorge Katz, and Giovanni Stumpo. "New Problems and Opportunities for Industrial Development in Latin America." *Oxford Development Studies* 25, no. 3 (1997): 261–77.

Benson, David C. "Why the Internet Is Not Increasing Terrorism." *Security Studies* 23, no. 2 (2014): 293–328.

Bergman, Manfred Max. "Combining Different Types of Data for Quantitative Analysis." In *The SAGE Handbook of Social Research Methods*, edited by Pertti Alasuutari, Leonard Bickman, and Julia Brannen, 585–601. London: Sage, 2008.

Berkowitz, Bruce. *The New Face of War: How War Will Be Fought in the 21st Century.* New York: Free Press, 2003.

Betts, Richard K. "The Lost Logic of Deterrence: What the Strategy That Won the Cold War Can—and Can't—Do Now." *Foreign Affairs* 92, no. 2 (2013): 87–99.

Betz, David J., and Tim Stevens. *Cyberspace and the State: Towards a Strategy for Cyber-Power.* London: Routledge, 2012.

Betz, David J., and Tim Stevens. "Analogical Reasoning and Cyber Security." *Security Dialogue* 44, No. 2 (2013): 147–64.

Bevir, Mark. *Democratic Governance.* Princeton: Princeton University Press, 2010.

Bhuiyan. Ali. *Internet Governance and the Global South: Demand for a New Framework.* Basingstoke: Palgrave Macmillan, 2016.

Bianculli, Andrea. "Latin America." In *The Oxford Handbook of Comparative Regionalism*, edited by Tanja A. Börzel and Thomas Risse, 154–77. Oxford: Oxford University Press, 2016.

Biddle, Stephen, and Robert Zirkle. "Technology, Civil-Military Relations, and Warfare in the Developing World." *Journal of Strategic Studies* 19, no. 2 (1996): 171–212.

Bildt, Carl, and Gordon Smith "The One and Future Internet." *Journal of Cyber Policy* 1, no. 2 (2016): 142–56.

Binnendijk, Hans. *Friends, Foes, and Future Directions: U.S. Partnerships in a Turbulent World: Strategic Rethink.* Santa Monica: RAND Corporation, 2016.

Boadle, Anthony. "Huawei Role in Brazil 5G Up to National Security Chief: Regulator." *Reuters*, February 18, 2020. https://uk.reuters.com/article/us-brazil-telecoms/huawei-role-in-brazil-5g-up-to-national-security-chief-regulator-idUKKBN20C1U1.

Boeke, Sergei, and Dennis Broeders. "The Demilitarisation of Cyber Conflict." *Survival* 60, no. 6 (2018): 73–90.

Boulanin, Vincent. "Cybersecurity: A Precondition to Sustainable Information and Communication Technology-Enabled Human Development." In *SIPRI*

Yearbook 2015: Armaments, Disarmament and International Security. https://www.sipri.org/yearbook/2015.
Boulanin, Vincent. "Mapping Key Actors and Efforts in Cybersecurity for Human Development." In *SIPRI Yearbook 2015: Armaments, Disarmament and International Security.* https://www.sipri.org/yearbook/2015.
Bowers, Ian. "The Use and Utility of Hybrid Warfare on the Korean Peninsula." *The Pacific Review* 31, No. 6 (2018): 762–86.
Bradshaw, Samantha, and Philip N. Howard. *The Global Disinformation Order: 2019 Global Inventory of Organised Social Media Manipulation.* Oxford: Project on Computational Propaganda, 2019.
Brands, Hal. "American Grand Strategy in Tthe Post–Cold War Era." In *New Directions in Strategic Thinking 2.0*, edited by Russell W. Glenn, 133–48. Canberra: ANU Press, 2018.
Brands, Hal. "Crime, Irregular Warfare, and Institutional Failure in Latin America: Guatemala as a Case Study." *Studies in Conflict & Terrorism* 34, no. 3 (2011): 228–47.
Brands, Hal. *American Grand Strategy in the Age of Trump.* Washington, DC: Brookings Institution Press, 2018.
Brantly, Aaron Franklin. "The Cyber Losers." *Democracy and Security* 10, no. 2 (2014): 132–55.
Braw, Elisabeth, and Gary Brown. "Personalised Deterrence of Cyber Aggression." *RUSI Journal* 165, no. 2 (2020): 48–54.
Brenner, Joel F. 2013. "Eyes Wide Shut: The Growing Threat of Cyber Attacks on Industrial Control Systems." *Bulletin of the Atomic Scientists* 69 (5): 15–20, 16.
Brenner, Susan. *Cyber Threats: The Emerging Fault Lines of the Nation State.* Oxford: Oxford University Press, 2009.
Bromley, Mark, and Alfredo Malaret. "ATT-Related Activities in Latin America and the Caribbean: Identifying Gaps and Improving Cooperation." February, 2017. https://www.sipri.org/sites/default/files/ATT-related-activities-Latin-America-and-Caribbean.pdf.
Bromley, Mark, Kolja Brockmann, and Giovanna Maletta. "Controls on Intangible Transfers of Technology and Additive Manufacturing." In *SIPRI Yearbook 2018: Armaments, Disarmament, and International Security.* https://www.sipriyearbook.org/view/9780198821557/sipri-9780198821557.xml.
Brooks, Rosa. *How Everything Became War and the Military Became Everything: Tales from the Pentagon.* New York: Simon and Schuster, 2016.
Bruneau, Thomas C. *Patriots for Profit: Contractors and the Military in U.S. National Security.* Stanford: Stanford University Press, 2011.
Bruneau, Thomas C. "Civilians and the Military in Latin America: The Absence of Incentives." *Latin American Politics and Society* 55, no. 4 (2013): 143–60.

Bruneau, Thomas C., and Florina Cristiana Matei. "Towards a New Conceptualization of Democratization and Civil-Military Relations." *Democratization* 15, no. 5 (2008): 909–29.

Bruneau, Thomas C., and Aurel Croissant. *Civil-Military Relations: Control and Effectiveness across Regimes.* Boulder: Lynne Rienner, 2019.

Bruneau, Thomas C., and Aurel Croissant. "Civil-Military Relations: Why Control Is Not Enough." In *Civil Military Relations: Control and Effectiveness across Regimes*, edited by Thomas C. Bruneau and Aurel Croissant, 1–18. Boulder: Lynne Rienner, 2019.

Brysk, Alison. *Global Good Samaritans: Human Rights as Foreign Policy.* Oxford: Oxford University Press, 2009.

Buchanan, Ben. *The Cybersecurity Dilemma.* London: Hurst, 2016.

Buchanan, Ben. *The Hacker and the State: Cyber Attacks and the New Normal of Geopolitics.* Cambridge: Harvard University Pres, 2020.

Burns, William J. "The Demolition of U.S. Diplomacy." *Foreign Affairs*, October 14, 2019. https://www.foreignaffairs.com/articles/2019-10-14/demolition-us-diplomacy.

Burton, Joe. "Small States and Cyber Security: The Case of New Zealand." *Political Science* 65, no. 2 (2013): 216–38.

Business Monitor International. "Latin America Security Risk Ratings." July 7, 2014. EBSCOhost.

Business Monitor International. "Defence and Security Report, Americas." July 7, 2015. EBSCOhost.

Business Monitor International. "Latin America Security Overview." July 7, 2015. EBSCOhost.

Business Monitor International. "BMI Defence and Security Report, Brazil." February 3, 2015. EBSCOhost.

Business Monitor International. "BMI Defence and Security Report, Colombia." January 1, 2017. EBSCOhost.

Business Monitor International. "BMI Defence and Security Report, Chile." January 1, 2018. EBSCOhost.

Business Monitor International. "BMI Defence and Security Report, Mexico." January 1, 2018. EBSCOhost.

Business Monitor International. "BMI Defence and Security Report, Venezuela." January 1, 2018. EBSCOhost.

Business Wire. "Future of the Argentina Defense Industry to 2022." July 13, 2017. https://www.businesswire.com/news/home/20170713005850/en/Future-Argentina-Defense-Industry-2022—Market.

Cadwalladr, Carole, and Emma Graham-Harrison. "Revealed: 50 Million Facebook Profiles Harvested from Cambridge Analytica in Major Data Breach." *Guardian*, March 17, 2018. https://www.theguardian.com/news/2018/mar/17/cambridge-analytica-facebook-influence-us-election.

Campbell, Kurt M., and Jake Sullivan. "Competition Without Catastrophe: How America Can Both Challenge and Coexist with China." *Foreign Affairs* 98, no. 5 (2019): 96–110.

Carr, Jeffrey. "The Misunderstood Acronym: Why Cyber Weapons Aren't WMD." *Bulletin of the Atomic Scientists* 69, No. 5 (2013): 32–37.

Carr, Madeline. *US Power and the Internet in International Relations.* Basingstoke: Palgrave Macmillan, 2016.

Carrapico, Helena, and Andre Barrinha. "European Union Cyber Security as an Emerging Research and Policy Field." *European Politics and Society* 19, no. 3 (2018): 299–303.

Casanova, Lourdes. "The Challenges of Chinese Investment in Latin America." *World Economic Forum*, March 15, 2018. https://www.weforum.org/agenda/2018/03/latin-america-china-investment-brazil-private-public/.

Catota, Frankie E., M. Granger Morgan, and Douglas C. Sicker. "Cybersecurity Education in a Developing Nation: The Ecuadorian Environment." *Journal of Cybersecurity* 5, no. 1 (2019): 1–19.

Centeno, Miguel, Atul Kohli, and Deborah Yashar. *States in the Developing World.* Cambridge: Cambridge University Press, 2017.

Cerami, Joseph, and John David Young. "Military Science." *Oxford Bibliographies.* DOI: 10.1093/OBO/9780199743292-0052.

Chan, Kelvin. "Apple CEO Backs Privacy Laws, Warns Data Being 'Weaponized.'" *Washington Post*, October 24, 2018. https://www.washingtonpost.com/business/technology/apple-ceo-backs-privacy-laws-warns-data-being-weaponized/2018/10/24/f7485a08-d773-11e8-8384-bcc5492fef49_story.html?utm_term=.cdabba2a4c36.

Chang, Lennon Y. C., and Peter Grabosky. "The Governance of Cyberspace." In *Regulatory Theory: Foundations and Applications*, edited by Peter Drahos, 533–51. Canberra: ANU Press, 2017.

Chateerjee, Samprit, and Alis S. Hadi. *Regression Analysis by Example.* London: John Wiley, 2012.

Cheresky, Isidoro. *El Nuevo Rostro de la Democracia en América Latina: Ciudadanía, Liderazgos E Instituciones.* Buenos Aires: Fondo de Cultura Económica, 2015.

Chertoff, Michael, and John Michael McConnell. "A Path Forward for Cyber Defense and Security." In *Securing Cyberspace: A New Domain for National Security*, edited by Nicholas Burns and Jonathan Price, 191–202. Washington, DC: Aspen Institute, 2012.

Cheung, Tai Ming. "The Rise of China as a Cybersecurity Industrial Power: Balancing National Security, Geopolitical, and Development Priorities." *Journal of Cyber Policy* 3, no. 3 (2018): 306–26.

Chipman, John, and James Lockhart Smith. "South America: Framing Regional Security." *Survival* 51, no. 6 (2009): 77–104.

Choucri, Nazli. *Cyberpolitics in International Relations.* Cambridge: MIT Press, 2012.

Choucri, Nazli, Stuart Madnick, and Jeremy Ferwerda. "Institutions for Cyber Security: International Responses and Global Imperatives." *Information Technology for Development* 20, no. 2 (2014): 96–121.

Christou, George, and Seamus Simpson. "The European Union, Multilateralism, and the Global Governance of the Internet." *Journal of European Public Policy* 18, no. 2 (2011): 241–57.

Christou, George. *Cybersecurity in the European Union: Resilience and Adaptability in Governance Policy*. Basingstoke: Palgrave Macmillan, 2015.

Chung, Jae Ho. "How America and China See Each Other: Charting National Views and Official Perceptions." *The Pacific Review* 32, no. 2 (2019): 188–209.

Clark, J. P. "Military Operations and the Defense Department." In *The Oxford Handbook of U.S. National Security*, edited by Derek S. Reveron, Nikolas K. Gvosdev, and John A. Cloud, 291–305. Oxford: Oxford University Press, 2018.

Clarke, Richard A., and Robert K. Knake. "The Internet Freedom League: How to Push Back Against the Authoritarian Assault on the Web." *Foreign Affairs* 98, no. 5 (2019): 184.

Clarke, Richard A., and Robert K. Knake. *Cyber War: The Next Threat to National Security and What to Do about It*. New York: Harper Collins, 2010.

Clement, Andrew, and Jonathan A. Obar. "Canadian Internet 'Boomerang' Traffic and Mass NSA Surveillance: Responding to Privacy and Network Sovereignty Challenges." In *Law, Privacy and Surveillance in Canada in the Post-Snowden Era*, edited by Michael Geist, 13–44. Ottawa: University of Ottawa Press, 2015.

Clemente, Dave. "International Security: Cyber Security as a Wicked Problem." *The World Today* 67, no. 10 (2011): 15–17.

Clough, Jonathan. *Principles of Cybercrime*. Cambridge: Cambridge University Press, 2015.

Cockburn, Andrew. *Kill Chain: Drones, and the Rise of High-Tech Assassins*. London: Verso, 2015.

Cohen, Ariel, and Ray Walser. "The Russia-Venezuela Axis: Using Energy for Geopolitical Advantage." July 21, 2008. https://www.heritage.org/europe/report/the-russia-venezuela-axis-using-energy-geopolitical-advantage

Cohen, Jacob, Patricia Cohen, Stephen G. West, and Leonna S. Aiken., *Applied Multiple Regression/Correlation Analysis for the Behavioral Sciences*. New York: Routledge, 2013.

Coleman, Stephen, and Jay Blumler. *The Internet and Democratic Citizenship: Theory, Practice, and Policy*. Cambridge: Cambridge University Press, 2009.

Colgan, Jeff. "Venezuela and Military Expenditure Data," *Journal of Peace Research* 48, no. 4 (2011): 547–56.

Comando Estratégico Operacional Fuerza Armada Nacional Bolivariana. "Ciberdefensa." Accessed December 20, 2019. https://ceofanb.mil.ve/direcciones/direccion-conjunta-de-ciberdefensa/.

Comando Estratégico Operacional Fuerza Armada Nacional Bolivariana. "FANB Desplegada en el marco del Ejercicio de Acción Integral 'Ana Karina Rote

2019.'" March 17, 2019. https://ceofanb.mil.ve/fanb-desplegada-en-el-marco-de-los-ejercicios-de-accion-integral-ana-karina-rote-2019/.
Connolly, Richard, and Cecilie Sendstad. "Russia's Role as an Arms Exporter: The Strategic and Economic Importance of Arms Exports for Russia." *Chatham House*, March 20, 2017. http://www.chathamhouse.org/sites/default/files/publications/research/2017-03-20-russia-arms-exporter-connolly-sendstad.pdf.
Contreras, Carolina. "Chilean Armed Forces Support Peace in Colombia." *Dialogo*, September 15, 2016. https://dialogo-americas.com/en/articles/chilean-armed-forces-support-peace-colombia.
Cooper, Andrew F. "Testing Middle Power's Collective Action in a World of Diffuse Power." *International Journal* 71, no. 4 (2016): 529–44.
Cooper, Andrew F. *Niche Diplomacy: Middle Powers after the Cold War*. New York: St. Martin's Press, 1997.
Cooper, Andrew F., and Jeremie Cornut. "The Changing Nature of Diplomacy." *Oxford Bibliographies*. DOI: 10.1093/OBO/9780199743292-0180.
Copeland, Dale C. "Realism and Neorealism in the Study of Regional Conflict." In *International Relations Theory and Regional Transformation*, edited by T. V. Paul, 49–72. Cambridge: Cambridge University Press, 2012.
Copp, Tara. "Deployed Border Troops Are Preparing for Militias Stealing Their Gear, Protester Violence, Documents Show." *Military Times*, October 31, 2018. https://www.militarytimes.com/news/your-military/2018/10/31/deployed-border-troops-are-preparing-for-militias-stealing-their-gear-protester-violence-documents-show/.
Cornish, Paul. "Governing Cyberspace through Constructive Ambiguity." *Survival* 57, no. 3 (2015): 153–76.
Cornish, Paul, and Kingsley Donaldson. *2020 World of War*. London: Hodder and Stoughton, 2017.
Cottey, Andrew, and Anthony Forster. *Reshaping Defence Diplomacy: New Roles for Military Cooperation and Assistance*. New York: Oxford University Press, 2004.
Council on Foreign Relations. "Cyber Operations Tracker." Accessed October 24, 2018, https://www.cfr.org/interactive/cyber-operations#Takeaways.
Cowie, Sam. "Brazilian Prosecutors Charge Journalist Glenn Greenwald with Cybercrimes." *Guardian*, January 21, 2020. https://www.theguardian.com/world/2020/jan/21/glenn-greenwald-charged-cybercrime-brazil.
Cox, Michael. "Power Shifts, Economic Change, and the Decline of the West?" *International Relations* 26, no. 4 (2012): 369–88.
Coyer, Paul. "China's Pivot to Latin America: Beijing's Growing Security Presence in America's Backyard." *Forbes*, February 20, 2016. https://www.forbes.com/sites/paulcoyer/2016/02/20/chinas-pivot-to-latin-america-beijings-growing-security-presence-in-americas-backyard/2/#70eb48951426.
Craig, Anthony. "Understanding the Proliferation of Cyber Capabilities." *Council on Foreign Relations*, October 18, 2018. https://www.cfr.org/blog/understanding-proliferation-cyber-capabilities.

Crandall, Matthew, and Collin Allan. "Small States and Big Ideas: Estonia's Battle for Cybersecurity Norms," *Contemporary Security Policy* 36, no. 2 (2015): 346–68.

Crandall, Russell. *The United States and Latin America after the Cold War*. New York: Cambridge University Press, 2008.

Crawford, Susan. "Net Neutrality Is Just a Gateway to the Real Issue: Internet Freedom." *Wired*, May 18, 2018. https://www.wired.com/story/net-neutrality-is-just-a-gateway-to-the-real-issue-internet-freedom/.

Creese, Sadie, Ruth Shillair, Maria Bada, and William H. Dutton. "Building the Cybersecurity Capacity of Nations." In *Society and the Internet: How Networks of Information and Communication are Changing Our Lives*, edited by Mark Graham and William H. Dutton, 166–79. Oxford: Oxford University Press, 2019.

Crowther, Glenn Alexander. "The Cyber Domain." *The Cyber Defense Review* 2, no. 3 (2017): 63–78.

Cuadros, Alex. "Open Talk of a Military Coup Unsettles Brazil." *New Yorker*, October 13, 2017. https://www.newyorker.com/news/news-desk/open-talk-of-a-military-coup-unsettles-brazil.

D'Elia, Danilo. "Industrial Policy: The Holy Grail of French Cybersecurity Strategy?" *Journal of Cyber Policy* 3, no. 3 (2018): 385–406.

Dargent, Eduardo. *Technocracy and Democracy in Latin America: The Experts Running Government*. New York: Cambridge University Press, 2015.

Das, Hari. "Statistical Analysis." In *Encyclopedia of Case Study Research*, edited by Albert J. Mills, Gabrielle Durepos, and Elden Wiebe, 889–892. Thousand Oaks: Sage, 2012.

Davis, Diane E., and Anthony W. Pereira. *Irregular Armed Forces and Their Role in Politics and State Formation*. New York: Cambridge University Press, 2003.

Davis, Ian. "Annex C. Chronology 2017." In *SIPRI Yearbook 2018: Armaments, Disarmament, and International Security*. https://www.sipriyearbook.org/view/9780198821557/sipri-9780198821557-chapter-13.xml.

Davis, Ian. "Annex C. Chronology 2018." In *SIPRI Yearbook 2019: Armaments, Disarmament, and International Security*. https://www.sipriyearbook.org/view/9780198821557/sipri-9780198821557-chapter-13.xml.

Deibert, Ronald J., and Masashi Crete-Nishihata, "Global Governance and the Spread of Cyberspace Controls." *Global Governance* 18, no. 3 (2012): 339–61.

Deibert, Ronald J. "Toward a Human-Centric Approach to Cybersecurity." *Ethics and International Affairs* 32, no. 4 (2018): 411–24.

Deibert, Ronald. "Trajectories for Future Cybersecurity Research." In *The Oxford Handbook of International Security*, edited by Alexandra Gheciu and William C. Wholforth, 532–46. New York: Oxford University Press, 2018.

Delgado-Ramos, Gian Carlo, and Silvina María Romano. "Political-Economic Factors in U.S. Foreign Policy: The Colombia Plan, the Mérida Initiative, and the Obama Administration." *Latin American Perspectives* 38, no. 4 (2011): 93–108.

Denoon, David B. H. *China, The United States, and the Future of Latin America.* New York: New York University Press, 2017.
DeRouen Jr., Karl R. "Defense Spending and Economic Growth in Latin America: The Externalities Effects." *International Interactions* 19, No. 3 (1994): 193–212.
Deutsche Welle. "Argentine and Russian Presidents Macri and Putin Sign Strategy Accord." January 23, 2018. http://www.dw.com/en/argentine-and-russian-presidents-macri-and-putin-sign-strategy-accord/a-42274790.
Deutsche Welle. "Venezuela Cyberattack Targets Government Websites." August 8, 2017. https://www.dw.com/en/venezuela-cyberattack-targets-government-websites/a-40002475.
DeVore, Marc R. "Commentary on the Value of Domestic Arms Industries: Security of Supply or Military Adaptation?" *Defence Studies* 17, no. 3 (2017): 242–59.
Diario Oficial de la República de Chile. "Ministerio De Defensa Nacional Aprueba Política De Ciberdefensa." March 3, 2018. https://www.diariooficial.interior.gob.cl/publicaciones/2018/03/09/42003/01/1363153.pdf.
Diaz, Fabio Andres, and Syed Mansoob Murshed. "'Give War a Chance': All-Out War as a Means of Ending Conflict in the Cases of Sri Lanka and Colombia." *Civil Wars* 15, no. 3 (2013): 281–305.
DiMarco, Louis. "Urban Warfare," *Oxford Bibliographies.* DOI: 10.1093/OBO/9780199791279-0171.
Dueck, Colin. *The Obama Doctrine: American Grand Strategy Today.* Oxford: Oxford University Press, 2015.
Dunn Cavelty, Myriam. "Cyber-Terror—Looming Threat or Phantom Menace? The Framing of the US Cyber-Threat Debate." *Journal of Information Technology & Politics* 4, no. 1 (2008): 19–36.
Dunn Cavelty, Myriam. *Cyber-Security and Threat Politics: US Efforts to Secure the Information Age.* Abingdon: Routledge, 2008.
Dunn Cavelty, Myriam. "Breaking the Cyber-Security Dilemma: Aligning Security Needs and Removing Vulnerabilities." *Science and Engineering Ethics* 20, no. 3 (2014): 701–15.
Dunn Cavelty, Myriam. "Europe's Cyber-Power." *European Politics and Society* 19, no. 3 (2018): 304–20.
Dunne, J. Paul, Sam Perlo-Freeman, and Aylin Soydan. "Military Expenditure and Debt in South America." *Defence and Peace Economics* 15, no. 2 (2004): 173–87.
Dunne, J. Paul, Sam Perlo-Freeman, and Ron P. Smith. "The Demand for Military Expenditure in Developing Countries: Hostility Versus Capability." *Defence and Peace Economics* 19, no. 4 (2008): 293–302.
Dupont, Benoit. "The Cyber-Resilience of Financial Institutions: Significance and Applicability." *Journal of Cybersecurity* 5, no. 1 (2019): 1–17.
Dutton, William H. "The Internet's Gift to Democratic Governance: The Fifth Estate." In *Can the Media Serve Democracy?* edited by Stephen Coleman, Giles Moss, and Katy Parry, 164–73. London: Palgrave Macmillan, 2015.

Eaglen, Mackenzie. "Pentagon Should Take the Lead on Tech." *Wall Street Journal*, September 4, 2019. https://www.wsj.com/articles/pentagon-should-take-the-lead-on-tech-11567636730.

Egloff, Florian J. "Contested Public Attributions of Cyber Incidents and the Role of Academia." *Contemporary Security Policy* 41, no. 1 (2020): 55–81.

El Tiempo. "Colombia También es Víctima del Ataque Global Informático." May 13, 2017. http://www.eltiempo.com/tecnosfera/novedades-tecnologia/ataque-cibernetico-afecta-redes-de-74-paises-87390.

El Universal. "El Gobierno Crea Comando Militar Para Garantizar Servicios Públicos Tras Apagón." March 15, 2019. http://www.eluniversal.com/economia/35501/el-gobierno-crea-comando-militar-para-garantizar-servicios-publicos-tras-apagon.

Ellis, R. Evan. *China in Latin America: The Whats and Wherefores*. Boulder: Lynne Rienner, 2009.

Ellis, R. Evan. *China on the Ground in Latin America: Challenges for the Chinese and Impacts on the Region*. New York: Palgrave Macmillan, 2014.

Erikson, Daniel P. "Make the Summit of the Americas Great Again." *Foreign Policy*, April 20, 2018. http://foreignpolicy.com/2018/04/20/make-the-summit-of-the-americas-great-again/.

Eriksson, Johan, and Giampiero Giacomello. "Who Controls the Internet? Beyond the Obstinacy or Obsolescence of the State." *International Studies Review* 11, no. 1 (2009): 205–30.

Fajardo Cabello, Daniel. "No Solo Amazon: Huawei Estudia Instalar En Chile Su Primer Data Center Para América Latina." *La Tercera*, February 21, 2019. https://www.latercera.com/pulso/noticia/no-solo-amazon-huawei-tambien-estudia-instalar-chile-gran-data-center-america-latina/537873/.

Farrell, Henry, and Martha Finnemore. "The End of Hypocrisy: American Foreign Policy in the Age of Leaks." *Foreign Affairs* 92, no. 6 (2013): 22–26.

Ferdinando, Lisa. "U.S. Wants to Remain 'Partner of Choice' in Latin America." U.S. Southern Command, April 6, 2017. https://www.defense.gov/Explore/News/Article/Article/1144536/us-wants-to-remain-partner-of-choice-in-latin-america/.

Fernandez-Osorio, Andres Eduardo, Fabio Nelson Cufiño-Gutierrez, Cesar Andres Gomez-Diaz, and Gustavo Andres Tovar-Cabrera. "Dynamics of State Modernization in Colombia: The Virtuous Cycle of Military Transformation." *Democracy and Security* 15, no. 1 (2018): 75–104.

Fernandez, Manny, David E. Sanger, and Marina Trahan Martinez. "Ransomware Attacks Are Testing Resolve of Cities across America." *New York Times*, August 22, 2019, Section A, page 1.

Financial Times. "Latin America Resists US Pressure to Exclude Huawei." June 9, 2019. https://www.ft.com/content/38257b66-83c5-11e9-b592-5fe435b57a3b.

Flournoy, Michèle, and Michael Sulmeyer. "Battlefield Internet: A Plan for Securing Cyberspace." *Foreign Affairs* 97, no. 5 (2018): 40–46.

Foreign Ministry of the People's Republic of China. "China's Defence Expenditure." April 7, 2011. http://www.fmprc.gov.cn/mfa_eng/wjb_663304/zzjg_663340/ jks_665232/ kjlc_665236/gjjk_665238/t410723.shtml.

Foreign Policy. "Can State's New Cyber Bureau Hack It?" January 18, 2019. https:// foreignpolicy.com/2019/01/18/state-department-cyber-security-cyber-threats-russia-china-diplomacy-capitol-hill-lawmakers-pompeo/.

Franceschi-Bicchierai, Lorenzo. "Venezuela's Government Appears to Be Trying to Hack Activists with Phishing Pages." *Vice*, February 15, 2019. https://motherboard. vice.com/en_us/article/d3mdxm/venezuela-government-hack-activists-phishing.

Fraundorfer, Markus. "Brazil's Organization of the NETmundial Meeting: Moving Forward in Global Internet Governance." *Global Governance* 23, no. 3 (2017): 503–21.

Freedman, Lawrence. "Deterrence: A Reply." *Journal of Strategic Studies* 28, no. 5 (2005): 789–801.

Fridman, Ofer. *Russian "Hybrid Warfare": Resurgence and Politicisation*. London: Hurst, 2018.

Fukuyama, Francis. "What Is Governance?" *Governance* 26, no 3 (2013): 347–68.

Galeotti, Mark. "The Cyber Menace." *The World Today* 68, no. 7 (2012): 32–35.

Gallagher, Kevin P., and Roberto Porzecanski. "China and the Future of Latin American Economic Development." In *The Oxford Handbook of Latin American Economics*, edited by José Antonio Ocampo and Jaime Ros, 461–87. Oxford: Oxford University Press, 2011.

Gao, Charlotte. "China Says Latin America 'Eager' To Join Belt and Road." *The Diplomat*, January 24, 2018. https://thediplomat.com/2018/01/ china-says-latin-america-eager-to- join-belt-and-road/.

Gardiner, C. H. "Rising Brazil Border Seizures Troubling Sign for the Region." *Insight Crime* March 5, 2018. https://www.insightcrime.org/news/brief/ brazil-frontier-region-busts/.

Garrison, Cassandra. "On South America's Largest Solar Farm, Chinese Power Radiates." *World Economic Forum*, April 2019. https://www.weforum.org/ agenda/2019/04/on-south-americas-largest-solar-farm-chinese-power-radiates/.

Gates, Robert M. "Helping Others Defend Themselves." *Foreign Affairs* 89, no. 3 (2010): 2–6.

Gehrke, Joel. "Pompeo to Latin America: 'We Love You, China Doesn't.'" *Washington Examiner*, April 12, 2019. https://www.washingtonexaminer.com/policy/ defense-national-security/pompeo-to-latin-america-we-love-you-china-doesn't.

Georgieva, Ilina. "The Unexpected Norm-Setters: Intelligence Agencies in Cyberspace." *Contemporary Security Policy* 41, no. 1 (2020): 33–54.

Gerring, John. "Case Selection for Case-Study Analysis: Qualitative and Quantitative Techniques." In *The Oxford Handbook of Political Methodology*, edited by Janet M. Box-Steffensmeier, Henry E. Brady, and David Collier, 646–84. New York: Oxford University Press, 2008.

Gill, Peter, and Mark Phythian, "What Is Intelligence Studies?" *The International Journal of Intelligence, Security, and Public Affairs* 18, no. 1 (2016): 5–19.

Gilley, Bruce, and Andrew O'Neil. "China's Rise through the Prism of Middle Powers." In *Middle Powers and the Rise of China*, edited by Bruce Gilley and Andrew O'Neil, 1–22. Washington, DC: Georgetown University Press, 2014.

Gilmard, Sean. *Statistical Modeling and Inference for Social Science*. New York: Cambridge University Press, 2014.

Girod, Desha. "How to Win Friends and Influence Development: Optimising US Foreign Assistance." *Survival* 61, no. 6 (2019): 99–114.

Glickhouse, Rachel. "Explainer: Fighting Cybercrime in Latin America." *Americas Society/Council of the Americas*, November 14, 2013. http://www.as-coa.org/articles/explainer-fighting-cybercrime-latin-america.

Gobierno de México. "Estrategia Nacional de Ciberseguridad, 2017." https://www.gob.mx/cms/uploads/attachment/file/271884/Estrategia_Nacional_Ciberseguridad.pdf.

Godoy, Emilio. "Fuerzas Armadas, Vulnerable a Ciberataque." *Proceso*, June 10, 2017. https://www.proceso.com.mx/490521/fuerzas-armadas-vulnerables-a-ciberataques.

Goldman, Emily O. *Power in Uncertain Times: Strategy in the Fog of Peace*. Stanford: Stanford University Press, 2010.

Goldman, Emily O., and Leslie C. Eliason. *The Diffusion of Military Technology and Ideas*. Stanford: Stanford University Press, 2003.

Gompert, David C., and Libicki, Martin. "Waging Cyber War the American Way." *Survival* 57, no. 4 (2015): 7–28.

Gouvea, Raul. "Brazil's New Defense Paradigm." *Defense & Security Analysis* 31, no. 2 (2015): 137–51.

Gouvea, Raul. "US-Latin America's Security: Moving Through an Inflection Point?" *Defense & Security Analysis* 33, no. 3 (2017): 223–41.

Graham, Stephen. *Cities under Siege: The New Military Urbanism*. New York: Verso, 2010.

Green, James A. *Cyber Warfare: A Multidisciplinary Analysis*. New York: Routledge, 2015.

Grigsby, Alex. "The Brazil-U.S. Cyber Relationship Is Back on Track." *Council on Foreign Relations*, July 1, 2015. https://www.cfr.org/blog/brazil-us-cyber-relationship-back-track.

Grinbaum, Mikael, and Susan T. Jackson. "Arms Production and Military Services." In *SIPRI Yearbook 2012: Armaments, Disarmament and International Security*. https://www.sipriyearbook.org/view/9780199650583/sipri-9780199650583-div1-30.xml.

Grindle, Merilee S. "Good Enough Governance Revisited." *Developmental Policy Review* 25, no. 5 (2007): 533–74.

Grindle, Merilee S. *Good Governance: The Inflation of an Idea*. HKS Faculty Research Working Paper Series, RWP10-023. John F. Kennedy School of Government, Harvard University, 2010.
Grindle, Merilee S. "Governance Reform: The New Analytics of Next Steps." *Governance* 24, No. 3 (2011): 415–18.
Grohe, Edwin. "The Cyber Dimensions of the Syrian Civil War: Implications for Future Conflict." *Comparative Strategy* 34, no. 2 (2015): 133–48.
Groll, Elias. "Many U.S. Weapons Systems Are Vulnerable to Cyberattack." *Foreign Policy*, October 9, 2018, https://foreignpolicy.com/2018/10/09/many-u-s-weapons-systems-are-vulnerable-to-cyberattack/.
Groll, Elias. "Spy vs. Spy, El Chapo Edition." *Foreign Policy*, January 11, 2019. https://foreignpolicy.com/2019/01/11/spy-vs-spy-el-chapo-edition-guzman-narco-flexispy-hacking/.
Groll, Elias. "The U.S.-Iran Standoff Is Militarizing Cyberspace." *Foreign Policy*, June 27, 2019. https://foreignpolicy.com/2019/09/27/the-u-s-iran-standoff-is-militarizing-cyberspace/.
Guéhenno, Jean-Marie. "10 Conflicts to Watch in 2017." *Foreign Policy*, May 10, 2017. http://foreignpolicy.com/2017/01/05/10-conflicts-to-watch-in-2017/.
Guitton, Clement. "Cyber Insecurity as a National Threat: Overreaction from Germany, France, and the UK?" *European Security* 22, no. 1 (2013): 21–35.
Gurney, Kyra. "Infiltration of Chile Air Force Emails Highlights LatAm Cyber Threats." *Insight Crime*, August 15, 2014. http://www.insightcrime.org/news-briefs/chile-air-force-peru-hackers-cyber-threat.
Gutmann, Ethan. "Hacker Nation: China's Cyber Assault." *World Affairs* 173, no. 1 (2010): 70–79.
Gvosdev, Nikolas K. "The Bear Goes Digital: Russia and Its Cyber Capabilities." In *Cyberspace and National Security: Threats, Opportunities, and Power in a Virtual World*, edited by Derek S. Reveron, 173–89. Washington, DC: Georgetown University Press, 2012.
Hague, William. "Arming the Information Highway Patrol." *The World Today* 68, no. 7 (2012): 35–36.
Hansen, Lene, and Helen Nissenbaum. "Digital Disaster, Cyber Security, and the Copenhagen School." *International Studies Quarterly* 53, no. 4 (2009): 1155–75.
Harknett, Richard J., and James A. Stever. "The New Policy World of Cybersecurity." *Public Administration Review* 71, no. 3 (2011): 455–60.
Harknett, Richard J., and Max Smeets. 2020. "Cyber Campaigns and Strategic Outcomes." *Journal of Strategic Studies* 00 (00): 1–34. https://doi.org/10.1080/01402390.2020.1732354.
Harold, Scott Warren, Martin C. Libicki, and Astrid Stuth Cevallos. *Getting to Yes with China in Cyberspace*. Santa Monica: RAND Corporation, 2016.

Harrison Dinniss, Heather. *Cyber Warfare and the Laws of War*. Cambridge: Cambridge University Press, 2012.
Hayes, Anna. "Human (In)security." In *The Oxford Handbook of US National Security. Oxford University Press*, edited by Derek S. Reveron, Nikolas K. Gvosdev, John A. Cloud, John A, 457–73. New York: Oxford University Press, 2018.
Healey, Jason. "Eight Questions and Answers on U.S. Cyber Statecraft." In *Securing Cyberspace: A New Domain for National Security*, edited by Nicholas Burns and Jonathan Price, 97–110. Washington, DC: Aspen Institute, 2012.
Heickerö, Roland. "Cyber Terrorism: Electronic Jihad." *Strategic Analysis* 38, no. 4 (2014): 554–65.
Helmke, Gretchen, and Steven Levitsky, *Informal Institutions, and Democracy: Lessons from Latin America*. Baltimore: Johns Hopkins University Press, 2006.
Herbst, John E., and Jason Marczak. "Russia's Intervention in Venezuela: What's at Stake?" *Atlantic Council*, September 12, 2019. https://www.atlanticcouncil.org/in-depth-research-reports/report/russias-intervention-in-venezuela-whats-at-stake/.
Hernández, Rafael. "Commentary: Alliances and Dis-alliances between the United States and Latin America and the Caribbean." *Latin American Perspectives* 38, no. 4 (2011): 131–36.
Herrera, Monica, and Ron Matthews. "Latin America in Step with Global Defence Offset Phenomenon." *The RUSI Journal* 159, no. 6 (2014): 50–57.
Herz, Monica. "Concepts of Security in South America." *International Peacekeeping* 17, no. 5 (2010): 598–612.
Hincks, Joseph. "Here's What a Top General Thinks Is the Next Big Threat to the U.S." *Time*, September 27, 2017. http://time.com/4958679/general-dunford-military-threat/.
Hindman, Matthew. *The Myth of Digital Democracy*. Princeton: Princeton University Press, 2008.
Holbraad, Carsten. *Middle Powers in International Politics*. London: Macmillan, 1984.
Holbraad, Carsten. "The Role of Middle Powers." *Cooperation and Conflict* 6, no. 1 (1991): 77–90.
Hollis, Duncan B., and Jens David Ohlin, "What if Cyberspace Were for Fighting?" *Ethics and International Affairs* 32, no. 4 (2018): 441–56.
Hood, Christopher. "The Tools of Government in the Information Age." In *The Oxford Handbook of Public Policy*, edited by Robert E. Goodin, Michael Moran, and Martin Rein, 469–81. Oxford: Oxford University Press, 2008.
Howard, Philip N. *Pax Technica: How the Internet of Things May Set Us Free or Lock Us Up*. New Haven: Yale University Press, 2015.
Howell O'Neill, Patrick. "Previously Unknown Cyber-Espionage Group Has Successfully Hacked in South America Since 2015." *CyberScoop*, November 7, 2017. https://www.cyberscoop.com/previously-unknown-cyber-espionage-group-successfully-hacked-south-america-since-2015/.

Huawei. "A New Era for Brazil's Digital Economy Powered by ICT Innovation." Accessed May 15, 2020. https://www.huawei.com/en/about-huawei/executives/articles/brazil-digital-economy-ict-innovation.

Huawei. "Huawei Branch Office." Accessed May 15, 2020. https://e.huawei.com/uk/branch-office-query.

Huntington, Samuel P. "Conventional Deterrence and Conventional Retaliation in Europe." *International Security* 8, no. 3 (1983/1984): 32–56.

Hurel, Louise Marie, and Luisa Cruz Lobato. "A Strategy for Cybersecurity Governance in Brazil." Igarapé Institute, Strategic Note 20, September 2018, https://igarape.org.br/wp-content/uploads/2019/01/A-Strategy-for-Cybersecurity-Governance-in-Brazil.pdf.

Ikenberry, G. John. "Between the Eagle and the Dragon: America, China, and Middle State Strategies in East Asia." *Political Science Quarterly* 131, no. 1 (2015): 9–43.

Indra Company. "Indra and CODALTEC to Develop the First Air Defense System in Colombia, Fully Manufactured in Latin America." https://www.indracompany.com/en/noticia/indra-codaltec-develop-first-air-defense-system-colombia-fully-manufactured-latin-america.

Infobae. "Hackearon la Página del Ejército Argentino con Supuestas Amenazas de ISIS." June 19, 2017. http://www.infobae.com/politica/2017/06/19/hackearon-la-pagina-del-ejercito-argentino-con-supuestas-amenazas-de-isis/.

Inkster, Nigel, "China in Cyberspace." In *Cyberspace, in Cyberspace and National Security: Threats, Opportunities, and Power in a Virtual World*, edited by Derek S. Reveron, 191–205. Washington, DC: Georgetown University Press, 2012.

Inkster, Nigel. "The Huawei Affair and China's Technology Ambitions." *Survival* 61, no. 1 (2019): 105–11.

Institute for Economics & Peace. "Global Peace Index 2018: Measuring Peace in a Complex World." Accessed December 5, 2018. http://visionofhumanity.org/reports.

Institute for Economics & Peace. "Global Peace Index 2019." Accessed December 10, 2019. https://www.visionofhumanity.org/maps/#/.

Instituto Federal de Telecomunicaciones. "En México 71.3 millones de usuarios de internet y 17.4 millones de hogares con conexión a este servicio: ENDUTIH 2017 (Comunicado 015/2018)." http://www.ift.org.mx/comunicacion-y-medios/comunicados-ift/es/en-mexico-713-millones-de-usuarios-de-internet-y-174-millones-de-hogares-con-conexion-este-servicio.

Inter-American Development Bank and Organization of American States. *Are We Ready in Latin America and the Caribbean?* March, 2016. https://publications.iadb.org/en/cybersecurity-are-we-ready-latin-america-and-caribbean.

International Institute for Strategic Studies. "Chapter Eight: Latin America and the Caribbean." *The Military Balance* 114, no. 1 (2014): 355–410.

International Institute for Strategic Studies. "Chapter Eight: Latin America and the Caribbean." *The Military Balance* 117, no. 1 (2017): 417–78.
International Institute for Strategic Studies. "Chapter Three: North America." *The Military Balance* 118, no. 1 (2018): 27–64.
International Institute for Strategic Studies. "Chapter Five: Russia and Eurasia." *The Military Balance*, 118, no. 1 (2018) 169–218.
International Institute for Strategic Studies. "Chapter Six: Asia." *The Military Balance* 118, no. 1 (2018): 219–314.
International Institute for Strategic Studies. "Chapter Eight: Latin America and the Caribbean." *The Military Balance* 118, no. 1 (2018): 375–428.
International Institute for Strategic Studies. "Chapter Eight: Latin America and the Caribbean." *The Military Balance* 119, no. 1 (2019): 380–437.
International Institute for Strategic Studies. "Chapter Ten: Military Cyber Capabilities." *The Military Balance* 120, No. 1 (2020): 515–18.
International Monetary Fund. "Real GDP Growth." Accessed December 10, 2019. https://www.imf.org/external/datamapper/NGDP_RPCH@WEO/OEMDC/ADVEC/WEOWORLD.
International Telecommunications Union. "Statistics." Accessed November 28, 2019. https://www.itu.int/en/ITU-D/Statistics/Pages/stat/default.aspx.
Internet Traffic Report. "Global Traffic Index." http://www.internettrafficreport.com/faq.htm
Internet World Stats. "World Internet Usage and Population Statistics." https://www.internetworldstats.com/stats.htm.
Interpol. "Hackers Reportedly Linked to 'Anonymous' Group Targeted in Global Operation Supported." February 28, 2012. https://www.interpol.int/News-and-media/News/2012/PR014.
Isacson, Adam, and Sara Kinosian. "A First Glance at the 2017 Foreign Aid Request for Latin America." *WOLA*, February 9, 2016. https://www.wola.org/analysis/a-first-glance-at-the-2017-foreign-aid-request-for-latin-america/.
Jackson, Susan T. "Arms Production." In *SIPRI Yearbook 2010: Armaments, Disarmament, and International Security*. https://www.sipriyearbook.org/view/9780199581122/sipri-9780199581122-div1-53.xml.
Janczewski, Lech J., and William Caelli. *Cyber Conflicts and Small States*. New York: Routledge, 2016.
Jindra, Cristoph, and Ana Vaz. "Good Governance and Multidimensional Poverty: A Comparative Analysis of 71 Countries." *Governance* 32, No. 4 (2019): 657–75.
Johnson, Rebecca. "Arms Control and Disarmament Diplomacy." In *The Oxford Handbook of Modern Diplomacy*, edited by Andrew F. Cooper, Jorge Heine, and Ramesh Thakur, 593–609. New York: Oxford University Press, 2015.
Jones, Sam. "Russian Government Behind Cyber Attacks, Says Security Group." *Financial Times*, October 28, 2014. https://www.ft.com/content/93108ba0-5ebe-11e4-a807-00144feabdc0.

Jones, Sam. "Timeline: How the Wannacry Cyber Attack Spread." *Financial Times*, May 14, 2017. https://www.ft.com/content/82b01aca-38b7-11e7-821a-6027b8a20f23.

Jones, Sam. "Venezuela Blackout: What Caused It and What Happens Next?" *Guardian*, March 13, 2019. https://www.theguardian.com/world/2019/mar/13/venezuela-blackout-what-caused-it-and-what-happens-next.

Kaljurand, Marina, and Michael Miklaucic. "An Interview with Marina Kaljurand, Former Minister of Foreign Affairs of Estonia." *PRISM* 7, no. 2 (2017): 116–20.

Kamarck, Elaine Ciulla, and Joseph S. Nye. *Governance.com: Democracy in the Information Age*. Washington, DC: Brookings Institution Press, 2002.

Kamarck, Elaine Ciulla, and Joseph S. Nye. *Visions of Governance in the 21st Century*. Washington, DC: Brookings Institution Press, 2002.

Kamel, Maha S. "China's Belt and Road Initiative: Implications for the Middle East." *Cambridge Review of International Affairs* 31, no. 1 (2018): 76–95.

Kaplan, Fred. *Dark Territory: The Secret History of Cyber War*. New York: Simon and Schuster, 2016.

Karatzogianni, Athina. *Cyber-Conflict and Global Politics*. London and New York: Routledge, 2009.

Karlin, Mara E., and Alice H. Friend. "Military Worship Hurts U.S. Democracy." *Foreign Policy*, September 21, 2018. https://foreignpolicy.com/2018/09/21/military-worship-hurts-us-democracy-civilian-trump/.

KasperskyLab. "El Machete." *SecureList*, August 20, 2014. https://securelist.com/el-machete/66108/.

Kassab, Hanna Samir, and Jonathan D. Rosen. *The Obama Doctrine in the Americas*. Lanham, MD: Lexington Books, 2017.

Kaufmann, Daniel, Aart Kraay, and Massimo Mastruzzi. "The Worldwide Governance Indicators: Methodology and Analytical Issues." World Bank Policy Research Working Paper No. 5430, 2010. https://ssrn.com/abstract=1682130.

Kelion, Leo. "Huawei Set for Limited Role in UK 5G Networks." *BBC News*, January 28, 2020. https://www.bbc.co.uk/news/technology-51283059.

Keller, John. "2017 DOD Budget to Roll Out Next Week: Has Pentagon Spending Finally Bottomed Out?" *Military & Aerospace Electronics*, February 2, 2016. https://www.militaryaerospace.com/computers/article/16715133/2017-dod-budget-to-roll-out-next-week-has-pentagon-spending-finally-bottomed-out.

Kello, Lucas. "The Meaning of the Cyber Revolution: Perils to Theory and Statecraft," *International Security* 38, no. 2 (2013): 7–40.

Kello, Lucas. "Cyber Security." In *Beyond Gridlock*, edited by Thomas Hale and David Hale, 206–28. Cambridge: Polity, 2017.

Keohane, Robert O., and Joseph S. Nye Jr. "Power and Interdependence in the Information Age." *Foreign Affairs* 77, no. 5 (1998): 81–94.

Khazaeli, Susan, and Daniel Stockemer. "The Internet: A New Route to Good Governance." *International Political Science Review* 34, no. 5 (2013): 463–82.

Khoo, Nicholas. "Interstate Rivalry in East Asia." In *The Oxford Handbook of U.S. National Security*, edited by Derek S. Reveron, Nikolas K. Gvosdev, and John A. Cloud, 573–90. Oxford: Oxford University Press, 2018.

Kile, Shannon N., and Hans M. Kristensen. "World Nuclear Forces." In *SIPRI Yearbook 2019: Armaments, Disarmament and International Security*. https://www.sipriyearbook.org/view/9780198839996/sipri-9780198839996-chapter-6-div1-034.xml.

King, Gary, Robert O. Keohane, and Sidney Verba. *Designing Social Inquiry: Scientific Inference in Qualitative Research*. Princeton: Princeton University Press, 1994.

Kissinger, Henry. *Diplomacy*. New York: Simon and Schuster, 1994.

Kissinger, Henry. "How the Enlightenment Ends." *The Atlantic*, June 2018. https://www.theatlantic.com/magazine/archive/2018/06/henry-kissinger-ai-could-mean-the-end-of-human-history/559124/.

Klimburg, Alexander. *The Darkening Web: The War for Cyberspace*. London: Penguin Press, 2017.

Kooiman, Jan. *Governing as Governance*. Thousand Oaks: SAGE, 2003.

Kornbluh, Karen. 2018. "The Internet's Lost Promise." *Foreign Affairs*, August 13, 2018. https://www.foreignaffairs.com/articles/world/2018-08-13/internets-lost-promise.

Krahmann, Elke. "Conceptualizing Security Governance." *Cooperation and Conflict* 38, no. 1 (2003): 5–26.

Krishnan, Aravind. "LAAD 2017 Defence & Security: The Biggest Arms Fair in Latin America." *Medium*, April 9, 2017. https://medium.com/@aravind17/laad-2017-defence-security-the-biggest-arms-fair-in-latin-america-81b9e268156d.

Kshetri, Nir. *Cybercrime and Cybersecurity in the Global South*. London: Palgrave Macmillan, 2013.

Kurç, Çağlar, and Stephanie G. Neuman. "Defence Industries in the 21st Century: A Comparative Analysis." *Defence Studies* 17, no. 3 (2017): 219–27.

La Tercera. "Chadwick señala que 'no está contemplado por ahora' visita de Piñera a oficinas de Huawei." *La Tercera*, April 14, 2019. https://www.latercera.com/politica/noticia/chadwick-senala-no-esta-contemplado-ahora-visita-pinera-oficinas-huawei/614168/.

Lacey, Michelle. "Concept of Significance Level." In *Encyclopedia of Research Design*, edited by Neil J. Salkind. Thousand Oaks: Sage, 2012.

Lambach, Daniel. "The Territorialization of Cyberspace." *International Studies Review* 22, no. 3 (2019): 482–506.

Lampton, David M. "China: Challenger or Challenged?" *The Washington Quarterly* 39, no. 3 (2016): 107–19.

Latin American Post. "Wannacry Cyberattack Numbers in Latin America." May 16, 2017. http://latinamericanpost.com/index.php/living-people/living-future/15225-wannacry-cyberattack-numbers-in-latin-america.

Latin American Public Opinion Project (LAPOP). "AmericasBarometer Data Sets 2016/17." The AmericasBarometer. Accessed May 10, 2018. https://www.vanderbilt.edu/lapop/raw-data.php.

Latin American Public Opinion Project (LAPOP). "AmericasBarometer Data Sets 2018/19." The AmericasBarometer. Accessed November 29, 2020. https://www.vanderbilt.edu/lapop/raw-data.php.

Lee, John. "Cyber Kleptomaniacs: Why China Steals Our Secrets." *World Affairs* 176, no. 3 (2013): 73–79.

Lemercier, Claire, and Claire Zalc. *Quantitative Methods in the Humanities: An Introduction*. Charlottesville: University of Virginia Press, 2019.

Lengyel, Joseph L. "Securing the Nation One Partnership at a Time." *Strategic Studies Quarterly* 12, No. 3 (2018): 3–9.

Lewis, James Andrew. "Advanced Experiences in Cybersecurity Policies and Practices: An Overview of Estonia, Israel, South Korea, and the United States." Discussion Paper No IDB-DP-457, Inter-American Development Bank, 2016.

Lewis, James Andrew. "Five Myths about Chinese Hackers." *Washington Post*, March 22, 2013. http://articles.washingtonpost.com/2013-03-22/opinions/37923854_1_chinese-hackers-cyberattacks-cold-war.

Lewis, James Andrew. *Rethinking Cybersecurity: Strategy, Mass Effect, and States*. Lanham, MD: Rowman and Littlefield, 2018.

Libicki, Martin C. *Cyberdeterrence and Cyberwar*. Santa Monica: RAND Corporation, 2009.

Libicki, Martin C. "Cyberspace Is not a Warfighting Domain." *I/S: A Journal of Law and Policy for the Information Society* 8, no. 2 (2012): 321–36.

Libicki, Martin C. "Is There a Cybersecurity Dilemma?" *The Cyber Defense Review* 1, no. 1 (2016): 129–40.

Lijphart, Arend. "The Comparable Cases Strategy in Comparative Research." *Comparative Political Studies* 8, no. 2 (1975): 158–77.

Limmnéll, Jarno. "The Cyber Arms Race Is Accelerating—What Are the Consequences?" *Journal of Cyber Policy* 1, no. 1 (2016): 50–60.

Lin, Herbert. "A Virtual Necessity: Some Modest Steps toward Greater Cybersecurity." *Bulletin of the Atomic Scientists* 68, no. 5 (2012): 75–87.

Lin, Herbet, and Amy Zegart. *Bytes, Bombs, and Spies: The Strategic Dimension of Offensive Cyber Operations*. Washington, DC: Brookings, 2018.

Lindsay, John R., and Eric Gartzke. "Politics by Many Other Means: The Comparative Strategic Advantages of Operations Domains." *Journal of Strategic Studies*. 1–34. https://doi.org/10.1080/01402390.2020.1768372.

Lindsay, Jon R. "The Impact of China on Cybersecurity Fiction and Friction." *International Security* 39, no. 3 (2014): 7–47.

Lindsay, Jon R., Tai Ming Cheung, and Derek S. Reveron. *China and Cybersecurity: Espionage, Strategy, and Politics in the Digital Domain*. Oxford: Oxford University Press, 2015.

Lopez-Lucia, Elisa. "Regional Powers and Regional Security Governance: An Interpretive Perspective on the Policies of Nigeria and Brazil." *International Relations* 29, no. 3 (2015): 348–62.

Lowenthal, Abraham F., Ted Piccone, Laurence Whitehead. *Shifting the Balance: Obama and the Americas*. Washington, DC: Brookings Institution Press, 2010.

Lowenthal, Abraham F., and Sergio Bitar. "Getting to Democracy: Lessons from Successful Transitions." *Foreign Affairs* 95, no. 1 (2016): 134–44.

Lynch, Justin. "North Korea 'Hacking the Hell Out of Latin America.'" *Fifth Domain*, June 29, 2018. https://www.fifthdomain.com/critical-infrastructure/2018/06/29/north-korea-hacking-the-hell-out-of-latin-america/.

Mahoney, James. "Qualitative Methodology and Comparative Politics." *Comparative Political Studies* 40, no. 2 (2007): 122–44.

Malcomson, Scott. "The Real Fight for the Future of 5G: Who Will Patrol the Borders of a New Network?" *Foreign Affairs*, November 14, 2019. https://www.foreignaffairs.com/articles/2019-11-14/real-fight-future-5g.

Malis, Christian. "Unconventional Forms of War." In *The Oxford Handbook of War*, edited by Yves Boyer and Julian Lindley-French, 186–98. New York: Oxford University Press, 2012.

Maness, Ryan C., and Brandon Valeriano. "The Impact of Cyber Conflict on International Interactions." *Armed Forces & Society* 42, no. 2 (2015): 301–23.

Manicom, James, and Jeffrey Reeves. "Locating Middle Powers in International Relations Theory and Power." In *Middle Powers and the Rise of China*, edited by Bruce Gilley and Andrew O'Neil, 23–44. Washington, DC: Georgetown University Press, 2014.

Manwaring, Max G. *Gangs, Pseudo-Militaries, and Other Modern Mercenaries: New Dynamics in Uncomfortable Wars*. Norman: University of Oklahoma Press, 2010.

Marchant-Shapiro, Theresa. *Statistics for Political Analysis: Understanding the Numbers*. London: Sage, 2017.

Mares, David R. *Latin America and the Illusion of Peace*. London: Routledge, 2012.

Mares, David, and Arie Kacowicz. *Routledge Handbook of Latin American Security*. London: Routledge, 2016.

Mares, David, and Francisco Rojas Aravena. *The United States and Chile: Coming in From the Cold*. New York: Routledge, 2001.

Margolis, Mac. "Obama's Low-Key Legacy of Success in Latin America." *Bloomberg*, January 3, 2017. https://www.bloomberg.com/view/articles/2017-01-03/obama-s-low-key-legacy-of-success-in-latin-america.

Marguelas, Oliver. "Venezuela Cybersecurity Profile: Seeking Stability on an Unstable Medium." *News*, The Henry M. Jackson School of International Studies, University of Washington. https://jsis.washington.edu/news/venezuela-cybersecurity-profile-seeking-stability-unstable-medium/.

Marks, Paul. "The Cybersecurity 202: International Cooperation on China Hacking Could Be Signal for 2019." *Washington Post*, December 21, 2018. https://www.washingtonpost.com/news/powerpost/paloma/the-cybersecurity-202/2018/12/21/the-cybersecurity-202-international-cooperation-on-china-hacking-could-be-signal-for-2019/5c1be9491b326b6a59d7b21c/.

Marks, Paul. "U.S. to Try New Approach to Punish Hacking Nations: Working with Allies." *Washington Post*, March 7, 2019. https://www.washingtonpost.com/news/powerpost/paloma/the-cybersecurity-202/2019/03/07/the-cybersecurity-202-u-s-to-try-new-approach-to-punish-hacking-nations-working-with-allies/5c80132b1b326b2d177d5ff1.

Marquis, Jefferson P., Jennifer D. P. Moroney, Justin Beck, Derek Eaton, Scott Hiromoto, David R. Howell, Janet Lewis, Charlotte Lynch, Michael J. Neumann, and Cathryn Quantic Thurston. *Developing an Army Strategy for Building Partner Capacity for Stability Operations*. Santa Monica: Rand Corporation, 2010.

Marshton, Hunter. "The U.S.-China Cold War Is a Myth." *Foreign Policy*, September 6, 2019. https://foreignpolicy.com/2019/09/06/the-u-s-china-cold-war-is-a-myth/.

Mattern, Benjamin. "Cyber Security and Hacktivism in Latin America: Past and Future." July 24, 2014. http://www.coha.org/cyber-security-and-hacktivism-in-latin-america-past-and-future/.

Maurer, Tim. *Cyber Mercenaries: The State, Hackers, and Power*. Cambridge: Cambridge University Press, 2017.

Mazanec, Brian M., and Bradley A. Thayer. *Deterring Cyber Warfare: Bolstering Strategic Stability in Cyberspace*. Basingstoke: Palgrave Macmillan, 2014.

McGraw, Gary. "Cyber War Is Inevitable (Unless We Build Security In)." *Journal of Strategic Studies* 36, no. 1 (2013): 109–19.

McGuffin, Chris, and Paul Mitchell. "On Domains: Cyber and the Practice of Warfare." *International Journal* 69, no. 3 (2014): 394–412.

Mearsheimer, John J., and Stephen M. Walt. "Leaving Theory Behind: Why Simplistic Hypothesis Testing Is Bad for International Relations." *European Journal of International Relations* 19, no. 3 (2013): 427–57.

Mello, Gabriela. "Huawei Partners with Local Retailers to Sell Flagship Smartphones in Brazil." *Reuters*, April 30, 2019. https://uk.reuters.com/article/uk-huawei-tech-brazil/huawei-partners-with-local-retailers-to-sell-flagship-smartphones-in-brazil-idUKKCN1S62TN.

Menard, Scott. "Logistic Regression." In *Encyclopedia of Research Design*, edited by Neil J. Salkind, 731–35. Thousand Oaks: Sage, 2012.

Ministerio de Defensa Nacional de Chile. "Balance De Gestión Integral Año 2017, Subsecretaría De Defensa." http://www.dipres.gob.cl/597/articles-172590_doc_pdf.pdf.

Ministerio del Interior y Seguridad Publica de Chile. "Estrategia Nacional de Ciberseguridad." April, 2017. http://ciberseguridad.interior.gob.cl/media/2017/04/NCSP-ENG.pdf.

Ministry for Europe and Foreign Affairs of France. "Paris Call for Trust and Security in Cyberspace." November 12, 2018. https://www.diplomatie.gouv.fr/en/french-foreign-policy/digital-diplomacy/france-and-cyber-security/article/cybersecurity-paris-call-of-12-november-2018-for-trust-and-security-in.

Ministry of Defense. "Livro Blanco de Defesa Nacional." http://www.defesa.gov.br/arquivos/estado_e_defesa/livro_branco/lbdn_2013_ing_net.pdf.

Ministry of National Defense the People's Republic of China. "China's Military Strategy." May 26, 2015. http://eng.mod.gov.cn/Press/2015-05/26/content_4586805.htm.

Miller, Robert, and Ciaran Acton. *SPSS for Social Scientists*. London: Palgrave Macmillan, 2009.

Montgomery, Douglas C., Elizabeth A. Peck, G. Geoffrey Vining. *Introduction to Linear Regression Analysis* London: John Wiley, 2012.

Moore, Tyler, and Ross Anderson. "Internet Security." In *The Oxford Handbook of the Digital Economy*, edited by Martin Peitz and Joel Waldfogel, 572–600. New York: Oxford University Press, 2012.

Mora, Frank, and Brian Fonseca. "Defense Budgets Get a Boost from Science." *Americas Quarterly*. May 7, 2015. https://www.americasquarterly.org/content/latin-americas-high-tech-warriors.

Morehouse, Matthew Alan, Tomáš Lovecký, Hannah Read, Marcus Woodman. "Quantify? or Wanna Cry? Integrating Methods Training in the IR Classroom." *International Studies Perspectives* 18, no. 2 (2017): 225–45.

Morrison, Kevin M. "The Washington Consensus and the New Political Economy of Economic Reform." In *The Oxford Handbook of the Politics of Development*, edited by Carol Lancaster and Nicolas van de Walle, 73–87. Oxford: Oxford University Press, 2018.

Moury, Taciana. "Brazilian Army Invests in Cyber Defense." *Dialogo*, May 12, 2017. https://dialogo-americas.com/en/articles/brazilian-army-invests-cyber-defense.

Mueller, Milton. *Networks and States: The Global Politics of Internet Governance*. Cambridge: MIT Press, 2010.

Mueller, Milton, Andreas Schmidt, and Brenden Kuerbis. "Internet Security and Networked Governance in International Relations." *International Studies Review* 15, no. 1 (2013): 86–104.

Muggah, Robert, and Misha Glenny. "Why Brazil Put Its Military in Charge of Cyber Security," January 13, 2015. http://www.defenseone.com/technology/2015/01/why-brazil-put-its-military-charge-cyber-security/102756/.

Muggah, Robert, Ilona Szabo de Carvalho, and Katherine Aguirre. "Latin America Is the World's Most Dangerous Region. But There Are Signs It Is Turning a Corner." *World Economic Forum on Latin America*, March 14, 2018. https://www.weforum.org/agenda/2018/03/latin-america-is-the-worlds-most-dangerous-region-but-there-are-signs-its-turning-a-corner/.

Muggah, Robert, and Nathan B. Thompson. "Brazil's Critical Infrastructure Faces a Growing Risk of Cyberattacks." *Council on Foreign Relations Blog*, April 10, 2018. https://www.cfr.org/blog/brazils-critical-infrastructure-faces-growing-risk-cyberattacks.

Muggah, Robert, John P. Sullivan. "The Coming Crime Wars." *Foreign Policy*, September 21, 2018. https://foreignpolicy.com/2018/09/21/the-coming-crime-wars/.

Mujis, Daniel. *Doing Quantitative Research in Education with SPSS*. London: Sage, 2011.

Muñoz, Jairo. "Will China Make the Most of Colombia's Peace Deal with the FARC." *Diplomat*, September 13, 2016. https://thediplomat.com/2016/09/will-china-make-the-most-of-colombias-peace-deal-with-farc/.

Murphy, Margi, and Matthew Field, "How Russia Would Wage Cyber War on Britain." *Telegraph*, April 16, 2018, https://www.telegraph.co.uk/technology/2018/04/16/russia-would-wage-cyber-war-britain/.

Naím, Moisés. "Mafia States: Organized Crime Takes Office." *Foreign Affairs* 91, no. 3 (2012): 100–11.

Naím, Moisés. "Why Democracies Are at a Disadvantage in Cyber Wars." *Journal of International Affairs*, The Next World Order: Special 70th Anniversary Issue (2017): 85–91.

Naím, Moisés, Katherine Pollock, and Sabin Ray. "The Decay of Power in Domestic and International Politics." *The Brown Journal of World Affairs* 21, no. 1 (2014), 241–51.

Nakashima, Ellen. "Chinese Hackers Who Breached Google Gained Access to Sensitive Data, U.S. Officials Say." *Washington Post*, May 20, 2013. https://www.washingtonpost.com/world/national-security/chinese-hackers-who-breached-google-gained-access-to-sensitive-data-us-officials-say/2013/05/20/51330428-be34-11e2-89c9-3be8095fe767_story.html?utm_term=.a19c779acaad.

Natalevich, Martín. "Uruguay Sufrió 15 Ciberataques De Alta Severidad Durante 2016." *El Observador*, January 23, 2017. http://www.elobservador.com.uy/uruguay-sufrio-15-ciberataques-alta-severidad-2016-n1022628.

Naughton, John. "The Evolution of the Internet: From Military Experiment to General Purpose Technology." *Journal of Cyber Policy* 1, no. 1 (2016): 5–28.

Navy Live. "Building Partnerships During 56th Annual UNITAS Exercise." October 15, 2015. http://navylive.dodlive.mil/2015/10/16/building-partnerships-during-56th-annual-unitas-exercise/.

Newman, Lily Hay. "The US Gives Cyber Command the Status It Deserves." *Wired*, August 19, 2017. https://www.wired.com/story/cyber-command-elevated/.

Newman, Lily Hay. "How Facebook Hackers Compromised 30 Million Accounts," *Wired*, October 12, 2018, https://www.wired.com/story/how-facebook-hackers-compromised-30-million-accounts/.

Nisbet, Erik C., Elizabeth Stoycheff, Katy E. Pearce. "Internet Use and Democratic Demands: A Multinational, Multilevel Model of Internet Use and Citizen Attitudes about Democracy." *Journal of Communication* 62, no. 2 (2012): 249–65.

Nixon, Allan. "China's Growing Arms Sales to Latin America." *Diplomat*, August 24, 2016. https://thediplomat.com/2016/08/chinas-growing-arms-sales-to-latin-america/.

Norden, Deborah L. *Argentina and the United States: Changing Relations in a Changing World*. New York: Routledge Press, 2002.

Norden, Deborah L. "Sowing Conflict in Venezuela: Political Violence and Economic Policy." In *Economic Development Strategies and the Evolution of Violence in Latin America*, edited by William Ascher and Natalia S. Mirovitskaya, 153–80. New York: Palgrave Macmillan, 2012.

Norden, Deborah L. "The Making of Socialist Soldiers: Radical Populism and Civil-Military Relations in Venezuela, Ecuador, and Bolivia." In *Debating Civil-Military Relations in Latin America*, edited by David R. Mares and Rafael Martinez, 155–80. Sussex: Sussex Academic Press, 2014.

Norden, Deborah L. "Latin American Militaries in the 21st Century: Civil-Military Relations in the Era of Disappearing Boundaries." In *Routledge Handbook of Latin American Security*, edited by David Mares and Arie Kacowicz, 242–53. London: Routledge, 2015.

Nordin, Astrid H. M., and Mikael Weissmann. "Will Trump Make China Great Again? The Belt and Road Initiative and International Order." *International Affairs* 94, no. 2 (2018): 231–49.

Nossel, Suzanne. "Google Is Handing the Future of the Internet to China." *Foreign Policy*, September 10, 2018. https://foreignpolicy.com/2018/09/10/google-is-handing-the-future-of-the-internet-to-china/ (accessed 19 October 2018).

Notimex. "Latinoamérica Está Expuesta a Ataques Cibernéticos." June 3, 2017. http://www.excelsior.com.mx/hacker/2017/06/03/1167503.

Nouwens, Meia. "Global Defence-Industry League: Where Is China?" *Military Balance Blog*, August 28, 2018. https://www.iiss.org/blogs/military-balance/2018/08/china-global-defence-industry-league.

Nye Jr., Joseph S. "Nuclear Lessons for Cyber Security?" *Strategic Studies Quarterly* 5, no. 4 (2011): 18–38.

Nye Jr., Joseph S. "Hard, Soft, and Smart Power." In *The Oxford Handbook of Modern Diplomacy*, edited by Andrew F. Cooper, Jorge Heine, and Ramesh Thakur, 559–75. New York: Oxford University Press, 2013.

Nye Jr., Joseph S. "Deterrence and Dissuasion in Cyberspace." *International Security* 41, No, 3 (2016): 44–71.

O'Connor, Tom. "Russian Military Sells New Missile Defense System to Brazil." *Newsweek*, April 4, 2017. http://www.newsweek.com/russia-military-influence-expand-brazil-missile-system-579156.

O'Donnell, Guillermo. "On the State, Democratization, and Some Conceptual Problems: A Latin American View with Glances at Some Postcommunist Countries." *World Development* 21, no. 8 (1993): 1355–69.

O'Donnell, Guillermo. *Democracy, Agency, and the State: Theory with Comparative Intent*. Oxford: Oxford University Press, 2010.

O'Neil, Shannon K. "Argentina's IMF Package Could Trigger Ugly Blowback." *Council on Foreign Relations*, June 14, 2018. https://www.cfr.org/blog/argentinas-imf-package-could-trigger-ugly-blowback?sp_mid=56809101&sp_rid=Y2Flc29sYXJAZ21haWwuY29tS0.

Observatory of Cybersecurity in Latin America and the Caribbean. "Home." Accessed November 29, 2019. https://observatoriociberseguridad.org/#/home.

Ohlheiser, Abby. "Brazil and Mexico Ask the U.S. to Explain Reports of NSA Spying on Their Leaders." *Atlantic*, September 2, 2013. https://www.theatlantic.com/international/archive/2013/09/brazil-and-mexico-want-us-explain-reports-nsa-spying-their-presidents/311546/.

Organization of American States. "Report on Cybersecurity and Critical Infrastructure in the Americas, 2015." https://www.sites.oas.org/cyber/Documents/2015%20-%20OAS%20Trend%20Micro%20Report%20on%20Cybersecurity%20and%20CIP%20in%20the%20Americas.pdf.

Organization of American States and Symantec. "Tendencias de Seguridad Cibernetica en America Latina y el Caribe, 2014." http://www.symantec.com/content/es/mx/enterprise/other_resources/b-cyber-security-trends-report-lamc.pdf.

Ortega, Miriam. "Colombian Armed Forces Counter Cybercrime." *Dialogo*, November 9, 2017. https://dialogo-americas.com/en/articles/colombian-armed-forces-counter-cybercrime.

Ortis, Cameron, and Paul Evans. "The Internet and Asia-Pacific Security: Old Conflicts and New Behaviour." *The Pacific Review* 16, No. 4 (2003): 549–50.

Oxford Analytica. "Huawei's Fate Is Inseparable from Europe's Geostrategy." March 27, 2019. https://dailybrief.oxan.com/Analysis/GA242262/Huaweis-fate-is-inseparable-from-Europes-geostrategy.

Panizza, Francisco. *Contemporary Latin America: Development and Democracy beyond the Washington Consensus*. London: Zed Books, 2009.

Patrick, Stewart. *Weak Links: Fragile States, Global Threats, and International Security*. New York: Oxford University Press, 2011.

Pawlak, Patryk, and Panagiota-Nayia Barmpaliou. "Politics of Cybersecurity Capacity Building: Conundrum and Opportunity." *Journal of Cyber Policy* 2, no. 1 (2017): 123–44.

Payne, Geoff, and Judy Payne. *Key Concepts in Social Research*. London: Sage, 2011.

Pearl, Judea. "The Foundations of Causal Inference." *Sociological Methodology* 40, no. 1 (2010): 75–149.

Pérez, Orlando J. "Can Nicaragua's Military Prevent a Civil War?" *Foreign Policy*, July 3, 2018. https://foreignpolicy.com/2018/07/03/can-nicaraguas-military-prevent-a-civil-war/.

Peters, B. Guy, and Jon Pierre. *Comparative Governance: Rediscovering the Functional Dimension of Governing*. New York: Cambridge University Press, 2016.

Peters, B. Guy. "What Is so Wicked About Wicked Problems? A Conceptual Analysis and a Research Program." *Policy and Society* 36, no. 3 (2017): 385–96.

Pierce, Roger. *Research Methods in Politics*. London: Sage, 2011.

Pigman, Lincoln. "Russia's Vision of Cyberspace: A Danger to Regime Security, Public Safety, and Societal Norms and Cohesion." *Journal of Cyber Policy* 4, no. 1 (2019): 22–34.

Pion-Berlin, David, and Rafael Martínez. *Soldiers, Politicians, and Citizens: Reforming Civil-Military Relations in Latin America.* New York: Cambridge University Press, 2017.

Pion-Berlin, David. *Military Missions in Democratic Latin America.* New York: Palgrave Macmillan, 2016.

Platt, Victor. "Still the Fire-Proof House? An Analysis of Canada's Cyber Security Strategy." *International Journal* 67, no. 1 (2011): 155–67.

Polonski, Vyacheslav. "The Biggest Threat to Democracy? Your Social Media Feed." World Economic Forum, August 4, 2016. https://www.weforum.org/agenda/2016/08/the-biggest-threat-to-democracy-your-social-media-feed/.

Pomerleau, Mark. "Here's How Cyber Command Is Using 'Defend Forward.'" *Fifth Domain*, November 12, 2019. https://www.fifthdomain.com/smr/cybercon/2019/11/12/heres-how-cyber-command-is-using-defend-forward/?utm_source=clavis.

Pomerleau, Mark. "In Era Of 'Defend Forward,' What Does Success Look Like?" *Fifth Domain*, April 24, 2019. https://www.fifthdomain.com/dod/2019/04/24/in-era-of-defend-forward-what-does-success-look-like

Pomerleau, Mark. "Two Years In, How Has a New Strategy Changed Cyber Operations?" *Fifth Domain*, November 11, 2019. https://www.fifthdomain.com/dod/2019/11/11/two-years-in-how-has-a-new-strategy-changed-cyber-operations/.

Poujol, Mathieu. "Key Cyber Security Issues in the Aerospace and Defense Industry." *Teknowlogy*, September 11, 2018. https://www.pac-online.com/key-cyber-security-issues-aerospace-defense-industry.

Presidency of the Republic of Brazil. "Estratégia Nacional de Defesa, Decreto N. 6.703." http://www.planalto.gov.br/ccivil_03/_ato2007-2010/2008/Decreto/D6703.htm.

Privacy International. "State of Privacy: Argentina." January 23, 2019. https://privacyinternational.org/state-privacy/57/state-privacy-argentina.

Pyman, Mark. "Addressing Corruption in Military Institutions." *Public Integrity* 19, no. 5 (2017): 513–28.

Radu, Roxana. *Negotiating Internet Governance.* Oxford: Oxford University Press, 2019.

Reichmann, Deb. "President Trump Reduces Restrictions on U.S. Responding Offensively to Cyberattacks." *Time*, October 31, 2018. http://time.com/5441193/trump-offensive-cyberattacks-bolton/.

Reid, Michael. "Obama and Latin America: A Promising Day in the Neighborhood." *Foreign Affairs* 94, no. 5 (2015): 45–53.

Renard, Thomas. "EU Cyber Partnerships: Assessing the EU Strategic Partnerships with Third Countries in the Cyber Domain." *European Politics and Society* 19, no. 3 (2018): 321–37.

Reporters Without Borders "Egypt: New Media Law Aims to Silence Independent Online Media, RSF Says." November 3, 2018, https://rsf.org/en/news/egypt-new-media-law-aims-silence-independent-online-media-rsf-says.

Reuters. "Russia Hacked Danish Defense for Two Years, Minister Tells Newspaper." April 23, 2017. http://www.reuters.com/article/us-denmark-security-russia/russia-hacked-danish-defense-for-two-years-minister-tells-newspaper-idUSK BN17P0NR.

Reuters. "Argentina Buys Five Used Fighter Jets from French Navy." May 11, 2018, https://www.reuters.com/article/us-argentina-military-france-idUKKBN1IC1OP.

Reuters. "China's Chinalco Starts $1.3 Billion Expansion of Peru Copper Mine." June 4, 2018. https://www.reuters.com/article/us-peru-copper-china/chinas-chinalco-starts-1-3-billion-expansion-of-peru-copper-mine-idUSKCN1J0OCI.

Reuters. "Chinas' COSCO to build $2 Billion Shipping Port in Peru: Ambassador." June 1, 2018. https://www.reuters.com/article/us-peru-china/chinas-cosco-to-build-2-billion-shipping-port-in-peru-ambassador-idUSKCN1IX5WW.

Reuters. "Factbox: Venezuela's Jailed, Exiled or Barred Opposition Politicians." February 19, 2018. https://www.reuters.com/article/us-venezuela-politics-factbox/factbox-venezuelas-jailed-exiled-or-barred-opposition-politicians-idUSKCN1 G31WU.

Reuters. "Pompeo to Visit Four Latin American Nations in Security, Migration Push." July 16, 2019. https://uk.reuters.com/article/uk-usa-pompeo/pompeo-to-visit-four-latin-american-nations-in-security-migration-push-idUKKCN1UB2Q7?il=0.

Reveron, Derek S. "An Introduction to National Security and Cyberspace." In *Cyberspace and National Security: Threats, Opportunities, and Power in a Virtual World*, edited by Derek S. Reveron, 3–19. Washington, DC: Georgetown University Press, 2012.

Reveron, Derek S., and Nikolas K. Gvosdev. "National Interests and Grand Strategy." In *The Oxford Handbook of U.S. National Security*, edited by Derek S. Reveron, Nikolas K. Gvosdev, and John A. Cloud, 35–56. Oxford: Oxford University Press, 2018.

Reveron, Derek S., Nikolas K. Gvosdev, and John A. Cloud, "Introduction: Shape and Scope of U.S. National Security." In *The Oxford Handbook of U.S. National Security*, Derek S. Reveron, Nikolas K. Gvosdev, and John A. Cloud, 2–12. Oxford: Oxford University Press, 2018.

Reyers, Gerardo. "Growing Scandal in Latin America over 'Pegasus' Spy-hacking Program." *Univision*, June 21, 2017. http://www.univision.com/univision-news/latin-america/growing-scandal-in-latin-america-over-pegasus-spy-hacking-program.

Reynolds, Matthew W. "Bivariate Regression." In *Encyclopedia of Research Design*, edited by Neil J. Salkind, 87–91. Thousand Oaks: Sage, 2012.

Rice, Condoleezza, and Amy Zegart. "Managing 21st-Century Political Risk." *Harvard Business Review*, May-June, 2018. https://hbr.org/2018/05/managing-21st-century-political-risk.

Rid, Thomas. "Cyber War Will Not Take Place." *Journal of Strategic Studies* 35, no. 1 (2012): 5–32.

Rid, Thomas. *Cyber War Will Not Take Place*. London: Hurst, 2013.
Rid, Thomas, and Peter McBurney. "Cyber-Weapons." *The RUSI Journal* 157, no. 1 (2012): 6–13.
Robertson, Jordan, Michael Riley, and Andrew Willis. "How to Hack an Election." *Bloomberg*, March 31, 2016. https://www.bloomberg.com/features/2016-how-to-hack-an-election/.
Rodríguez, Israel. "Reporta Condusef Récord de Presuntos Fraudes." *Jornada*, July 18, 2017. http://www.jornada.com.mx/ultimas/2017/07/18/reporta-condusef-record-de-presuntos-fraudes.
Roett, Riordan, and Guadalupe Paz. *Latin America and the Asian Giants: Evolving Ties with China and India*. Washington, DC: Brookings Institute Press, 2016.
Rojas Aravena, Francisco. "Williamsburg: ¿Un Giro Definitivo En Las Relaciones Hemisféricas De Seguridad?" *Estudios Internacionales* 29, no. 114 (1996): 139–64.
Rojas Aravena, Francisco. *The Difficult Task of Peace: Crisis, Fragility, and Conflict in an Uncertain World*. New York: Palgrave, 2020.
Rolland, Nadège. "China's "Belt and Road Initiative": Underwhelming or Game-Changer?" *The Washington Quarterly* 40, no. 1 (2017): 127–42.
Rosenberg, Matthew, and Maggie Haberman, "When Trump Phones Friends, the Chinese and the Russians Listen and Learn." *New York Times*, October 24, 2018. https://www.nytimes.com/2018/10/24/us/politics/trump-phone-security.html.
Rotberg, Robert I. "Good Governance Means Performance and Results." *Governance* 27, no. 3 (2014): 511–18.
Rothstein, Bo. *The Quality of Government: Corruption, Social Trust, and Inequality in International Perspective*. Chicago: University of Chicago Press, 2011.
Rovner, Joshua, and Tyler Moore. "Does the Internet Need an Hegemon?" *Journal of Global Security Analysis* 2, no. 3 (2017): 184–203.
Rovner, Joshua. "Cyber War as an Intelligence Contest." *War on the Rocks*, September 16, 2019. https://warontherocks.com/2019/09/cyber-war-as-an-intelligence-contest/.
RT. "Argentina, Brazil Agree on a Cyber-Defense Alliance against US Espionage." September 15, 2013. https://www.rt.com/news/brazil-argentina-cyber-defense-879/.
RT. "Venezuela Crea Un Comando Cibernético Para Defenderse Contra La 'Conjura Mediática.'" June 7, 2014. https://actualidad.rt.com/actualidad/view/130410-maduro-comando-cibernetico-venezuela.
Saalman, Lora. "New Domains of Crossover and Concern in Cyberspace." *SIPRI Commentary*, July 26, 2017. https://www.sipri.org/commentary/topical-backgrounder/2017/new-domains-crossover-and-concern-cyberspace.
Sachs, Jeffrey S. "America's War on Chinese Technology." *Project Syndicate*, Nov 7, 2019. https://www.project-syndicate.org/commentary/cheney-doctrine-us-war-on-chinese-technology-by-jeffrey-d-sachs-2019-11.
Salomon, Feliz. "The Thai Junta Looks Set to Tighten Control of the Internet Even Further." *Time*, December 16, 2016, http://time.com/4604521/thailand-junta-cyber-crime-internet-rights-expression/.

Sambuli, Nanjira. "Challenges and Opportunities for Advancing Internet Access in Developing Countries While Upholding Net Neutrality." *Journal of Cyber Policy* 1, no. 1 (2016): 61–74.

Sanchez Nieto, W. Alejandro. "Whatever Happened to South America's Splendid Little Wars?" *Small Wars & Insurgencies* 22, no. 2 (2011): 329–35.

Sancho, Carolina. "Ciberseguridad: Presentación del Dossier." *Urvio* 20 (2018): 8–18.

Sandler, Todd, and Justin George. "Military Expenditure Trends for 1960–2014 and What They Reveal." *Global Policy* 7, no. 2 (2016): 174–84.

Saunders, Jamie. "Tackling Cybercrime—the UK Response." *Journal of Cyber Policy* 2, no. 1 (2017): 4–15.

Schmidt, Eric, and Jared Cohen. "The Digital Disruption: Connectivity and the Diffusion of Power." *Foreign Affairs* 89, no. 6 (2010): 75–85.

Schneider, Jacquelyn. "The Capability/Vulnerability Paradox and Military Revolutions: Implications for Computing, Cyber, and the Onset of War." *Journal of Strategic Studies* 42, no. 6 (2019): 841–63.

Schulte, Heinz. "Industry and War." In *The Oxford Handbook of War*, edited by Julian Lindley-French and Ives Boyer, 517–30. New York: Oxford University Press, 2012.

Sechser, Todd S., Neil Narang, and Caitlin Talmadge. "Emerging Technologies and Strategic Stability in Peacetime, Crisis, and War." *Journal of Strategic Studies* 42, no. 6 (2019): 727–35.

Segal, Adam. "Cyber Week in Review: August 23, 2019." *Council on Foreign Relations*, August 23, 2019. https://www.cfr.org/blog/cyber-week-review-august-23-2019.

Seib, Philip M. *Real-Time Diplomacy: Politics and Power in the Social Media Era*. New York: Palgrave Macmillan, 2012.

Seiff, Abby, and Phnom Penh. "Chinese State-Linked Hackers in Large Scale Operation to Monitor Cambodia's Upcoming Elections, Report Says." *Time*, July 10, 2018. http://time.com/5334262/chinese-hackers-cambodia-elections-report/.

Seligman, Lara. "U.S. Military Wary of China's Foothold in Venezuela." *Foreign Policy*, April 8, 2019. https://foreignpolicy.com/2019/04/08/us-military-wary-of-chinas-foothold-in-venezuela-maduro-faller-guaido-trump-pentagon/.

Semana. "Maduro Contra la Pared." September 17, 2017. https://www.semana.com/china-y-rusia-quieren-demostrar-que-amenazas-de-trump-son-vacias/540741/.

Serrano, Mónica. "States of Violence: State-crime Relations in Mexico." In *Violence, Coercion, and State Making in Twentieth Century Mexico*, edited by Wil G. Pansters, 135–58. Stanford, Stanford: University Press, 2012.

Serrano, Mónica. "¿Promesas Idealistas? América Latina en Tiempos de la Responsabilidad de Proteger." *Foro Internacional* 57, no. 1 (2017): 55–108.

Serrano, Mónica. "Del 'Momento Mexicano' A la Realidad de la Violencia Político-Criminal." *Foro Internacional* 60, no. 2 (2020): 791–852.

Shaheen, Jeanne. "The Russian Company That Is a Danger to Our Security," *New York Times*, September 4, 2017. https://www.nytimes.com/2017/09/04/opinion/kapersky-russia-cybersecurity.html?_r=0.

Shane, Scott. "To Sway Vote, Russia Used Army of Fake Americans." *New York Times*, September 8, 2017. https://www.nytimes.com/2017/09/07/us/politics/russia-facebook-twitter-election.html.

Sharpe, Kenneth E. "The Military, The Drug War, and Democracy in Latin America: What Would Clausewitz Tell Us?" *Small Wars & Insurgencies* 4, no. 3 (1993): 56–86.

Shifter, Michael. "The US and Latin America through the Lens of Empire." *Current History* 103, no. 670 (2004): 61–67.

Shirreff, Richard. "Conducting Joint Operations." In *The Oxford Handbook of War*, edited by Julian Lindley-French and Ives Boyer, 373–86. New York: Oxford University Press, 2012.

Shuya, Mason. "Russian Influence in Latin America." *Journal of Strategic Security* 12, no. 2 (2019): 17–41.

Silva, Paulina. "Chile: Presidential Instructive on Cybersecurity." November 5, 2018. https://www.mondaq.com/security/751016/presidential-instructive-on-cybersecurity.

Simon, Luis, Alexander Lanoszka, and Hugo Meijer. "Nodal Defence: The Changing Structure of U.S. Alliance Systems in Europe and East Asia." *Journal of Strategic Studies* 44, no. 3 (2019): 360–88.

Singer, P. W. *Wired for War: The Robotics Revolution and Conflict in the 21st Century.* New York: Penguin, 2009.

Singer, P. W., and Allan Friedman. *Cybersecurity and Cyberwar*. Oxford: Oxford University Press, 2014.

Slatta, Richard W. "Time Line of US-Latin American Relations." Accessed May 15, 2020. https://faculty.chass.ncsu.edu/slatta/hi216/hi453time.htm.

Slaughter, Anne-Marie. "America's Edge: Power in the Networked Century." *Foreign Affairs* 88, no. 1 (2009): 94–113.

Slayton, Rebecca. "What Is the Cyber Offense-Defense Balance? Conceptions, Causes, and Assessment." *International Security* 41, no. 3 (2016): 72–109.

Smith, Dan. "Introduction: International Stability and Human Security in 2018." In *SIPRI Yearbook 2019: Armaments, Disarmament, and International Security*. https://www.sipriyearbook.org/view/9780198839996/sipri-9780198839996-chapter-1-div1-006.xml.

Smith, Dan. "Dual-Use and Arms Trade Controls." In *SIPRI Yearbook 2018: Armaments, Disarmament and International Security*. https://www.sipriyearbook.org/view/9780198821557/sipri-9780198821557-chapter-10-div1-019.xml.

Smith, Martin. *Pressure, Power, and Policy*. Pittsburgh: Pittsburgh University Press.

Solar, Carlos. "State, Violence, and Security in Mexico: Developments and Consequences for Democracy." *Mexican Studies* 30, no. 1 (2014): 241–55.

Solar, Carlos. "Governance of Defense and Policymaking in Chile." *Latin American Policy* 6, no. 2 (2015): 205–25.

Solar, Carlos. "Police Bribery: Is Corruption Fostering Dissatisfaction with the Political System?" *Democracy and Security* 11, no. 4 (2015): 373–94.

Solar, Carlos. "The Inter-Institutional Governance of Money Laundering: An In-Depth Look at Chile following Re-Democratisation." *Global Crime* 16, no. 4 (2015): 328–50.

Solar, Carlos. *Government and Governance of Security: The Politics of Organised Crime in Chile*. New York: Routledge, 2018.

Solar, Carlos. "Chile's Peacekeeping and the Post-UN Intervention Scenario in Haiti." *International Studies* 56, no. 4 (2019): 272–91.

Solar, Carlos. "Civil-Military Relations and Human Security in a Post-Dictatorship." *Journal of Strategic Studies* 42, no, 3–4 (2019): 507–31.

Solar, Carlos. "Reassessing Chilean International Security." *International Politics* 56, no. 5 (2019): 569–84.

Sørensen, Eva. "Democratic Network Governance." In *Handbook on Theories of Governance*, edited by Christopher Ansell and Jacob Torfing, 419–27. Cheltenham: Edward Elgar, 2016.

Sotomayor, Arturo. *The Myth of the Democratic Peacekeeper: Civil-Military Relations and the United Nations*. Baltimore: Johns Hopkins University Press, 2014.

Spanakos, Anthony Peter, and Joseph Marques. "Brazil's Rise as a Middle Power: The Chinese Contribution." In *Middle Powers and the Rise of China*, edited by Bruce Gilley and Andrew O'Neil, 213–36. Washington: Georgetown University Press, 2014.

Sperling, James, and Mark Webber, "Security Governance in Europe: A Return to System." *European Security* 23, no. 2 (2014): 126–44.

Spetalnick. Matt. "Russian Deployment in Venezuela includes 'Cybersecurity Personnel': U.S. Official." *Reuters*, March 26, 2019. https://www.reuters.com/article/us-venezuela-politics-russians/russian-deployment-in-venezuela-includes-cybersecurity-personnel-u-s-official-idUSKCN1R72FX?feedType=RSS&feedName=worldNews.

Stallings, Barbara. "Does Asia Matter? The Political Economy of Latin America's International Relations." In *The Oxford Handbook of Latin American Political Economy*, edited by Javier Santiso and Jeff Dayton-Johnson, 210–32. Oxford: Oxford University Press, 2012.

Stevens, Tim, and Kevin O'Brien. "Brexit and Cyber Security." *RUSI Journal* 164, no. 3 (2019): 22–30.

Stevens, Tim. "A Cyberwar of Ideas? Deterrence and Norms in Cyberspace." *Contemporary Security Policy* 33, no. 1 (2012): 148–70.

Stewart, Phil, and Valerie Volcovichi and Reuters. "Trump Backtracks on His Idea for a Joint Cyber Security Unit with Russia after Harsh Criticism." *Time*, July 10, 2017. http://time.com/4850902/trump-russia-cyber-security-putin-criticism/.

Stockholm International Peace Research Institute (SIPRI). *Yearbook 2019: Armaments, Disarmament, and International Security.* Accessed December 30, 2019. https://www.sipri.org/sites/default/files/2019-08/yb19_summary_eng_1.pdf.

Stockholm International Peace Research Institute (SIPRI). "SIPRI Military Expenditure Database." Accessed November 30, 2020, and July 12, 2022. https://www.sipri.org/databases/milex.

Stokes, Mark. "The Chinese People's Liberation Army Computer Network Operations Infrastructure." In *China and Cybersecurity: Espionage, Strategy, and Politics in the Digital Domain*, edited by Jon R. Lindsay, Tai Ming Cheung, and Derek S. Reveron, 164–77. Oxford: Oxford University Press, 2015.

Stoycheff, Elizabeth, and Erik C. Nisbet. "What's the Bandwidth for Democracy? Deconstructing Internet Penetration and Citizen Attitudes about Governance." *Political Communication* 31, no. 4 (2014): 628–46.

Stuenkel, Oliver. "Huawei Heads South: The Battle over 5G Comes to Latin America" *Foreign Affairs*, May 10, 2019. https://www.foreignaffairs.com/articles/brazil/2019-05-10/huawei-heads-south.

Sullivan, Andrew. "Avoiding Lamentation: To Build a Future Internet." *Journal of Cyber Policy* 2, no. 3 (2017): 323–37.

Sullivan, John P., and Robert J. Bunker. "Mexican Cartel Strategic Note No. 18: Narcodrones on the Border and Beyond." *Small Wars Journal.* https://smallwarsjournal.com/jrnl/art/mexican-cartel-strategic-note-no-18-narcodrones-on-the-border-and-beyond.

Surana, Kavitha. "Spyware Sold to Mexican Government Was Used to Target Experts Investigating Missing Students." *Foreign Policy*, July 10, 2017. https://foreignpolicy.com/2017/07/10/spyware-sold-to-mexican-government-was-used-to-target-experts-investigating-missing-students-cyberwar-ayotzinapa-hacking/.

Teufel Dreyer, June. "Managing Sino-American Relations." *Orbis* 61, no. 1 (2017): 83–90.

The Business Year. "Connecting Country: Colombia, 2018." Accessed May 15, 2020. https://www.thebusinessyear.com/colombia-2018/connecting-country/interview.

The Cylance Threat Research Team. "El Machete's Malware Attacks Cut through LATAM." *ThreatVector*, March 22, 2017. https://threatvector.cylance.com/en_us/home/el-machete-malware-attacks-cut-through-latam.html.

The Economist. "Defence Companies Target the Cyber-Security Market." July 26, 2018. https://www.economist.com/business/2018/07/26/defence-companies-target-the-cyber-security-market.

The Economist. "The Tax Heaven at the End of the World." July 16, 2017. https://www.economist.com/the-americas/2016/07/16/the-tax-haven-at-the-end-of-the-world.

The Economist. "What Latin Americans Think of the United States." March 29, 2016. https://www.economist.com/graphic-detail/2016/03/29/what-latin-americans-think-of-the-united-states.

The Global Americans. "On Trump's Latin America Team: Tomás Regalado Appointed as New Head of Radio and TV Martí." May 13, 2018. https://theglobalamericans.org/2018/05/trumps-latin-america-team/.

The Observatory of Economic Complexity (OEC). "Argentina." https://oec.world/en/profile/country/arg.

The Washington Post. "Trump Has a Chance to Correct Obama's Mistake on Venezuela." March 17, 2017. https://www.washingtonpost.com/opinions/global-opinions/trumps-chance-to-correct-obamas-mistake-on-venezuela/2017/03/17/8389943a-09aa-11e7-b77c-0047d15a24e0_story.html?noredirect=on&utm_term=.d88d6583ea18.

The White House. "Advancing Our Interests: Actions in Support of the President's National Security Strategy." May 27, 2010. https://obamawhitehouse.archives.gov/the-press-office/advancing-our-interests-actions-support-presidents-national-security-strategy.

The White House. "Strategy to Combat Transnational Organized Crime." July 2011. https://obamawhitehouse.archives.gov/sites/default/files/microsites/2011-strategy-combat-transnational-organized-crime.pdf.

The White House. "Fact Sheet: Cybersecurity National Action Plan." February 9, 2016. https://obamawhitehouse.archives.gov/the-press-office/2016/02/09/fact-sheet-cybersecurity-national-action-plan.

The White House. "National Security Strategy of the United States of America, December 2017." December 18, 2017. https://www.whitehouse.gov/wp-content/uploads/2017/12/NSS-Final-12-18-2017-0905.pdf.

Thompson, Robert. "The Cyber Domains: Understanding Expertise for Network Security." In *The Oxford Handbook of Expertise*, edited by Paul Ward, Jan Maarten Schraagen, Julie Gore, and Emilie M. Roth, 718–39. New York: Oxford University Press, 2019.

Tian, Nan. "Military Expenditure," In *SIPRI Yearbook 2017: Armaments, Disarmament and International Security*. https://www.sipriyearbook.org/view/9780198811800/sipri-9780198811800-chapter-9-div1-45.xml.

Tian, Nan, and Diego Lopes da Silva. "Improving South American Military Expenditure Data." September 4, 2017. https://www.sipri.org/commentary/topical-backgrounder/2017/improving-south-american-military-expenditure-data.

Tian, Nan, Alexandra Kuimova, Diedo Lopes da Silva, Pieter D Wezeman, and Siemon T. Wezeman. "Trends in World Military Expenditure, 2019." April, 2020. https://www.sipri.org/sites/default/files/2020-04/fs_2020_04_milex_0_0.pdf.

Tikk, Eneken, "Ten Rules for Cyber Security." *Survival* 53, no. 3 (2011): 119–32.

Torfing, Jacob, B. Guy Peters, Jon Pierre, and Eva Sorensen. *Interactive Governance: Advancing the Paradigm*. Oxford: Oxford University Press, 2012.

Torralba, Carlos. "Mexico Triples Its Arms Purchases in Space of Five Years." *El País*, February 21, 2017. https://elpais.com/elpais/2017/02/21/inenglish/1487673374_904061.html.

Tulchin, Joseph S. *Latin America in International Politics: Challenging US Hegemony.* Boulder: Lynne Rienner, 2016.

Tusalem, Rollin F. "Bringing the Military Back in: The Politicisation of the Military and Its Effect on Democratic Consolidation." *International Political Science Review* 35, no. 4 (2014): 482–501.

U. K. National Cyber Security Centre. "Annual Review 2018: Making the UK the Safest Place to Live and Work Online." http://www.ncsc.gov.uk/annual-review-2018.

U. K. Office for National Statistics. "Internet Users, UK: 2018." https://www.ons.gov.uk/businessindustryandtrade/itandinternetindustry/bulletins/internetusers/2018.

U.S. Department of Defense. "Press Briefing by Admiral Kurt Tidd." March 5, 2018. https://www.defense.gov/Newsroom/Transcripts/Transcript/Article/1458541/.

U.S. Department of Defense. "Remarks by Secretary Mattis on Cybersecurity in Santiago, Chile." August 16, 2018. https://www.defense.gov/Newsroom/Transcripts/Transcript/Article/1608504/remarks-by-secretary-mattis-on-cybersecurity-in-santiago-chile/.

U.S. Department of Defense. "Summary: Department of Defense Cyber Strategy." September 18, 2018. https://media.defense.gov/2018/Sep/18/2002041658/-1/-1/1/CYBER_STRATEGY_SUMMARY_FINAL.PDF.

U.S. Department of Defense. "Southcom Chief Outlines Keys for Success in South America." May 24, 2019. https://www.defense.gov/explore/story/Article/1857725/southcom-chief-outlines-keys-for-success-in-south-america/.

U.S. Department of State. "Secretary of State Michael R. Pompeo and Argentine Foreign Minister Jorge Faurie at a Press Availability at the Western Hemisphere Counterterrorism Ministerial Plenary." July 19, 2019. https://www.state.gov/secretary-of-state-michael-r-pompeo-and-argentine-foreign-minister-jorge-faurie-at-a-press-availability-at-the-western-hemisphere-counterterrorism-ministerial-plenary/.

U.S. Department of State. "Joint Statement on U.S.-Argentina Partnership on Cyber Policy." April 27, 2017. https://www.state.gov/r/pa/prs/ps/2017/04/270496.htm.

U.S. Department of State. "U.S-Brazil Bilateral Cooperation on Cyber and Internet Policy." May 28, 2018. https://www.state.gov/r/pa/prs/ps/2018/05/282074.htm.

U.S. Department of State. "Recommendations to the President on Protecting American Cyber Interests through International Engagement, Office of the Coordinator for Cyber Issues." May 31, 2018. https://www.state.gov/wp-content/uploads/2019/04/Recommendations-to-the-President-on-Protecting-American-Cyber-Interests-Through-International-Engagement.pdf.

U.S. Department of State. "U.S.-Chile Executive Cyber Consultation." September 18, 2018. https://www.state.gov/r/pa/prs/ps/2018/09/285999.htm.

U.S. Department of State. "Remarks Kimberly Breier, Assistant Secretary Bureau of Western Hemisphere Affairs. China's New Road in the Americas: Beyond Silk and Silver." April 26, 2019. https://www.state.gov/chinas-new-road-in-the-americas-beyond-silk-and-silver/.

U.S. Department of State. "Joint Statement on the Inaugural U.S.-Dutch Cyber Dialogue." May 20, 2019. https://www.state.gov/joint-statement-on-the-inaugural-u-s-dutch-cyber-dialogue/.

U.S. Department of State. "Secretary Pompeo's Meeting with Argentine President Macri." July 19, 2019. https://www.state.gov/secretary-pompeos-meeting-with-argentine-president-macri/.

U.S. Department of State. "Remarks of Deputy Assistant Secretary Robert L. Strayer at a Joint U.S.-Mexico Reception at the OAS for Participants of the UN GGE Regional Consultation." August 15, 2019. https://www.state.gov/remarks-of-deputy-assistant-secretary-robert-l-strayer-at-a-joint-u-s-mexico-reception-at-the-oas-for-participants-of-the-un-gge-regional-consultation/.

U.S. Department of State. "U.S. Policy on 5G Technology." August 28, 2019. https://www.state.gov/US-Policy-On-5g-Technology.

U.S. Department of State. "The United States Co-Hosts Second Ministerial Meeting on Advancing Responsible State Behavior in Cyberspace." September 23, 2019. https://www.state.gov/the-united-states-co-hosts-second-ministerial-meeting-on-advancing-responsible-state-behavior-in-cyberspace/.

U.S. Department of State. "Co-Chairs' Statement on the Inaugural ASEAN-U.S. Cyber Policy Dialogue." October 3, 2019. https://www.state.gov/co-chairs-statement-on-the-inaugural-asean-u-s-cyber-policy-dialogue/.

U.S. Department of State. "U.S. Relations with Venezuela." July 6, 2020. https://www.state.gov/u-s-relations-with-venezuela/.

U.S. Department of the Navy. "US 4th Fleet Participates in SIFOREX 2018." April 19, 2018. https://www.southcom.mil/MEDIA/NEWS-ARTICLES/Article/1501727/us-4th-fleet-participates-in-siforex-2018/.

U.S. Government Accountability Office (GAO), "Weapon Systems Cybersecurity: DOD Just Beginning to Grapple with Scale of Vulnerabilities." October, 2018. https://www.gao.gov/assets/700/694913.pdf.

U.S. Immigration and Customs Enforcement. "Russian Cyber-Criminal Pleads Guilty to Role in Multimillion Dollar Online Identity Theft Ring." August 9, 2017. https://www.ice.gov/news/releases/russian-cyber-criminal-pleads-guilty-role-multimillion-dollar-online-identity-theft.

U.S. Senate. "Statement of General Paul M. Nakasone Commander United States Cyber Command before the Senate Committee on Armed Services" February 14, 2019. https://www.armed-services.senate.gov/imo/media/doc/Nakasone_02-14-19.pdf.

U.S. Southern Command. "Areas of Responsibility." http://www.southcom.mil/About/Area-of-Responsibility/.

U.S. Southern Command. "Theater Strategy." May 19, 2017. http://www.southcom.mil/Portals/7/Documents/USSOUTHCOM_Theater_Strategy_Final.pdf?ver=2017-05-19-120652-483.

U.S. Southern Command. "Posture Statement of Admiral Kurt W. Tidd Commander, United States Southern Command before the 115th Congress Senate Armed

Services Committee." February 15, 2018. http://www.southcom.mil/Portals/7/Documents/Posture%20Statements/SOUTHCOM_2018_Posture_Statement_FINAL.PDF?ver=2018-02-15-090330-243.

U.S. Southern Command. "United States Southern Command Strategy: Enduring Promise for the Americas." May 8, 2019. https://www.southcom.mil/Portals/7/Documents/SOUTHCOM_Strategy_2019.pdf?ver=2019-05-15-131647-353.

UNESCO, "World Trends in Freedom of Expression and Media Development, Global Report 2017/2018." http://unesdoc.unesco.org/images/0026/002610/261065e.pdf, 109–10.

UNIDIR. "Cybersecurity, and Cyberwarfare: Preliminary Assessment of National Doctrine and Organization." October 5, 2011. https://unidir.org/publication/cybersecurity-and-cyberwarfare-preliminary-assessment-national-doctrine-and.

UNIDIR. "Cyber Policy Portal." Accessed October 25, 2020. https://unidir.org/publication/cybersecurity-and-cyberwarfare-preliminary-assessment-national-doctrine-and.

UNISDR. *Poverty and Death: Disaster Mortality 1996–2015, 2016*. Geneva: UNISDR.

United Nations Secretary-General, "Address at the Opening Ceremony of the Munich Security Conference," February 16, 2018. https://www.un.org/sg/en/content/sg/speeches/2018-02-16/address-opening-ceremony-munich-security-conference.

Uriegas, Fernanda. "The Chinese Brands Latin Americans Love." *Americas Quarterly*, April 15, 2019. https://www.americasquarterly.org/content/chinese-brands-latin-americans-love.

Valeriano, Brandon, and Ryan C. Maness. "The Dynamics of Cyber Conflict between Rival Antagonists, 2001–11." *Journal of Peace Research* 51, no. 3 (2014): 347–60.

Valeriano, Brandon, and Ryan C. Maness. *Cyber War versus Cyber Realities: Cyber Conflict in the International System*. Oxford and New York: Oxford University Press, 2015.

Van Puyvelde, Damien, and Aaron F. Brantley. *Cybersecurity: Politics, Governance, and Conflict in Cyberspace*. Cambridge: Polity, 2019.

Vargas Vásquez, Suldery. "Ejercito Pone en Marcha el Primer Batallón de Prevención y Atención de Desastres." January, 2009. https://www.ejercito.mil.co/?idcategoria=246906.

Velediaz, Juan. "Crecen Operaciones Conjuntas entre Marina, Ejército y Fuerza Aérea." *EstadoMayor.mx*, September 7, 2017. https://www.estadomayor.mx/76656.

Vidal, John. "Mexico's Zapatista Rebels, 24 Years on and Defiant in Mountain Strongholds." *Guardian*, February 17, 2018. https://www.theguardian.com/global-development/2018/feb/17/mexico-zapatistas-rebels-24-years-mountain-strongholds.

Villar Gertner, Andrés. "The Beagle Channel Frontier Dispute between Argentina and Chile: Converging Domestic and International Conflicts." *International Relations* 28, no. 2 (2014): 207–27.

Volkert, Zachary. "Microsoft to Bring ICT Innovation to Colombia." *BN Americas*, December 6, 2013. http://www.bnamericas.com/news/technology/microsoft-to-bring-ict-innovation-to-colombia?lang=en.
Voltz, Dustin, and Timothy Gardner, "In a First, U.S. Blames Russia for Cyber Attacks on Energy Grid." *Reuters*, March 15, 2018. https://www.reuters.com/article/us-usa-russia-sanctions-energygrid/in-a-first-u-s-blames-russia-for-cyber-attacks-on-energy-grid-idUSKCN1GR2G3.
Votel, Joseph L., David J. Julazadeh, and Weilun Lin. "Operationalizing the Information Environment: Lessons Learned from Cyber integration in the USCENTCOM AOR." *The Cyber Defense Review* 3, no. 3 (2018): 15–20.
VR World "Regin: Stuxnet's Best Spying Malware Cousin." November 24, 2014, http://vrworld.com/2014/11/24/regin-stuxnets-best-spying-malware-cousin/.
Walt, Stephen M. "Alliance Formation and the Balance of World Power." *International Security* 9, no. 4 (1985): 3–43.
Walt, Stephen M. "Alliances in a Unipolar World." *World Politics* 61, no. 1 (2009): 86–120.
Walt, Stephen. *The Hell of Good Intentions: America's Foreign Policy Elite and the Decline of U.S. Primacy* New York: Farrar, Straus and Giroux, 2019.
Wang, Hongying. "The Missing Link in Sino-Latin American Relations." *Journal of Contemporary China* 24, no. 95 (2015): 922–42.
Wang, Yong. "Offensive for Defensive: The Belt and Road Initiative and China's New Grand Strategy." *The Pacific Review* 29, no. 3 (2016): 455–63.
Washington Examiner. "China Selling High-Tech Tyranny to Latin America, Stoking US Concern." April 10, 2019. https://www.washingtonexaminer.com/policy/defense-national-security/china-selling-high-tech-tyranny-to-latin-america-stoking-us-concern.
Wazeman, Peter, Aude Fleurant, Alexandra Kuimova, Nan Tian, and Siemon T. Wezeman. "Trends in international Arms Transfers, 2017." *SIPRI Fact Sheet*, March 2018. https://www.sipri.org/sites/default/files/2018-03/fssipri_at2017_0.pdf.
Weare, Christopher. "The Internet and Democracy: The Causal link Between Technology and Politics." *International Journal of Public Administration* 25, no. 5 (2002): 659–91.
Weaver, Jay. "Will Ex-President Be Sent Home to Panama to Face Charges? His Fate up to Miami Judge." *Miami Herald*, August 3, 2017. http://www.miamiherald.com/news/nation-world/article165290732.html.
Webber, Mark, Stuart Croft, Jolyon Howorth, Terry Terriff, and Elke Krahmann. "The Governance of European Security." *Review of International Studies* 30, no. 1 (2004): 3–26.
Weeks, Gregory B. *U.S. and Latin American Relations*. Chichester: Wiley Blackwell, 2015.
Weiffen, Brigitte, Leslie Wehner, and Detlef Nolte. "Overlapping Regional Security Institutions in South America: The Case of OAS and UNASUR." *International Area Studies Review* 16, no. 4 (2013): 370–89.

Weimann, Gabriel. "Cyberterrorism: The Sum of All Fears?" *Studies in Conflict & Terrorism* 28, no. 2 (2005): 129–49.

Weiss, Margaret. "Assad's Secretive Cyber Force." *Washington Institute*, April 12, 2012. https://www.washingtoninstitute.org/policy-analysis/view/assads-secretive-cyber-force.

Weiss, Moritz, and Vytautas Jankauskas. "Securing Cyberspace: How States Design Governance Arrangements." *Governance* 32, no. 2 (2019): 259–75.

Wells, Miriam. "Brazil Investigating Hack of Military Police Data." *Insight Crime*, September 17, 2013. http://www.insightcrime.org/news-briefs/hackers-publish-personal-details-of-50000-rio-police.

Wesley, Michael. *Global Allies: Comparing US Alliances in the 21st Century*. Canberra: Australian National University Press, 2017.

White, Adam. "Introduction: A State-in Society Agenda." In *The Everyday Life of the State*, edited by Adam White, 3–12. Seattle: University of Washington Press, 2013.

Whyte, Christopher, and Brian Mazanec. *Understanding Cyber Warfare: Politics, Policy, and Strategy*. New York: Routledge, 2018.

Williams, Phil, and Sandra Quincoses. *The Evolution of Threat Networks in Latin America*. Accessed February 20, 2022. https://gordoninstitute.fiu.edu/research/publications/evolution-of-threat-networks-in-latam.pdf.

Willett, Marcus. "Assessing Cyber Power." *Survival* 61, no. 1 (2019): 85–90.

Wolf, Cristof, and Helen Best. "Linear Regression." In *The SAGE Handbook of Regression Analysis and Causal Inference*, edited by Henning Best and Christof Wolf, 57–82. London: Sage, 2013.

Wolfgang, Ben. "U.S., To Provide Military Security for the G-20 Summit in Buenos Aires." *Washington Post*, August 16, 2018. https://www.washingtontimes.com/news/2018/aug/16/us-provide-military-security-g-20-summit-argentina/.

Wong, Wendy H., and Peter A. Brown. "E-Bandits in Global Activism: WikiLeaks, Anonymous, and the Politics of No One." *Perspectives on Politics* 11, no. 4 (2013): 1015–33.

World Bank. "World Bank Open Data." Accessed December 30, 2019, and July 12, 2022. www.data.worldbank.org.

World Bank. "World Development Indicators: States and Markets, 2019." Accessed November 29, 2020. http://datatopics.worldbank.org/world-development-indicators/themes/states-and-markets.html#science-and-innovation_1.

World Bank. "Ease of Doing Business." Accessed November 29, 2020. https://www.doingbusiness.org/content/dam/doingBusiness/pdf/db2020/Doing-Business-2020_rankings.pdf.

World Bank. "World Governance Indicators." Accessed January 15, 2021. http://www.govindicators.org.

Wortzel, Larry. "Change Partners: Who Are America's Military and Economic Allies in the 21st Century?" *Heritage Foundation*, June 6, 2005. https://www.

heritage.org/defense/report/change-partners-who-are-americas-military-and-economic-allies-the-21st-century.

Xinhua. "Ambassadors Expect More China, Latin America Cooperation." October 26, 2016. http://english.gov.cn/news/international_exchanges/2016/10/26/content_281475475859018.htm.

Xinhua. "Full Text of China's Policy Paper on Latin America and the Caribbean." November 24, 2016. http://english.gov.cn/archive/white_paper/2016/11/24/content_281475499069158.htm.

Xinhua. "Xi, Morales Hold Talks, Agree to Establish China-Bolivia Strategic Partnership." June 19, 2018. http://www.xinhuanet.com/english/2018-06/19/c_137265305.htm.

Xu, Yanran. *China's Strategic Partnerships in Latin America: Case Studies of China's Oil Diplomacy in Argentina, Brazil, Mexico, and Venezuela, 1991–2015.* London: Lexington Books, 2017.

Yau, Hon-min. "A Critical Strategy for Taiwan's Cybersecurity: A Perspective from Critical Security Studies." *Journal of Cyber Policy* 4, no. 1 (2019): 35–55.

Zeng, Jinghan, Tim Stevens, Yaru Chen. "China's Solution to Global Cyber Governance: Unpacking the Domestic Discourse of 'Internet Sovereignty.'" *Politics & Policy* 45, no. 3 (2017): 432–64.

Zhang, Jian. "China's New Foreign Policy under Xi Jinping: Towards 'Peaceful Rise 2.0'?" *Global Change, Peace & Security* 27, no. 1 (2015): 5–19.

Zhao, Suisheng. "A New Model of Big Power Relations? China-US Strategic Rivalry and Balance of Power in the Asia-Pacific." *Journal of Contemporary China* 24, no. 93 (2015): 377–97.

Zhou, Weifeng, and Mario Esteban. "Beyond Balancing: China's Approach towards the Belt and Road Initiative." *Journal of Contemporary China* 27, no. 112 (2018): 487–501.

Index

Abebe, Daniel, 58
Afghanistan, 15, 33, 80, 169
Africa, 28, 30, 48, 59, 140, 150, 171, 228
African Union, 1
Al-Qaeda, 72, 169
Alcatel, 126
Alliances, 14, 79, 85, 108, 122, 124, 132, 140, 165, 167, 212, 219
Alliance of the Pacific, 153
Amorim, Celso, 117
António Guterres, 2
Apple, 35, 126, 219
Argentina, 13, 16, 19, 20, 35, 39, 49, 57, 113, 114, 146, 168
Artificial Intelligence, 71, 206
Asia, 25, 28, 37–40, 54, 88, 102, 115, 137, 149, 150, 175
ASEAN, 137, 142, 173
Astore, William, 33
Australia, 7, 25, 40, 57, 142, 180
  Australia Group, 96
  Signals Directorate, 21
Austria, 25

Bachelet, Michelle, 121
Baird, Zoe, 26
Bangladesh, 40
Bashar Al-Assad, 40
Berners Lee, Tim, 25

Belgium, 26, 140, 241, 283
Belize, 68, 232
Betts, Richard, 91
Bitcoin, 203
Boeing, 155
Bolivia, 35, 59, 72, 101, 108, 120, 148, 159, 173
Bolivarian Alliance (ALBA), 65, 138, 149
Bolsonaro, Eduardo, 208
Bolsonaro, Jair, 143, 167
Bolton, John, 81, 133
Brantly, Aaron, 78
Brazil, 10, 13, 19, 20, 26, 30, 32, 39, 40, 46, 49, 57, 113, 118, 145
  Cyber Defense Command, 17, 118, 121, 178
  Cybersecurity strategy, 204
  Federal police, 17
  First Capital Command (PCC), 108
  National Defense Strategy, 17, 118
  Planalto Palace, 101
  Rio de Janeiro, 100
Breier, Kimberly, 147, 180
Brenner, Suzanne W., 53
Brown, Peter A., 73
Brunei, 39
Bush, George W., 152

Cambodia, 30

CAN, 138
Canada, 25, 35, 40, 57, 113, 147, 167, 175
Calderon, Felipe, 32
Caribbean, 13
　Caribben Community (CARICOM), 138
Carr, Jeffrey, 81
Carr, Madeleine, 213
CELAC, 138
Central America, 13, 64, 84, 101, 109, 151, 179, 200, 232
Central African Republic, 32
Cerami, Joseph, 99
Chile, 13, 19, 20, 26, 49, 57, 113, 121, 138, 141, 164, 181, 204, 212, 228, 232
　ASMAR, 122
　Endesa, 67
　FACH, 73
　FAMAE, 122
　General Secretariat of the Presidency, 16, 123
　Ministry of Foreign Affairs, 16, 123
　Ministry of Interior, 16, 123
　National Intelligence Agency, 16, 123
　Public Prosecutor's Office, 16, 123
　War of the Pacific, 72
Chang, Lennon, 16
China, 1–2, 6, 23, 53
　Beijing, 14
　Belt and Road, 40, 138, 147, 210
　Companies, 154
　Cyberspace Administration, 21
　Government, 30
　Military, 150
　People's Liberation Army's, 10, 42, 154, 175
　Sino-U.S. rivalry, 137, 139, 147, 163
　Xi Jinping, 70
　White paper, 141

Chadwick, Andrés, 143
Choucri, Nazli, 59
Cisco, 126, 203
Clark, J. P., 99
Clarke, Richard, 209
Climate change, 17, 19, 93, 106, 110, 124, 134, 169
Cohen, Jared, 78
Colombia, 13, 19, 26, 30, 32, 35, 46, 49, 57, 113, 145, 164 174, 180, 191, 200, 212, 217
　BACRIM, 32
　CODALTEC, 125
　COTECMAR, 125
　Ejército de Liberación Nacional, 124
　Fuerzas Armadas Revolucionarias de Colombia, 32, 72, 124, 136, 152
　INDUMIL, 125
　Joint Cyber Command, 74, 127
　Ministry of Defense, 74, 127
　Plan Colombia, 152
　War College, 74, 127
Cold War, 15, 34, 36, 84, 163, 205, 208
Cook, Tim, 35
Computer Emergency Response Team, 25, 60, 110, 180, 205
COVID-19, 109
CSIS, 156
Craig, Anthony, 37
Crime
　Criminal violence, 9, 13
　Criminology, 3
Cryptocurrency, 203
Cotton, Tom, 206
Council on Foreign Relations, 37, 155
Cuba, 35, 65, 127, 131, 151, 169
Cyber, 7, 25, 26–28
　Alliances, 167, 177
　Attack, 23, 39, 42, 43, 65, 68, 105
　Conflict, 34, 82, 88
　Crime, 8, 21, 193, 202

Index

Defense, 21, 75, 192, 201
Deterrence, 14, 54
Espionage, 8
Incidents, 12, 43, 99, 165, 203
Military units, 37
Militia, 10
Partnership, 19, 166
Power, 90
Terrorism, 8, 72
Threats, 7, 190
War, 8, 28, 31, 34, 58, 88
Weapons, 11, 34, 81, 91
Cyberpsychology, 2, 9
Cyberspace, 1–4, 7, 25, 136, 211
Cybersecurity
  Capacity building, 13, 28, 39, 95, 205
  Geopolitics, 54
  Governance, 21, 23, 36, 58, 61, 182, 192, 209, 213, 2017
  Legislation, 198
  Policy communities, 26
  Policymaking, 28, 57, 74, 185, 198, 201
  Preparedness, 44, 134
  Technology, 198
  Training, 200
Czech Republic, 26, 140, 180

Data
  Destruction, 23
De Morais, Leonardo, 207
Defacement, 23
Deibert, Ronald, 15
Dell, 126, 208
Democracy, 21, 31, 62, 78, 98, 141, 150, 166
  Advanced democracies, 8
  Democratic states, 44
  Government, 42, 63
  Interests, 21
Deterrence, 14, 91

Denmark, 26, 70, 140, 170
DeRouen, Karl, 97
Developing countries, 7
Digital pax Latin Americana, 18, 64, 73, 80, 203, 208
Distributed Denial of Service, 23, 27
Dominican Republic, 35, 68, 151, 185, 201, 232
Doxing, 23
Dreyer, June Teufel, 164
Dunford, Joseph, 24
Duque, Iván, 167

Ecuador, 35, 39, 59, 101, 113, 142, 173, 228
Egypt, 29, 140
El Salvador, 32, 69, 108, 142, 151, 174
Espionage, 23
Estonia, 10, 11, 170, 181
Europe, 12, 15, 25, 28, 35, 59, 71, 98, 102, 121, 140, 171, 191, 206, 216
  European Union, 2, 7
Export liberalization, 19

Facebook, 12, 35, 69, 133, 203
Faller, Craig, 124, 166, 218
Fernández, Cristina, 138
Financial Intelligence Units and FATF, 76
FireEye, 126
Five Eyes, 144, 172, 178
Fonseca, Brian, 119
Fox, Vicente, 32
France, 12, 40, 68, 84, 98, 115, 117, 154, 167, 181, 191
Friedman, Allan, 54
Fukuyama, Francis, 46

Gardner, Cory, 171
Gates, Robert, 15
Geneva Convention, 2

336 Index

Germany, 35, 40, 68, 87, 98, 117, 129, 140, 154, 168, 170
  Federal Office for Information Security, 21
Georgia, 10, 82, 170, 179
Google, 29, 35, 70, 78, 146
Gouvea, Raul, 97
Governance, 1, 5–6, 7, 21, 45
  Collaborative, 16
  Cybersecurity governance, 10, 216
  Networked, 44
  Quality, 46
Gingrich, Newt, 207
Global Peace Index, 12
Grabosky, Peter, 16
Graham, Stephen, 32
Greenwald, Glenn, 204
Gmail, 133
Greece, 15
Groll, Elias, 36
Guatemala, 35, 40, 69, 80, 142, 151
Guyana, 113, 120, 188
Guzmán, Joaquín, 130

Hackers, 5, 7, 40
Hacking, 27
Haiti, 110, 131, 148, 152
Haugh, Timothy, 169
Heleno, Augusto, 207
Helmke, Gretchen, 43
Herbst, John E., 179
Hezbollah, 30
Hood, Christopher, 43
Hong Kong, 12, 40, 71, 155
Honduras, 30, 32, 69, 108, 141, 151, 228, 232
Hongying Wang, 163
Huawei, 143, 182, 206, 208
Human security, 59
Human trafficking, 13, 61, 109, 136, 142, 176, 211
Hybrid warfare, 98, 99, 134, 136, 155

India, 40
Indonesia, 131, 140
Indo-Pacific, 179, 228
Indra, 126
Information and communication technologies (ICTs), 1–4, 28, 34, 43, 48, 54, 168, 178, 190
Inkster, Nigel, 147
Institute for Economics and Peace, 31
International Institute for Strategic Studies, 137, 176, 212, 213
International Telecommunications Union, 49
Internet
  5G, 180, 203, 207
  Balkanization, 189
  E-government, 48, 201
  Governance, 1, 14
  Penetration, 37, 45, 188, 192, 198, 221
  Usage, 162
  Weaponized, 48, 73
Interstate conflict, 9, 14, 19, 31. 83, 93, 101, 121, 134, 142
Inter-American Development Bank, 64
Iran, 6, 11, 30, 140, 169
Irregular warfare, 33, 99, 107
Israel, 11, 59, 60, 73, 85, 117, 121, 129
Iraq, 15, 32, 80

Jamaica, 32, 60
Japan, 40, 140, 155, 173, 180

Kaljurand, Marina, 11
Kaspersky, 66, 126
Kello, Lucas, 82
Kenya, 40
Khashoggi, Jamal, 29
Kirchner, Néstor, 138
Kissinger, Henry, 206
Kuczynski, Pedro Pablo, 152
Kurç, Çaglar, 102

LAPOP Americas Barometer, 62, 69, 157
La Tercera, 145
League of Arab States, 1
Lenovo, 147, 208
Levitsky, Steven, 43
Lewis, James Andrew, 156
Libicki, Martin, 7
López, Vladimiro, 133
López Obrador, Andrés Manuel, 212

Macri, Mauricio, 139
Macron, Emmanuele, 191, 203
Maness, Ryan, 3, 34, 87, 92, 110, 135
Mahoney, James, 113
Malaysia, 39, 189
Malware, 11, 27, 35, 41, 66, 82, 103, 166, 203
Marguleas, Oliver, 133
Marczak, Jason, 179
Markle Foundation, 26
Martinelli, Ricardo, 73
Mattis, Jim, 124, 178
Maurer, Tim, 54
Mazanec, Brian, 55
McBurney, Peter, 11
Mexico, 13, 19, 20, 26, 32, 35, 49, 59, 127, 145
  Center for Cyberspace Operations, 129
  Ejercito Zapatistas de Liberación Nacional, 128
  Merida Initiative, 152
  Military, 128
  National Guard, 212
  Pemex, 203
  Sedena, 129, 130
Microsoft, 126, 133, 203, 208
Middle states, or powers, 140
Military, 3, 63
  Civil-military relations, 87
  Expenditure, 85, 110
  Demilitarization, 211–12

Diplomacy, 19
Industrialization, 94, 96, 98
Militarization, 21, 26
Personnel, 110
R&D, 97, 102
Roles and mission, 100
Transformation, 93
Mongolia, 40
Mora, Frank, 119
Morales, Evo, 138
Moro, Sérgio, 204
Mourao, Hamilton, 207
Muggah, Robert, 88
Myanmar, 32

Nakasone, Paul, 172
National security, 7
National Strategy to Secure Cyberspace, 11
Natural disasters, 19, 93, 106, 110, 124, 131, 179
Netherlands, The, 26, 84, 167, 181, 203
Neuman, Stephanie G., 102
New York Times, 5, 35
New Zealand, 59, 142, 173, 180
North American Free Trade Agreement (NAFTA), 137
Nicaragua, 31, 35, 60, 108, 180, 185
Nokia, 207
North Atlantic Treaty Organization, 2, 8, 26, 40, 126, 140, 167, 203
Non-state actors, 9, 120
North Korea, 12, 23, 40, 155, 169
Norway, 26, 41, 140
NSO Group, 129
Nye Jr., Joseph, 79, 90, 104

Obama, Barack, 151, 169
O'Donnell, Guillermo, 44
Organization for Security and Co-operation in Europe (OSCE), 41

Organization of American States (OAS), 1, 14, 20, 64, 76, 126, 159, 216
  Cyber Observatory, 173, 187, 192
  Cyber Security Program, 14, 60
  Insulza, José Miguel, 141
  Inter-American Committee against Terrorism (CICTE), 14, 181

Pakistan, 15, 164, 169
Panama, 57, 65 73, 142, 172
Paraguay, 30, 60, 72, 105, 113, 120, 173
  Asociación Campesina Armada, 108
  Ejército del Pueblo Paraguayo, 108
Paramilitaries, 19
Peru, 30, 35, 39, 50, 60 100, 113, 152, 164
Peña-Nieto, Enrique, 68, 129
Piñera, Sebastián, 124, 143, 167
Police, 63, 65
Political science, 3
Pompeo, Michael R., 165, 170
Protectionism, 19
Putin, Vladimir, 24, 81, 95, 131, 143, 145

Ransomware, 5–6
Realism, 140
Regin, 12
Ren Zhengfei, 207
Reveron, Dereck, 65
Rid, Thomas, 11, 58
Rossi, Agustin, 116
Rotberg, Robert I., 46
Rouseff, Dilma, 68, 170
Rubio, Marco, 218
Russia, 2, 10, 23, 35, 53, 70, 157, 163, 170, 189, 203, 215, 228
  Military, 74
  Moscow, 11, 14, 41
  RT, 64
  Security Council, 21

Saakashvili, Mikheil, 179
Saalman, Lora, 136
Sabotage, 23
Sachs, Jeffrey, 211
Saudi Arabia, 29, 85, 140, 157
Samsung, 207
Sancho, Carolina, 217
Schmidt, Eric, 78
Schulte, Heinz, 89
Security, 7
  Networks, 44
Shanghai Cooperation Organization, 1
Sharpe, Ken, 100
Shirreff, Richard, 103
Singapore, 155, 157, 175
Singer, P. W., 54
Slayton, Rebecca, 105
Smith, Martin, 216
Snowden, Edward, 80, 204
Spain, 67, 129, 150, 228
Somalia, 32
South Africa, 142
South Sudan, 33
South Korea, 15, 129, 137, 155, 175, 181, 207
Soviet Union, 20
Sowbug, 39
STATA, 187, 221
Stockholm International Peace Research Institute (SIPRI), 71, 85, 154
Stokes, Mark, 144
Strayer, Robert, 173, 182
Stuxnet, 11
St. Kitts, 32
Sullivan, John, 88, 177
Suriname, 113, 120, 188
Sweden, 140, 207
Syrian Electronic Army, 29

Taiwan, 144, 153, 175, 189
Tallin Manual, 18

Temer, Michael, 120
Terrorism, 13, 19
Thailand, 30, 40, 140
Thayer, Bradley, 55
Thompson, Robert, 5
Tidd, Kurt W., 162
Transnational crime, 13, 19, 27, 31 73, 83, 106, 130, 153, 174, 209, 212
Trans-Pacific Partnership, 153, 189
Trinidad and Tobago, 60, 180, 181
Triple Frontier, 72, 108
Trudeau, Justin, 147
Trump, Donald, 12, 24, 35, 124, 133, 137, 189, 212
Turkey, 32, 86, 140, 168
Twitter, 12, 133, 218

Union of South American Nations (UNASUR), 92, 138
United Nations, 2, 30
  Group of Governmental Experts, 181
  Institute for Disarmament Research (UNIDIR), 8
  Internet Governance Forum, 191
  Secretary-General, 2
  Security Council, 140, 148
  Stabilization Mission in Haiti (MINUSTAH), 110, 148
United Kingdom, 7, 40, 80, 84, 98, 140, 151, 167, 191, 215
  GCHQ, 68
  Ministry of Defense, 89
  National Cybersecurity Centre, 21, 23
United States, 1, 5, 8, 14, 20, 30, 35, 55, 57, 116, 136, 182
  Agency for International Development, 166
  Cyber Command, 15, 21, 38, 68, 165, 170, 172
  Democratic Party, 69

Department of Commerce, 16
Department of Defense 8, 16, 36, 42, 142, 157, 166, 190
Department of Homeland Security, 16, 41, 60
Department of Justice, 16, 41
Department of State, 16, 68, 74, 138, 147, 169, 172, 210
Department of the Treasury, 16, 169
European Command, 171
Federal Bureau of Investigation, 41, 70, 169
Government, 160
Government Accountability Office, 36
Immigration and Customs Enforcement, 67
National Guard, 170
National Security Agency, 18, 65
National Security Council, 16, 68
National Security Strategy, 156, 170, 172
Republican Party, 25
Secret Service, 67
Southern Command, 65, 124, 142, 148, 151, 162, 171
U.S.-Sino rivalry, 137, 139, 147, 163
Unmanned Aerial Vehicle (UAV), 120, 130
Ukraine, 10, 24, 32, 35, 41, 59, 170, 175
Uruguay, 57, 6178, 101, 111, 141, 152, 164, 173

Valeriano, Brandon, 3, 43, 87, 92, 110, 135
Venezuela, 13, 19, 20, 26, 32, 35, 49, 130, 157, 2017
  CAVIM, 131
  Chávez, Hugo, 131, 138, 142, 179
  Cyberdefense Directorate, 74
  Guaidó, Juan, 132, 179

Venezuela *(continued)*
  Joint Cyberdefense Directorate, 132
  Maduro, Nicolás, 131, 179, 2017
  Strategic Operations Command, 74, 132
Vietnam, 25, 150, 189
Villegas, Luis, 127

Washington Consensus, 153
Walt, Stephen, 165
Web 2.0, 32
World Bank, 28, 49, 86, 110

World Governance Indicators, 46, 221
World War I, 98
Wong, Wendy, 73
Wealth creation, 19
Wikileaks, 73, 80, 157
Willet, Markus, 208

Xi Jinping, 70, 138, 163

Yemen, 15, 32
Yong Duo, 145
Young, John, 99

www.ingramcontent.com/pod-product-compliance
Lightning Source LLC
Chambersburg PA
CBHW031433230426
43668CB00007B/517